Lecture Notes in Computer Science 7774

Commenced Publication in 1973
Founding and Former Series Editors:
Gerhard Goos, Juris Hartmanis, and Jan van Leeuv

Hideaki Takeda   Yuzhong Qu
Riichiro Mizoguchi   Yoshinobu Kitamura (Eds.)

Semantic Technology

Second Joint International Conference, JIST 2012
Nara, Japan, December 2-4, 2012
Proceedings

 Springer

Volume Editors

Hideaki Takeda
National Institute of Informatics
Tokyo, Japan
E-mail: takeda@nii.ac.jp

Yuzhong Qu
Nanjing University, China
E-mail: yzqu@nju.edu.cn

Riichiro Mizoguchi
Japan Advanced Institute of Science and Technology
Ishikawa, Japan
E-mail: mizo@jaist.ac.jp

Yoshinobu Kitamura
Osaka University, Japan
E-mail: kita@ei.sanken.osaka-u.ac.jp

ISSN 0302-9743   e-ISSN 1611-3349
ISBN 978-3-642-37995-6   e-ISBN 978-3-642-37996-3
DOI 10.1007/978-3-642-37996-3
Springer Heidelberg Dordrecht London New York

Library of Congress Control Number: 2013936007

CR Subject Classification (1998): I.2.4, I.2.6, H.3, H.2

LNCS Sublibrary: SL 3 – Information Systems and Application, incl. Internet/Web
and HCI

Typesetting: Camera-ready by author, data conversion by Scientific Publishing Services, Chennai, India

Printed on acid-free paper

Springer is part of Springer Science+Business Media (www.springer.com)

Preface

The Joint International Semantic Technology Conference (JIST) is a regional federation of semantic technology-related conferences. JIST 2012 was a joint effort between the JIST Conference and SIG on the Semantic Web and Ontology of Japanese Society for AI (JASI). The mission of JIST is to bring together researchers from the semantic technology community and other areas of semantic-related technologies to present their innovative research results or novel applications of semantic technologies. By semantic technology, we mean not only the Semantic Web but also ontology, metadata, linked open data, OWL/RDF languages, description logics, etc.

JIST 2012 consisted of main technical tracks including a regular paper track, an in-use track, two special tracks on linked data in practice and database integration, three workshops, and a tutorial. We had 90 submissions for the main technical tracks from 26 countries which include not only Asian countries but also many European countries. Statistics is shown below.

Regular papers	55
In-use track papers	16
Special track papers	15
Linked data in practice:	11
Database integration:	4
Total	86

Every paper was reviewed by three reviewers on average and then discussed by the Chairs. In these proceedings, there are 33 papers that were accepted and presented in the conference: 20 full papers for the research track, 7 short papers for the in-use track, and 6 short papers for the special tracks (linked data in practice: 4; database integration: 2).

The topics were well distributed, from theoretical studies on ontology and description logics to applications of semantic technology. In particular, many papers are related to linked data. The trend is encouraging to us since it indicates that semantic technology has been deployed in the real world.

Without the hard work and close cooperation of the Organizing Committee, the great success of JIST 2012 would not have been possible. We all thank the members of the Program Committee and local Organizing Committee for their constant support. Special thanks also go to the sponsors and supporting organizations for their financial support.

December 2012

Hideaki Takeda
Yuzhong Qu
Riichiro Mizoguchi
Yoshinobu Kitamura

Conference Organization

Organizing Committee

General Chair
Riichiro Mizoguchi   Japan Advanced Institute of Science and Technology, Japan

Program Chairs
Hideaki Takeda   National Institute of Informatics (NII), Japan
Yuzhong Qu   Nanjing University, China

In-Use Track Chair
Takahiro Kawamura   Toshiba Corp., Japan

Special Track Chair
Tetsuro Toyoda   RIKEN, Japan

Poster and Demo Chairs
Ryutaro Ichise   National Institute of Informatics (NII), Japan
Seokchan Yun   Daum Communications Corp., Korea

Workshop Chairs
Zhiqiang Gao   Southeast University, China
Haklae Kim   Samsung Electronics Co., Ltd., Korea

Tutorial Chairs
Ikki Ohmukai   National Institute of Informatics (NII), Japan
Hanmin Jung   Korea Institute of Science and Technology Information, Korea

Publicity Chair
Kouji Kozaki   Osaka University, Japan

Local Organizing Chair
Yoshinobu Kitamura   Osaka University, Japan

Program Committee

Chutiporn Anutariya
Huajun Chen
Ling Chen
Gong Cheng
Hong Cheng
Key-Sun Choi
Bernardo Cuenca
Ying Ding
Dejing Dou
Jianfeng Du
Zhiqiang Gao
Masahiro Hamasaki
Sungkook Han
Hyoil Han
Siegfried Handschuh
Koiti Hasida
Pascal Hitzler
Laura Hollink
Itaru Hosomi
Wei Hu
Hanmin Jung
Jaieun Jung
Takahiro Kawamura
Hong-Gee Kim
Pyung Kim
Yoshinobu Kitamura
Seiji Koide
Kouji Kozaki
Myungjin Lee
Seungwoo Lee
Juanzi Li
Diana Maynard
Takeshi Morita

Mitsuharu Nagamori
Shinichi Nagano
Hyun Namgoong
Ekawit Nantajeewarawat
Ikki Ohmukai
Yue Pan
Young-Tack Park
Dimitris Plexousakis
Guilin Qi
Yuzhong Qu
Marco Ronchetti
Riccardo Rosati
Hiroyuki Sato
Kai-Uwe Sattler
Stefan Schlobach
Yi-Dong Shen
Sa-Kwang Song
Giorgos Stoilos
Umberto Straccia
Thepchai Supnithi
Hideaki Takeda
Tetsuro Toyoda
Kewen Wang
Haofen Wang
Shenghui Wang
Chris Welty
Gang Wu
Vilas Wuwongse
Bin Xu
Takahira Yamaguchi
Mun Yong Yi
Yuting Zhao
Yi Zhou

Additional Reviewers

Marut Buranarach
Jun Deng
Bing He
Lei Hou
Ernesto Jimenez-Ruiz
Amit Joshi
Haridimos Kondylakis
Adila A. Krisnadhi
Zhixing Li
Chang Liu
Yuanchao Ma
Akshay Maan
Raghava Mutharaju
Jungyeul Park
Martin Rezk
Kunal Sengupta
Eleni Tsalapati
Tassos Venetis
Zhigang Wang
Erjia Yan

Table of Contents

Semantic Search

Knowledge Building

Semantic Web Application

In-Use Track

Social Semantic Web

Semantic Search

Special Track

Linked Data in Practice

Database Integration

A Resolution Procedure for Description Logics with Nominal Schemas

Cong Wang and Pascal Hitzler

Kno.e.sis Center, Wright State University, Dayton OH 45435, USA
{cong,pascal}@knoesis.org

Abstract. We present a polynomial resolution-based decision procedure for the recently introduced description logic $\mathcal{ELHOV}_n(\sqcap)$, which features nominal schemas as new language construct. Our algorithm is based on ordered resolution and positive superposition, together with a lifting lemma. In contrast to previous work on resolution for description logics, we have to overcome the fact that $\mathcal{ELHOV}_n(\sqcap)$ does not allow for a normalization resulting in clauses of globally limited size.

1 Introduction

Description Logic (DL) and rule-based formalism are two prominent paradigms for Knowledge Representation. Although both paradigms are based in classical logic, they provide different expressivity and neither of them contains the other one, i.e. there exist axioms in DL which are not expressible in the rules paradigm and viceversa. Despite significant research efforts [4,5,17], many integrations of the two paradigms lead to undecidability (see [14] for a survey).

Currently, the most notable language in DLs family is the W3C recommendation Web Ontology Language (OWL[1]). OWL can express many rules (see [14]), but it cannot express many others, such as

$$hasParent(x, z) \land hasParent(x, y) \land married(y, z) \rightarrow C(x) \quad (1)$$

which defines a class C of children whose parents are married.

One idea for retaining decidability is to restrict the applicability of rules to *named individuals*. Rules that are understood in this sense are called *DL-safe rules*, and the combination between OWL DL and DL-safe rules is decidable [21].

Very recently, this idea is found to be able to carry further to description logic paradigm. *Nominal schemas,* as a new element of description logic syntax construct, was introduced in this sense [16]. It does not only further generalize the notion of DL-safety, but also enables to express the DL-safe rules within the description logic syntax. Using nominal schemas, rule (1) could be represented as:

$$\exists hasParent.\{z\} \sqcap \exists hasParent.\exists married.\{z\} \sqsubseteq C \quad (2)$$

---

[1] http://www.w3.org/TR/owl2-overview/

H. Takeda et al. (Eds.): JIST 2012, LNCS 7774, pp. 1–16, 2013.

The expression $\{z\}$ in (2) is a nominal schema, which is a variable that only binds with known individuals in a knowledge base and the binding is the same for all occurrences of the same nominal schema in an axiom.

Consequently, a new description logic $\mathcal{SROIQV}_n$ was introduced, which indeed is a extension of OWL 2 DL $\mathcal{SROIQ}$ with nominal schemas $\mathcal{V}_n$. It is decidable and has the same worst-case complexity as $\mathcal{SROIQ}$ [16]. $\mathcal{SROELV}_n$ is a tractable fragment of $\mathcal{SROIQ}$. And extending $\mathcal{SROELV}_n$ with role conjunction on simple roles and concept product would not change its worst-case complexity [24]. The importance of $\mathcal{SROELV}_n(\sqcap_s, \times)$ is that it can incorporate OWL EL and OWL RL (two tractable fragments of OWL 2), and to allow restricted semantic interaction between the two. Also, it is more easy for ontology modelers to write rules in OWL syntax.

Although reasoning for description logic with nominal schema is theoretically feasible, the simple experiment in [19] shows that the naive approach, based on *full grounding* nominal schemas to all known individuals, is extremely slow. Therefore, it is really necessary to design a *smarter* algorithm. One idea is to ground nominal schemas in a more intelligent way, e.g. intelligent grounding, which is quite well known in *Answer Set Programming* (ASP) field [23]. In [13], the authors applied this strategy on $\mathcal{ALCO}$ with nominal schema, but it needs a very good heuristics for grounding choices.

Another idea is to find a procedure that could do reasoning without grounding. We apply this idea in the paper by using resolution procedure with a lifting lemma. In this paper, we restrict $\mathcal{SROELV}_n(\sqcap_s, \times)$ to $\mathcal{ELHOV}_n(\sqcap)$, which disallows self role, complex role inclusion (role chain) and concept product, but allows role conjunction even for complex roles. The reason of this restriction will be discussed in Section 6. We provide a tractable resolution procedure for $\mathcal{ELHO}(\sqcap)$, then show that the algorithm can also apply for $\mathcal{ELHOV}_n(\sqcap)$ via a lifting lemma.

The structure of this paper is as follows. Section 2 describes some preliminaries of description logic and resolution procedure. Section 3 presents a tractable, sound and complete resolution procedure for $\mathcal{ELHO}(\sqcap)$. Section 4 extends the algorithm to deal with $\mathcal{ELHOV}_n(\sqcap)$. We will provide an example to illustrate the resolution procedure in Section 5 and then briefly discuss some related works in Section 6. Finally we conclude.

2   Preliminaries

2.1   Description Logic

We start by introducing the description logic $\mathcal{ELHOV}_n(\sqcap)$.

A signature of $\mathcal{ELHOV}_n(\sqcap)$ is a tuple $\Sigma = \langle \mathsf{N}_I, \mathsf{N}_C, \mathsf{N}_R, \mathsf{N}_V \rangle$ of mutually disjoint finite sets of *individual names, concept names, role names,* and *variables.*

Table 1. Semantics of $\mathcal{ELHOV}_n(\sqcap)$

Name	Syntax	Semantics
concept name	A	$A^{\mathcal{I}} \subseteq \Delta^{\mathcal{I}}$
role name	R	$R^{\mathcal{I}} \subseteq \Delta^{\mathcal{I}} \times \Delta^{\mathcal{I}}$
individual name	a	$a^{\mathcal{I}} \in \Delta^{\mathcal{I}}$
variable	x	$\mathcal{Z}(x) \in \Delta^{\mathcal{I}}$
top	$\top$	$\Delta^{\mathcal{I}}$
bottom	$\bot$	$\emptyset$
nominal(schema)	$\{t\}$	$\{t^{\mathcal{I},\mathcal{Z}}\}$
existential restriction	$\exists R.C$	$\{\delta \mid$ there is ϵ with $\langle \delta, \epsilon \rangle \in R^{\mathcal{I},\mathcal{Z}}$ and $\epsilon \in C^{\mathcal{I},\mathcal{Z}}\}$
concept conjunction	$C \sqcap D$	$C^{\mathcal{I},\mathcal{Z}} \cap D^{\mathcal{I},\mathcal{Z}}$
role conjunction	$R \sqcap S$	$R^{\mathcal{I}} \cap S^{\mathcal{I}}$
concept assertion(Abox)	$A(t)$	$t^{\mathcal{I},\mathcal{Z}} \in A^{\mathcal{I},\mathcal{Z}}$
role assertion(ABox)	$R(t,u)$	$\langle t^{\mathcal{I},\mathcal{Z}}, u^{\mathcal{I},\mathcal{Z}} \rangle \in R^{\mathcal{I},\mathcal{Z}}$
TBox axiom	$C \sqsubseteq D$	$C^{\mathcal{I},\mathcal{Z}} \subseteq D^{\mathcal{I},\mathcal{Z}}$
RBox axiom	$R \sqsubseteq S$	$R^{\mathcal{I}} \subseteq S^{\mathcal{I}}$

Definition 1. *The role sets* **R** *of* $\mathcal{ELHOV}_n(\sqcap)$ *and the concept sets* **C** *of* $\mathcal{ELHOV}_n(\sqcap)$ *are defined by the following grammar:*

$$\mathbf{R} ::= \mathbf{R} \mid \mathbf{R} \sqcap \mathbf{R}$$
$$\mathbf{C} ::= \top \mid \bot \mid \mathsf{N}_C \mid \{\mathsf{N}_I\} \mid \{\mathsf{N}_V\} \mid \mathbf{C} \sqcap \mathbf{C} \mid \exists \mathbf{R}.\mathbf{C}$$

Definition 2. *An interpretation* $\mathcal{I} = (\Delta^{\mathcal{I}}, \cdot^{\mathcal{I}})$ *consists of a domain of discourse* $\Delta^{\mathcal{I}} \neq \emptyset$ *and a function* $\cdot^{\mathcal{I}}$ *which maps* $\mathsf{N}_C, \mathsf{N}_R,$ *and* N_I *to elements, sets and relations of* $\Delta^{\mathcal{I}}$ *as shown in Table 1. A variable assignment* $\mathcal{Z}$ *for an interpretation* $\mathcal{I}$ *is a function* $\mathcal{Z} : \mathsf{N}_V \to \Delta^{\mathcal{I}}$ *such that for each* $v \in \mathsf{N}_V$, $\mathcal{Z}(v) = a^{\mathcal{I}}$ *for some* $a \in \mathsf{N}_I$. *For any interpretation* $\mathcal{I}$, *assignments* $\mathcal{Z}$, *and* $C_{(i)} \in C$, $R_{(i)} \in \mathsf{N}_R$, $t_{(i)} \in T$, *the function* $\cdot^{\mathcal{I},\mathcal{Z}}$ *is defined as shown in Table 1.*

$\mathcal{I}$ *and* $\mathcal{Z}$ *satisfy a* $\mathcal{ELHOV}_n(\sqcap)$ *axiom* α, *written* $\mathcal{I}, \mathcal{Z} \models \alpha$, *if the corresponding condition shown in Table 1 holds.* $\mathcal{I}$ *satisfies* α, *written* $\mathcal{I} \models \alpha$, *if* $\mathcal{I}, \mathcal{Z} \models \alpha$ *for all variable assignments* $\mathcal{Z}$ *for* $\mathcal{I}$. $\mathcal{I}$ *satisfies a* $\mathcal{ELHOV}_n(\sqcap)$ *knowledge base* KB, *written* $\mathcal{I} \models KB$, *if* $\mathcal{I} \models \alpha$ *for all* $\alpha \in KB$, *and* KB *is satisfiable if such an* $\mathcal{I}$ *exists. The axiom* α *is entailed by* KB, *written* $KB \models \alpha$, *if all models of* KB *are also models of* α.

If α is a $\mathcal{ELHOV}_n(\sqcap)$ axiom, we call $\mathsf{ground}(\alpha)$ as the set of all axioms that can be obtained by uniformly replacing nominal schemas in α with individuals in N_I. Given a $\mathcal{ELHOV}_n(\sqcap)$ knowledge base KB, $\mathsf{ground}(KB) := \bigcup_{\alpha \in KB} \mathsf{ground}(\alpha)$.

Example 1. If a $\mathcal{ELHOV}_n(\sqcap)$ KB contains $\exists R.\{z_1\} \sqcap A \sqsubseteq \exists S.\{z_2\}$ and known individuals a and b. Then, $\mathsf{ground}(KB) = \{\exists R.\{a\} \sqcap A \sqsubseteq \exists S.\{a\}, \exists R.\{a\} \sqcap A \sqsubseteq \exists S.\{b\}, \exists R.\{b\} \sqcap A \sqsubseteq \exists S.\{a\}, \exists R.\{b\} \sqcap A \sqsubseteq \exists S.\{b\}\}$.

Table 2. Translating $\mathcal{ELHOV}_n(\sqcap)$ into First Order Logic

Translating Concepts into FOL
$\pi_x(\bot) = \bot$
$\pi_x(\top) = \top$
$\pi_x(A) = A(x)$
$\pi_x(C \sqcap D) = \pi_x(C) \wedge \pi_x(D)$
$\pi_x(\exists R.C) = \exists y.[R(x,y) \wedge \pi_y(C)]$
$\pi_x(\{z\}) = x \approx z$
Translating TBox Axioms without nominal schemas into FOL
$\pi(C \sqsubseteq D) = \forall x : [\pi_x(C) \to \pi_x(D)]$
$\pi(R \sqsubseteq S) = \forall x \forall y : [R(x,y) \to S(x,y)]$
$\pi(R \sqcap S \sqsubseteq T) = \forall x \forall y : [R(x,y) \wedge S(x,y) \to T(x,y)]$
Translating ABox Axioms without nominal schemas into FOL
$\pi(C(a)) = C(a)$
$\pi(R(a,b)) = R(a,b)$
Translating Axioms containing nominal schemas into FOL
$\pi(C \sqsubseteq D) = \forall x \forall z_1 \ldots \forall z_n : [O(z_1) \wedge \ldots \wedge O(z_n) \to [\pi_x(C) \to \pi_x(D)]]$
where $z_1, \ldots, z_n$ are variables in C or D (*)
Translating KB into FOL
$\pi(KB) = \bigwedge_{\alpha \in KB_T \cup KB_A \cup KB_R} \pi(\alpha)$

2.2 Translating $\mathcal{ELHOV}_n(\sqcap)$ into First Order Logic

The first step in deciding satisfiability by resolution procedure is to transform KB into a set of clauses in first order logic. We apply the well-known FOL translation for DLs in [6], and also show how to translate nominal schemas into first order logic.

Definition 3. *Let $O(x) = x \approx a_1 \vee x \approx a_2 \vee \ldots \vee x \approx a_n$ be a first order logic predicate symbol, where $a_k(1 \leq j \leq n) \in \mathsf{N}_I$. The $\approx$ symbol refers to the first order logic equality.*

Table 2 shows such DL-to-FOL translation. The translation from DL to FOL is straightforward based on the semantics of DL. We just make it clear for axioms containing nominal schemas, which the formula (*) in Table 2 states.

From Definition 2, an interpretation $\mathcal{I}$ *satisfies* α, if $\mathcal{I}, \mathcal{Z} \models \alpha$ for all variable assignments $\mathcal{Z}$ for $\mathcal{I}$. That is to say, if $\mathcal{I}$ satisfies an axiom α which contains nominal schema $z_1, \ldots, z_n$, $\mathcal{I}$ must satisfy ground(α). Suppose $a_j \in \mathsf{N}_I (1 \leqslant j \leqslant m)$ in a KB. $\mathcal{I}$ must satisfy $\bigwedge_{i=1}^{n}\bigwedge_{j=1}^{m}((z_i \approx a_j) \to \pi(\alpha))$. From Definition 3, $\bigwedge_{j=1}^{m}[z_i \approx a_j \to \pi(\alpha)] = O(z_i) \to \pi(\alpha)$. Therefore, $\bigwedge_{i=1}^{n}(O(z_i) \to \pi(\alpha)) = O(z_1) \wedge \ldots \wedge O(z_n) \to \pi(\alpha)$. So, we get the formula (*) $\pi(C \sqsubseteq D) = \forall x \forall z_1 \ldots \forall z_n : (O(z_1) \wedge \ldots \wedge O(z_n) \to (\pi_x(C) \to \pi_x(D)))$, where $z_1, \ldots, z_n$ are in C or D.

Example 2. Given two DL axioms containing three nominal schemas z_1, z_2 and z_3, $\alpha = \exists R.\{z\} \sqsubseteq C$, $\beta = \exists R.\{z_1\} \sqcap \exists S.\{z_2\} \sqsubseteq \exists T.\{z_3\}$. The corresponding first order logic translations of α and β according to Table 2 are, $\pi(\alpha) = O(z) \rightarrow [R(x, z) \rightarrow C(x)]$, $\pi(\beta) = [O(z_1) \wedge O(z_2) \wedge O(z_3)] \rightarrow [(R(x, z_1) \wedge S(x, z_2)) \rightarrow T(x, z_3)]$.

2.3   Ordered Resolution

Ordered resolution [3] is a widely used calculus for theorem proving in first order logic. The calculus has two parameters, an admissible ordering $\succ$ on literals and a selection function.

   An ordering $\succ$ on literals is admissible if (1) it is well-founded, stable under substitutions, and total on ground literals; (2) $\neg A \succ A$ for all ground atoms A; and (3) $B \succ A$ implies $B \succ \neg A$ for all atoms A and B. A literal L is (strictly) maximal with respect to a clause C if there is no other literal $L' \in C$ such that $(L' \succeq L)L' \succ L$. A literal $L \in C$ is (strictly) maximal in C if and only if L is (strictly) maximal with respect to $C \backslash L$. [6]

   A *selection function* S assigns to each clause C a subset of negative literals of C (empty possibly); the literals are said to be *selected* if they are in $S(C)$. No other restrictions are imposed on the selection function, i.e., any arbitrary functions mapping to negative literals are allowed.

   With $\mathcal{R}$ we denote the ordered resolution calculus, consisting of the following inference rules, where $D \vee \neg B$ is called the main premise. $C \vee A$ is called the side premise, and $C\sigma \vee D\sigma$ is called conclusion:

Ordered Resolution:
$$\frac{C \vee A \quad D \vee \neg B}{C\sigma \vee D\sigma}$$

where (1) $\sigma = mgu(A, B)$, (2) $A\sigma$ is strictly maximal with respect to $C\sigma$, and no literal is selected in $C\sigma \vee A\sigma$, (3) $\neg B\sigma$ is either selected in $D\sigma \vee \neg B\sigma$, or it is maximal with respect to $D\sigma$ and no literal is selected in $D\sigma \vee \neg B\sigma$.

   For general FOL, there is another rule needed, called Positive factoring. It resolves two positive literals in one clause. However, since the target logic language in the paper is a Horn logic, such that this rule is not required any more.

2.4   Superposition

Translation $\mathcal{ELHOV}_n(\sqcap)$ into FOL will produce equality symbol. In order to deal with equality, we also need superposition, a calculus for equational theorem proving.

Positive superposition:
$$\frac{(C \vee s \approx t) \cdot p \quad (D \vee w \approx v) \cdot p}{(C \vee D \vee w[t]_p \approx v) \cdot \theta}$$

where (i) $\sigma = mgu(sp, wp|_p)$ and $\theta = p\sigma$, (ii) $t\theta \not\succeq s\theta$ and $v\theta \not\succeq w\theta$, (iii) $(s \approx t) \cdot \theta$ is strictly eligible for superposition in $(C \vee s \approx t) \cdot \theta$, (iv) $(w \approx v) \cdot \theta$ is strictly eligible for superposition in $(D \vee w \approx v) \cdot \theta$, (v) $s\theta \approx t\theta \not\succeq w\theta \approx v\theta$, (vi) $w|_p$ is not a variable.

   Superposition [20] contains 4 rules, positive superposition, negative superposition, reflexivity resolution and equality factoring. However, due to the preprocess in Section 3, only positive superposition is needed. Since negative superposition and reflexivity resolution need clauses containing $\not\approx$, which will not occur in

Table 3. Normal forms of $\mathcal{ELHO}(\sqcap)$ axioms

$A \sqsubseteq \bot$	$\bot \sqsubseteq C$	$A \sqsubseteq C$	$A \sqcap B \sqsubseteq C$	$\exists R.A \sqsubseteq C$	$A \sqsubseteq \exists R.B$
$\exists R.\{a\} \sqsubseteq C$	$A \sqsubseteq \exists R.\{a\}$	$A \sqsubseteq \{a\}$	$\{a\} \sqsubseteq A$	$R \sqsubseteq T$	$R \sqcap S \sqsubseteq T$

$\mathcal{ELHOV}_n(\sqcap)$ clauses. Also, since $\mathcal{ELHOV}_n(\sqcap)$ is a Horn logic, therefore, equality factoring, which requires two positive literals in the premise, cannot be applied.

Ordered resolution and superposition are sound and complete algorithms for first order logic [3]. But, with different settings of the order and selection function, the procedure may be terminated or not.

3   Deciding Satisfiability of $\mathcal{ELHO}(\sqcap)$ by Resolution Procedure

In order to make the resolution procedure simpler, we first eliminate some equality literals such that the clauses contains only positive equality literals. Then we use the well-known structure transformation [20] to get $\mathcal{ELHO}(\sqcap)$ normal forms.

3.1   Eliminating Equality Literals

Superposition rules are designed to deal with equality in saturating first order logic clauses. However, some superposition inferences often make the resolution procedure very complicated. Directly translating of DL to FOL may contain negative equality literals. We use the following equivalent translation to eliminate negative equality literals, such that only positive superposition can be applied.

For DL concepts containing a nominal $\{a\}$, $\pi_x(\exists R.\{a\}) = \exists y.[R(x,y) \wedge y \approx a] = R(x,a)$, $\pi_x(\{a\} \sqcap C) = x \approx a \wedge \pi_x(C) = \pi_a(C)$. For DL axioms containing a nominal $\{a\}$, $\pi_x(\{a\} \sqsubseteq C) = C(a)$. For $C \sqsubseteq \{a\}$, we still directly translate it into FOL, i.e., $\neg C(x) \vee x \approx a$.

Similarly, for DL concepts containing a nominal schema $\{z\}$, $\pi_z(\exists R.\{z\}) = R(x,z)$, $\pi_x(\{z\} \sqcap C) = \pi_z(C)$. For DL axioms containing a nominal schema $\{z\}$, $\pi_x(\{z\} \sqsubseteq C) = O(z) \rightarrow (x \approx z \rightarrow \pi_x(C)) = (O(z) \wedge x \approx z) \rightarrow \pi_x(C) = O(x) \rightarrow \pi_x(C)$. For $C \sqsubseteq \{z\}$, C is either empty or the subconcept of each individual. Without losing generality, we assume that no concept C can be subconcept of each individual. Therefore, C has to be empty, $\pi_x(C \sqsubseteq \{z\}) = \pi_x(C \sqsubseteq \bot)$.

After such transformation, all negative equality literals can be eliminated.

Example 3. For DL axiom $\alpha = \exists R.\{z\} \sqcap \{z\} \sqsubseteq C$, where $\{z\}$ is a nominal schema, $\pi(\alpha) = O(z) \rightarrow [\pi_x(\exists R.\{z\} \sqcap \{z\}) \rightarrow \pi_x(C)] = [O(z) \wedge (x \approx z) \wedge \pi_x(\exists R.\{z\})] \rightarrow \pi_x(C) = \neg O(x) \vee \neg R(x,x) \vee C(x)$.

3.2   Preprocessing

All the $\mathcal{ELHO}(\sqcap)$ axioms can be translated into normal forms in Table 3 in polynomial time using the structure transformation [20]. Table 4 shows all possible clause types appearing in $\mathcal{ELHO}(\sqcap)$ saturation. We first give the definition of $\Xi(KB)$, which denotes the FOL clause set of a $\mathcal{ELHO}(\sqcap)$ KB.

Definition 4. *The set of clauses $\Xi(KB)$, encoding an $\mathcal{ELHO}(\sqcap)$ knowledge base KB in FOL, is defined as follows:*

- For each ABox or RBox axiom α in KB , $\pi(\alpha) \in \Xi(KB)$.
- For each TBox axiom $C \sqsubseteq D$ in KB , $\pi(C \sqsubseteq D) \in \Xi(KB)$.
- For each TBox axiom $C \equiv D$ in KB , $\pi(C \sqsubseteq D) \in \Xi(KB)$ and $\pi(D \sqsubseteq C) \subseteq \Xi(KB)$.

Theorem 1. *Let KB be an $\mathcal{ELHO}(\sqcap)$ knowledge base. Then, the following claims hold:*

- KB *is satisfiable if and only if $\Xi(KB)$ is satisfiable.*
- $\Xi(KB)$ *can be computed in polynomial time in the $|KB|$.*

Proof. From Definition 4, equisatisfiability of KB and $\Xi(KB)$ is trivial to check [6]. All the $\mathcal{ELHO}(\sqcap)$ axioms can be translated into the normal forms of Table 3 in polynomial time, and translating from DL normal forms into first order logic clauses is in polynomial time. Therefore, $\Xi(KB)$ can be computed in time polynomial in the $|KB|$.

3.3 Deciding $\mathcal{ELHO}(\sqcap)$

Now we are ready to show that the resolution procedure for $\mathcal{ELHO}(\sqcap)$ is in time polynomial in $|KB|$.

Definition 5. *Let $\mathcal{R}_{DL}$ denote the ordered resolution calculus $\mathcal{R}$ with positive superposition parameterized as follows:*

- *The literal ordering is an admissible ordering $\succ$ such that $f \succ c \succ R \succ A$, for all function symbol f, constant symbol c, binary predicate symbol R and unary predicate symbol A.*
- *The selection function selects every negative maximal binary literal in each clause.*

Next, we enumerate all $\mathcal{R}_{DL}$ inferences between clauses and show that every conclusion is one of clause types of Table 4. With [n, m] $\rightsquigarrow$ [k] we denote an inference between clause type n and m resulting in clause type k, where n, m, k are integers.

Lemma 1. *Each $\mathcal{R}_{DL}$ inference, when applied to $\mathcal{ELHO}(\sqcap)$ -clauses, produces a $\mathcal{ELHO}(\sqcap)$ -clause type in Table 4. The maximum length of each clause is 3. And the number of clauses different up to variable renaming is polynomial in $|KB|$.*

Table 4. $\mathcal{ELHO}(\sqcap)$ -clause types

(1) $\neg A(x)$	(11) $\neg A(x) \vee f(x) \approx a$
(2) $C(x)$	(12) $a \approx b$
(3) $\neg A(x) \vee C(x)$	(13) $\neg R(x,y) \vee S(x,y)$
(4) $\neg A(x) \vee \neg B(x) \vee C(x)$	(14) $\neg R(x,y) \vee \neg S(x,y) \vee T(x,y)$
(5) $\neg R(x,y) \vee \neg A(y) \vee C(x)$	(15) $(\neg)A(a)$
(6) $\neg A(x) \vee R(x,f(x))$	(16) $(\neg)R(a,b)$
(7) $\neg A(x) \vee B(f(x))$	(17) $\neg A(x) \vee \neg R(x,f(x)) \vee S(x,f(x))$
(8) $\neg A(x) \vee R(x,a)$	(18) $\neg A(x) \vee \neg B(f(x)) \vee C(f(x))$
(9) $\neg R(x,a) \vee A(x)$	(19) $\neg A(x) \vee \neg B(f(x)) \vee C(x)$
(10) $\neg A(x) \vee x \approx a$	

Proof. The ordered resolution inferences are possible between the following clauses. $[2,3] \rightsquigarrow [2]$, $[2,4] \rightsquigarrow [3]$. $[6,5] \rightsquigarrow [19]$, $[6,13] \rightsquigarrow [6]$, $[6,14] \rightsquigarrow [17]$. $[7,3] \rightsquigarrow [7]$, $[7,4] \rightsquigarrow [18]$, $[7,10] \rightsquigarrow [11]$, $[7,15] \rightsquigarrow [1]$. $[8,5] \rightsquigarrow [3]$ with unifying x to a, $[8,9] \rightsquigarrow [3]$, $[8,13] \rightsquigarrow [8]$, $[8,14] \rightsquigarrow [17]$ with unifying x to a, $[8,16] \rightsquigarrow [15]$. $[15,1] \rightsquigarrow \bot$, $[15,3] \rightsquigarrow [15]$, $[15,4] \rightsquigarrow [3]$ with unifying x to a, $[15,15] \rightsquigarrow \bot$. $[16,5] \rightsquigarrow [3]$ with unifying x to a and y to b, $[16,9] \rightsquigarrow [15]$, $[16,16] \rightsquigarrow [15]$.

The positive superposition inferences are possible between the following clauses. $[6,11] \rightsquigarrow [8]$, $[7,11] \rightsquigarrow [3]$ with unifying x to a, $[8,12] \rightsquigarrow [8]$. $[10,12] \rightsquigarrow [10]$.

(18) $\neg A(x) \vee \neg B(f(x)) \vee C(f(x))$ can only resolve with clause (7) $\neg A(x) \vee B(f(x))$ or (2) $B(x)$, and produce clause (7) $\neg A(x) \vee C(f(x))$. Since ordered resolution only resolves on maximal literals, thus literal $\neg A(x)$ in clause type (7) can never participate. In addition, due to that every function symbol is unique after skolemization, there is no other clauses in clause type (7) containing $B(f(x))$. Since $\neg B(f(x))$ in (18) has to resolve with $B(f(x))$ or $B(x)$, (18) $\neg A(x) \vee \neg B(f(x)) \vee C(f(x))$ can only resolve with clause $\neg A(x) \vee B(f(x))$ or $B(x)$. Similarly, (17) $\neg A(x) \vee \neg S(x,f(x)) \vee T(x,f(x))$ can only resolve with (6) $\neg A(x) \vee S(x,f(x))$ producing (6) $\neg A(x) \vee T(x,f(x))$, (19) $\neg A(x) \vee \neg B(f(x)) \vee C(x)$ can only resolve with (2) and (3) producing (3).

Any other inferences are not applicable. Therefore, every clause is one of the clause types of Table 4, and the maximum length of clauses is 3. Let c be the number of unary predicates, r the number of binary predicates, f the number of unary function symbols, and i the number of constants in the signature of $\Xi(KB)$. Then, trivially c, r, f and i are linear in $|KB|$. Consider now the maximal $\mathcal{ELHO}(\sqcap)$ -clause of type 5 in Table 4. There are possibly at most rc^2 clauses of type 5. The number of clauses is polynomial in $|KB|$. For other $\mathcal{ELHO}(\sqcap)$ -clause types, the bounds on the length and on the number of clauses can be derived in an analogous way. Therefore, the number of $\mathcal{ELHO}(\sqcap)$ -clauses different up to variable renaming is polynomial in $|KB|$.

Theorem 2. *For an $\mathcal{ELHO}(\sqcap)$ knowledge base KB, saturating $\Xi(KB)$ by $\mathcal{R}_{DL}$ decides satisfiability of KB and runs in time polynomial in $|KB|$.*

Proof. The number of clauses by translating KB is polynomial in $|KB|$. By Lemma 1, the length of every clauses derivable by $\mathcal{R}_{DL}$ is at most 3. And each inference can be performed polynomially. Hence, the saturation terminates in polynomial time. Since $\mathcal{R}_{DL}$ is sound and complete [3], therefore $\mathcal{R}_{DL}$ decides satisfiability of $\Xi(KB)$ in time polynomial in $|KB|$.

4   Deciding Satisfiability of $\mathcal{ELHOV}_n(\sqcap)$ by Resolution Procedure

$\mathcal{ELHOV}_n(\sqcap)$ axioms may contain several nominal schemas or one nominal schema appearing in different positions of an axiom. In such situation, normalization of axioms becomes difficult. For example, $\exists R.(C \sqcap \exists S.\{z\}) \sqsubseteq \exists R.\{z\}$, since $\{z\}$ binds to the same variable, the axiom can not be normalized. $\exists R.(C \sqcap \exists S.\{z\}) \sqsubseteq \exists R.\{z\}$ has to be translated into first order logic directly, which is $\neg O(z) \vee \neg R(x,y) \vee \neg C(y) \vee \neg S(y,z) \vee R(x,z)$. Hence, there are possibly very complex clauses. In order to solve such issue, we use a lifting lemma to show show the resolution procedure for $\mathcal{ELHOV}_n(\sqcap)$ is still polynomial in $|KB|$.

In general, the lifting lemma states that reasoning on a $\mathcal{ELHOV}_n(\sqcap)$ KB without grounding nominal schemas takes fewer steps or produces fewer clauses than reasoning on the grounding KB. Since after grounding all the nominal schemas to nominals, $\mathcal{ELHOV}_n(\sqcap)$ KB becomes actually $\mathcal{ELHO}(\sqcap)$ KB. And since we already showed that the resolution procedure for $\mathcal{ELHO}(\sqcap)$ is polynomial in Theorem 2, therefore reasoning on a $\mathcal{ELHOV}_n(\sqcap)$ KB is still polynomial.

At first, we need to define *safe environment* and $\mathsf{ground}^+(KB)$. The intuition behind of *safe environment* is to restrict the KB with tree-shaped dependencies in order to avoid exponential blow-up (see details in [16]). Then, we show that the size of $\mathsf{ground}^+(KB)$ is polynomial in $|KB|$.

Definition 6. *An occurrence of a nominal schema x in a concept C is safe if C has a sub-concept of the form $\{a\} \sqcap \exists R.D$ for some $a \in N_I$, such that D contains the occurrence of $\{x\}$ but no other occurrence of any nominal schema. In this case, $\{a\} \sqcap \exists R.D$ is a safe environment for this occurrence of $\{x\}$. $S(a,x)$ will sometimes be used to denote an expression of the form $\{a\} \sqcap \exists R.D$ within which $\{x\}$ occurs safe.*

Definition 7. *We define a $\mathcal{ELHOV}_n(\sqcap)$ knowledge base $\mathsf{ground}^+(KB)$ as follows. The RBox and ABox of $\mathsf{ground}^+(KB)$ are the same as the RBox and ABox of KB. For each TBox axiom $\alpha = C \sqsubseteq D \in KB$, the following axioms are added to $\mathsf{ground}^+(KB)$:*

1. *For each nominal schema $\{x\}$ safe for α, with safe occurrences in environments $S_i(a_i, x)$ for $i = 1, \ldots, l$, introduce a fresh concept name $O_{x,\alpha}$. For every individual $b \in N_I$ in KB, $\mathsf{ground}^+(KB)$ contains an axiom*

$$\bigcap_{i=1}^{l} \exists U.S_i(a_i, b) \sqsubseteq \exists U.(\{b\} \sqcap O_{x,\alpha}),$$

2. *A concept C' is obtained from C as follow. Initialize $C' := C$. For each nominal schema $\{x\}$ that is safe for α: (a) replace all safe occurrences $S(a, x)$ in C' by $\{a\}$; (b) replace the non-safe occurrence (if any) of $\{x\}$ in C' by $O_{x,\alpha}$; (c) set $C' := C' \sqcap \exists U.O_{x,\alpha}$. After these steps, C' contains only nominal schemas that are not safe for α, and neither for $C' \sqsubseteq D$.*

Now add axioms ground$(C' \sqsubseteq D)$ *to* ground$^+(KB)$.

Theorem 3. *Given a $\mathcal{ELHOV}_n(\sqcap)$ knowledge base KB, the size of* ground$^+$ (KB) *is exponential in n and polynomial in $|KB|$ [16].*

From Theorem 3, we know that deciding $\mathcal{ELHOV}_n(\sqcap)$ is in polynomial time. We also showed that resolution for $\mathcal{SROEL}(\sqcap_s, \times)$ is a polynomial algorithm in Section 3. Now, we are ready to bridge the gap by the lifting lemma. Before giving the lifting lemma, the ordered resolution parameters must be redefined.

Definition 8. *Let $\mathcal{R}^O_{DL}$ denote the resolution calculus $\mathcal{R}_{DL}$ parameterized as follows:*

- The literal ordering is an admissible ordering $\succ$ such that $f \succ c \succ R \succ O \succ A$, for all function symbol f, constant symbol c, binary predicate symbol R, unary predicate symbol A and first order logic predicate O for nominal schemas.
- The selection function selects every negative maximal binary literal in each clause.

Lemma 2 (lifting lemma). *For a $\mathcal{ELHOV}_n(\sqcap)$ knowledge base KB, clause $C \in \Xi(KB)$ and $D \in \Xi(KB)$, if C can resolve with D, then there must exist at least one resolution inference between a clause $C' \in$ ground(C) and $D' \in$ ground(D).*

Proof. We show this lemma by proving contradiction, which is to show the statement that if there is no clause $C' \in$ ground(C) can resolve with clause $D' \in$ ground(D), then C can not resolve with D. Without losing generality, we assume C in clause type (n) and D in clause type (m). We denote $[n \not\sim m]$ by clause type (n) cannot resolve with clause type (m). There are two possibilities.

- Clause type (n) and clause type (m) can resolve, but the resolved literals are not on the same predicate name. So no matter the clauses are grounded or not, they cannot resolve.
- Clause type (n) and clause type (m) cannot resolve. So we need to show they cannot resolve even before grounding. We enumerate all the impossible resolution cases. For ordered resolution inference, $[2 \not\sim 5]$, because before grounding, $\neg A(x)$ is not selected and not the maximal literal, so $C(x)$ in (2) cannot resolve with $\neg A(x)$ in (5). Similarly, we have $[2 \not\sim 6, 7, 8, 10, 11]$ and $[7 \not\sim 5, 6, 7, 8, 9, 10, 11]$. For positive superposition, $[6 \not\sim 10, 12]$, because violate the condition of positive superposition before grounding. Similarly, $[7 \not\sim 10, 12]$ and also $[10 \not\sim 11]$.

Therefore, if there is no clause $C' \in$ ground(C) which can resolve with clause $D' \in$ ground(D), then C can not resolve with D. Hence, as for clause $C \in \Xi(KB)$ and $D \in \Xi(KB)$, if C can resolve with D, then there must exist at least one resolution inference between a clause $C' \in$ ground(C) and $D' \in$ ground(D).

Theorem 4. *Given an $\mathcal{ELHOV}_n(\sqcap)$ knowledge base KB, saturating $\Xi(KB)$ by $\mathcal{R}^O_{DL}$ decides satisfiability of KB and runs in time polynomial in $|KB|$.*

Proof. By the lifting lemma, reasoning on clauses before grounding take fewer steps than the clauses after grounding. By theorem 3, we know that reasoning on ground$^+(KB)$ is in polynomial time in $|KB|$. We also know that the resolution procedure for $\mathcal{ELHO}(\sqcap)$ is in polynomial time by Theorem 2 and ground$^+(KB)$ is a $\mathcal{ELHO}(\sqcap)$ KB. Therefore, the resolution procedure for $\mathcal{ELHOV}_n(\sqcap)$ KB takes fewer steps than ground$^+(KB)$, and so $\mathcal{R}^O_{DL}$ decides satisfiability of $\Xi(KB)$ in time polynomial in $|KB|$.

The proof is closely relevant with the order and selection parameter of ordered resolution. If we change the setting of the parameters, the lifting lemma might not hold.

To the best of our knowledge, this parameter setting of the order and select function is best. If O has the highest order among all predicates, then the clauses which contain O cannot resolve with others unless $O(x)$ literals are resolved. Thus, it has no difference with *full grounding* method, because resolving $O(x)$ is actually grounding x with all known individuals. If $f \succ O \succ P$, where P denotes the DL predicate name, $\neg A(x) \vee R(x, f(x))$ and $\neg O(z) \vee \neg R(x, z) \vee S(x, z)$ cannot resolve. That is to say, we still need to ground $O(x)$ in some clauses. If we set the parameter as $f \succ R \succ O \succ A$, $\neg A(x) \vee R(x, f(x))$ and $\neg O(z) \vee \neg R(x, z) \vee S(x, z)$ can resolve. So it delays the grounding even later. However, if we force O to be the lowest order, there are undesired clauses violating termination. Therefore, we choose $f \succ R \succ O \succ A$ as the order.

There are several reasons that resolution procedure can be much more efficient than the naive *full grounding* method. First of all, the number of clauses translated from $\mathcal{ELHOV}_n(\sqcap)$ knowledge base KB is much fewer than the number of clauses of ground$^+(KB)$. Secondly, some clauses cannot do any further resolution, such that they can be seen as redundant clauses. For example, resolving $\neg A(x) \vee R(x, f(x))$ and $\neg O(z) \vee \neg R(x, z) \vee B(x)$ produces a clause containing $\neg O(f(x))$. However, $\neg O(f(x))$ cannot resolve with any others, because there are only positive literal of $O(a)$ in KB and thus $\neg O(f(x))$ cannot unify with others. For the resolution procedure, we can even apply the powerful redundancy technique to reduce the number of clauses [6], e.g., all the tautologies can be removed directly in the resolution procedure.

After saturation, we can reduce all the clauses into a disjunctive datalog program DD(KB). The program DD(KB) entails the same set of ground facts as KB, Thus, instance checking in KB can be performed as query answering in DD(KB). Database systems usually contain Datalog reasoning and compute query answers in one pass efficiently, either bottom-up or top-down. Especially, when KB containing nominal schemas, the decision procedure needs to do a lot of Instance

Checking implicitly. So resolution approach is particularly suitable for nominal schemas. Comparatively, tableau algorithms might need to run for each individual in ABox. (see more details about disjunctive datalog program in [6]).

5   Examples

We now present a rather simple example that points out how the resolution procedure works and why it's more efficient than *full grounding* approach in general. Intuitively, our approach delays grounding only when it's necessary to do so.

Consider the following clearly unsatisfiable KB containing nominal schema.

$KB = \{$

$\exists hasParent.\{z\} \sqcap \exists hasParent.\exists married.\{z\} \sqcap Teen \sqsubseteq Child,$

$hasParent(john, mark),$

$hasParent(john, mary),$

$married(mary, mark),$

$Teen(john),$

$\neg Child(john)\}$

We first translate all the DL axioms into first order logic clauses.

$\varXi(KB) = \{$

(1) $\neg O(z) \vee \neg hasParent(x, z) \vee \neg hasParent(x, y) \vee \neg married(y, z)$
$\vee \neg Teen(x) \vee Child(x)$

(2) $hasParent(john, mark),$

(3) $hasParent(john, mary),$

(4) $married(mary, mark),$

(5) $Teen(john),$

(6) $\neg Child(john)\}$

$\varXi(KB)$ also contains $O(john)$, $O(mary)$ and $O(mark)$ because they are known individuals. By saturating $\varXi(KB)$ we obtain the following clauses (the notation R(n,m) means that a clause is derived by resolving clauses n and m):

(7) $\neg O(mark) \vee \neg hasParent(john, y) \vee \neg married(y, mark) \vee \neg Teen(john)$
$\vee Child(john)$ R(1,2)

(8) $\neg O(mark) \vee \neg married(mary, mark) \vee \neg Teen(john)$
$\vee Child(john)$ R(7,3)

(9) $\neg O(mark) \vee \neg Teen(john) \vee Child(john)$ R(8,4)

Since O symbol has a higher order than unary predicate, (9) $\neg O(mark) \vee \neg Teen(john) \vee Child(john)$ will resolve with $O(mark)$ and produce (10) $\neg Teen(john) \vee Child(john)$.

(11) $Child(john)$ R(10,5)

(12) $\bot$ R(11,6)

Now, we can see that the O symbol literal is resolved at the very last. That is to say, the grounding of nominal schemas in such procedure has been delayed. However, if we use the *full grounding* approach, the KB will contain clause $\exists hasParent.\{john\} \sqcap \exists hasParent.\exists married.\{john\} \sqcap Teen \sqsubseteq Child$, which is absolutely unnecessary for inference. Consider KB has even more irrelevant known individuals, it will generate even more useless clauses. Therefore, clearly, our approach is much better than the *full grounding* one.

6   Related Work and Discussion

There are several algorithms particularly for $\mathcal{EL}$ family, but none of them can be easily to extend to DLs containing nominal schemas. For $\mathcal{SROEL}(\sqcap_s, \times)$, the only reasoning approach was proposed in [15]. Instead of using traditional tableau method, all DL axioms are rewritten into a set of datalog rules. However, it is unclear how to translate nominal schemas into such rules. Since all of the rewriting rules apply on the normal forms, but axioms containing nominal schemas are not able to be normalized to the best of our knowledge. Also, we do not know how to extend completion-rule based algorithm [1] and the recently concurrent algorithm [9,8], because they also need to normalize axioms at first. In [13], the author tried to apply a selective and delayed grounding technique to decide $\mathcal{ALCO}$ extended with nominal schemas. The advantage is the technique can easily extend to more expressive logic, such as $\mathcal{SROIQV}_n$. But it is hardly to say such algorithm is suitable for DLs with nominal schemas, because one has to find a very good heuristics for grounding choices.

When we want to extend our resolution procedure to capture general role chain, it becomes much more difficult. The possible cyclic role chain axioms can do self-resolution, which prevent termination. For example, a transitive relation S satisfies $S(a, b) \land S(b, c) \rightarrow S(a, c)$ can resolve with itself to yield a new clause $S(a, b) \land S(b, c) \land S(c, d) \rightarrow S(a, d)$ and so on. In [10], the problem was partially solved by eliminating transitive role with another equisatisfiable axiom containing universal quantifier. In [2], the authors developed a so-called ordered chaining rule based on standard techniques from term rewriting. It can deal with binary relations with composition laws of the form $R \circ S \sqsubseteq U$ in the context of resolution-type theorem proving. However, the approach applies even more restricted order on role predicates than the acyclic role chains in $\mathcal{SROIQ}$. So it can not solve the problem of general role chain neither. To the best of our knowledge, there is no resolution procedure that can deal with general role chain. And hence, it becomes to our next goal.

The extension of our algorithm to deal with cross-products becomes intractable for conjunction of roles. The reason is that using extended role hierarchies, it is possible to express inclusion axioms with universal value restrictions of form $C \sqsubseteq \forall R.D$, or equivalently, inclusion axioms with inverse roles of form $\exists R^-.C \sqsubseteq D$ which were shown to cause intractability in [1]. Indeed, these axioms are expressible using three inclusion axioms: $C \times \top \sqsubseteq S$, $S \sqcap R \sqsubseteq H$ and $H \sqsubseteq \top \times D$, where S and H are fresh role names [7].

The extension of $\mathcal{ELHO}(\sqcap)$ with self role should not affect tractability, although it may cause clauses have longer length. For example, consider the following $KB = \{C \sqsubseteq \exists R.\mathsf{Self}, D \sqsubseteq \exists S.\mathsf{Self}, R \sqcap S \sqsubseteq T\}$. After saturation, $\Xi(KB)$ contains $\neg C(x) \lor \neg D(x) \lor T(x, x)$, which may resolve with other role conjunction axioms and so forth. So it is possible to have clauses with longer length, like $\neg C_1(x) \vee \ldots \vee \neg C_n(x) \vee R(x,x)$. However, we conjecture that the number n in $\neg C_1(x) \vee \ldots \vee \neg C_n(x) \vee R(x,x)$ is linear to the number of concept names in KB. So the resolution procedure for $\mathcal{ELHO}(\sqcap)$ with self role should be still in polynomial time. Due to that we want to keep this paper easier to read, we disallow self role constructor.

More problems will occur when extending to more expressive description logic $\mathcal{SROIQV}_n$. When translating $\mathcal{SROIQV}_n$ into first order logic, since one nominal schema can appear at different positions in an axiom, such that the number of corresponding FOL clauses can be exponential blow-up. For example, for such a DL axiom, $(\exists R_1.\{z\} \sqcup \exists S_1.\{z\}) \sqcap \ldots \sqcap (\exists R_n.\{z\} \sqcup \exists S_n.\{z\}) \sqsubseteq C$, since we cannot normalize it into smaller axioms, it has to be translated into a number of clauses in conjunctive normal form, and the number of such clauses is exponential blow-up.

Although theoretically optimal, the resolution procedure may not be scalable in practice. The reason seems to be that, despite optimizations, resolution still produces many unnecessary clauses (see discussion in [25]). Another algorithm, called hypertableau, seems to be very potential to efficiently deal with nominal schemas. Hypertableau algorithm takes unnormalized *DL-clauses* to infer based on the hypertableau rule. It can avoid unnecessary nondeterminism and the construction of large models, which are two primary sources of inefficiency in the tableau-based reasoning calculi [22]. We believe that the idea of the lift lemma can also work for hypertableau method, such that we may use the similar way to prove the feasibility of hypertableau for nominal schemas.

Nominal schemas have even more good properties. In [11,12], the author describes nominal schemas allow not only for a concise reconciliation of OWL and rules, but also that the integration can in fact be lifted to cover established closed world formalisms on both the OWL and the rules side. More precisely, they endow $\mathcal{SROIQ}$ with both nominal schemas and a generalized semantics based on the logic of minimal knowledge and negation as failure (MKNF). The latter is non-monotonic and captures both open and closed world modeling.

7   Conclusion and Future work

In this paper, we provide a polynomial resolution procedure for the description logic language $\mathcal{ELHOV}_n(\sqcap)$. We show that the algorithm is sound, complete and tractable. For future work, the main task is to implement the algorithm and compare it with the tableau approach with selective grounding strategy. We will also look into the hypertableau method to see if it can be extended. In general, we hope to develop a more efficient algorithm to be applicable for $\mathcal{SROELV}_n(\sqcap_s, \times)$, $\mathcal{SROIQV}_n(\sqcap_s, \times)$ and even more powerful DL languages.

Acknowledgements. This work was supported by the National Science Foundation under award 1017225 "III: Small: TROn—Tractable Reasoning with Ontologies.

References

1. Baader, F., Brandt, S., Lutz, C.: Pushing the el envelope. In: Kaelbling, L.P., Saffiotti, A. (eds.) IJCAI, pp. 364–369. Professional Book Center (2005)
2. Bachmair, L., Ganzinger, H.: Ordered chaining for total orderings. In: Bundy, A. (ed.) CADE 1994. LNCS, vol. 814, pp. 435–450. Springer, Heidelberg (1994)
3. Bachmair, L., Ganzinger, H.: Resolution theorem proving. In: Robinson, J.A., Voronkov, A. (eds.) Handbook of Automated Reasoning, pp. 19–99. Elsevier, MIT Press (2001)
4. Boley, H., Hallmark, G., Kifer, M., Paschke, A., Polleres, A., Reynolds, D. (eds.): RIF Core Dialect. W3C Recommendation (June 22, 2010), `http://www.w3.org/TR/rif-core/`
5. Horrocks, I., Patel-Schneider, P., Boley, H., Tabet, S., Grosof, B., Dean, M.: SWRL: A Semantic Web Rule Language. W3C Member Submission (May 21, 2004), `http://www.w3.org/Submission/SWRL/`
6. Hustadt, U., Motik, B., Sattler, U.: Reasoning for Description Logics around SHIQ in a Resolution Framework. Tech. Rep. 3-8-04/04, FZI, Germany (2004)
7. Kazakov, Y.: Saturation-based decision procedures for extensions of the guarded fragment. Ph.D. thesis, Saarländische Universitäts- und Landesbibliothek, Postfach 151141, 66041 Saarbrücken (2005), `http://scidok.sulb.uni-saarland.de/volltexte/2007/1137`
8. Kazakov, Y., Krötzsch, M., Simančík, F.: Practical reasoning with nominals in the $\mathcal{EL}$ family of description logics. In: Brewka, G., Eiter, T., McIlraith, S.A. (eds.) Proceedings of the 13th International Conference on Principles of Knowledge Representation and Reasoning (KR 2012), pp. 264–274. AAAI Press (2012)
9. Kazakov, Y., Krötzsch, M., Simančík, F.: Concurrent classification of el ontologies (2011) (to appear)
10. Kazakov, Y., Motik, B.: A resolution-based decision procedure for $\mathcal{SHOIQ}$. In: Furbach, U., Shankar, N. (eds.) IJCAR 2006. LNCS (LNAI), vol. 4130, pp. 662–677. Springer, Heidelberg (2006)
11. Knorr, M., Alferes, J.J., Hitzler, P.: Local closed world reasoning with description logics under the well-founded semantics. Artif. Intell. 175(9-10), 1528–1554 (2011)
12. Knorr, M., Martínez, D.C., Hitzler, P., Krisnadhi, A.A., Maier, F., Wang, C.: Recent advances in integrating owl and rules (technical communication). In: Krötzsch, Straccia (eds.) [18], pp. 225–228
13. Krisnadhi, A., Hitzler, P.: A tableau algorithm for description logics with nominal schema. In: Krötzsch, Straccia (eds.) [18], pp. 234–237
14. Krisnadhi, A., Maier, F., Hitzler, P.: Owl and rules. In: Polleres, A., d'Amato, C., Arenas, M., Handschuh, S., Kroner, P., Ossowski, S., Patel-Schneider, P. (eds.) Reasoning Web 2011. LNCS, vol. 6848, pp. 382–415. Springer, Heidelberg (2011)
15. Krötzsch, M.: Efficient inferencing for OWL EL. In: Janhunen, T., Niemelä, I. (eds.) JELIA 2010. LNCS (LNAI), vol. 6341, pp. 234–246. Springer, Heidelberg (2010)
16. Krötzsch, M., Maier, F., Krisnadhi, A., Hitzler, P.: A better uncle for owl: nominal schemas for integrating rules and ontologies. In: Srinivasan, S., Ramamritham, K., Kumar, A., Ravindra, M.P., Bertino, E., Kumar, R. (eds.) WWW, pp. 645–654. ACM (2011)
17. Krötzsch, M., Rudolph, S., Hitzler, P.: ELP: Tractable rules for OWL 2. In: Sheth, A.P., Staab, S., Dean, M., Paolucci, M., Maynard, D., Finin, T., Thirunarayan, K. (eds.) ISWC 2008. LNCS, vol. 5318, pp. 649–664. Springer, Heidelberg (2008)

18. Krötzsch, M., Straccia, U. (eds.): RR 2012. LNCS, vol. 7497. Springer, Heidelberg (2012)
19. Martinezi, D.C., Krisnadhi, A., Maier, F., Sengupta, K., Hitzler, P.: Reconciling owl and rules. Tech. rep. Kno.e.sis Center, Wright State University, Dayton, OH, U.S.A. (2011), http://www.pascal-hitzler.de/
20. Motik, B.: Reasoning in Description Logics using Resolution and Deductive Databases. Ph.D. thesis, Universität Karlsruhe (TH), Karlsruhe, Germany (January 2006)
21. Motik, B., Sattler, U., Studer, R.: Query answering for owl-dl with rules. In: McIl-raith, S.A., Plexousakis, D., van Harmelen, F. (eds.) ISWC 2004. LNCS, vol. 3298, pp. 549–563. Springer, Heidelberg (2004)
22. Motik, B., Shearer, R., Horrocks, I.: Hypertableau reasoning for description logics. J. Artif. Intell. Res. (JAIR) 36, 165–228 (2009)
23. Dal Palù, A., Dovier, A., Pontelli, E., Rossi, G.: Answer set programming with constraints using lazy grounding. In: Hill, P.M., Warren, D.S. (eds.) ICLP 2009. LNCS, vol. 5649, pp. 115–129. Springer, Heidelberg (2009)
24. Rudolph, S., Krötzsch, M., Hitzler, P.: Cheap boolean role constructors for de-scription logics. In: Hölldobler, S., Lutz, C., Wansing, H. (eds.) JELIA 2008. LNCS (LNAI), vol. 5293, pp. 362–374. Springer, Heidelberg (2008)
25. Simancik, F., Kazakov, Y., Horrocks, I.: Consequence-based reasoning beyond horn ontologies. In: Walsh, T. (ed.) IJCAI, pp. 1093–1098. IJCAI/AAAI (2011)

Get My Pizza Right: Repairing Missing is-a Relations in $\mathcal{ALC}$ Ontologies

Patrick Lambrix, Zlatan Dragisic, and Valentina Ivanova

Department of Computer and Information Science
and Swedish e-Science Research Centre
Linköping University, 581 83 Linköping, Sweden
firstname.lastname@liu.se
http://www.ida.liu.se/

Abstract. With the increased use of ontologies in semantically-enabled applications, the issue of debugging defects in ontologies has become increasingly important. These defects can lead to wrong or incomplete results for the applications. Debugging consists of the phases of detection and repairing. In this paper we focus on the repairing phase of a particular kind of defects, i.e. the missing relations in the is-a hierarchy. Previous work has dealt with the case of taxonomies. In this work we extend the scope to deal with $\mathcal{ALC}$ ontologies that can be represented using acyclic terminologies. We present algorithms and discuss a system.

1 Introduction

Developing ontologies is not an easy task, and often the resulting ontologies are not consistent or complete. Such ontologies, although often useful, also lead to problems when used in semantically-enabled applications. Wrong conclusions may be derived or valid conclusions may be missed. Defects in ontologies can take different forms (e.g. [16]). Syntactic defects are usually easy to find and to resolve. Defects regarding style include such things as unintended redundancy. More interesting and severe defects are the modeling defects which require domain knowledge to detect and resolve, and semantic defects such as unsatisfiable concepts and inconsistent ontologies. Most work up to date has focused on debugging (i.e. detecting and repairing) the semantic defects in an ontology (e.g. [16,15,29,5]). Modeling defects have mainly been discussed in [2,20,19] for taxonomies, i.e. from a knowledge representation point of view, a simple kind of ontologies. The focus has been on defects regarding the is-a structure of the taxonomies. In this paper we tackle the problem of repairing the is-a structure of $\mathcal{ALC}$ ontologies that can be represented using acyclic terminologies.

In addition to its importance for the correct modeling of a domain, the structural information in ontologies is also important in semantically-enabled applications. For instance, the is-a structure is used in ontology-based search and annotation. In ontology-based search, queries are refined and expanded by moving up and down the hierarchy of concepts. Incomplete structure in ontologies

H. Takeda et al. (Eds.): JIST 2012, LNCS 7774, pp. 17–32, 2013.

influences the quality of the search results. As an example, suppose we want to find articles in the MeSH (Medical Subject Headings [25], controlled vocabulary of the National Library of Medicine, US) Database of PubMed [27] using the term *Scleral Diseases* in MeSH. By default the query will follow the hierarchy of MeSH and include more specific terms for searching, such as *Scleritis*. If the relation between *Scleral Diseases* and *Scleritis* is missing in MeSH, we will miss 738 articles in the search result, which is about 55% of the original result. The structural information is also important information in ontology engineering research. For instance, most current ontology alignment systems use structure-based strategies to find mappings between the terms in different ontologies (e.g. overview in [21]) and the modeling defects in the structure of the ontologies have an important influence on the quality of the ontology alignment results.

Debugging modeling defects in ontologies consists of two phases, detection and repair. There are different ways to detect missing is-a relations. One way is by inspection of the ontologies by domain experts. Another way is to use external knowledge sources. For instance, there is much work on finding relationships between terms in the ontology learning area [3]. In this setting, new ontology elements are derived from text using knowledge acquisition techniques. Regarding the detection of is-a relations, one paradigm is based on linguistics using lexico-syntactic patterns. The pioneering research conducted in this line is in [11], which defines a set of patterns indicating is-a relationships between words in the text. Another paradigm is based on machine learning and statistical methods. Further, guidelines based on logical patterns can be used [5]. When the ontology is part of a network of ontologies connected by mappings between the ontologies, knowledge inherent in the ontology network can be used to detect missing is-a relations using logical derivation [2,20,19]. However, although there are many approaches to detect missing is-a relations, these approaches, in general, do not detect *all* missing is-a relations. For instance, although the precision for the linguistics-based approaches is high, their recall is usually very low.

In this paper we assume that the detection phase has been performed. We assume that we have obtained a set of missing is-a relations for a given ontology and focus on the repairing phase. In the case where our set of missing is-a relations contains *all* missing is-a relations, the repairing phase is easy. We just add all missing is-a relations to the ontology and a reasoner can compute all logical consequences. However, when the set of missing is-a relations does not contain all missing is-a relations - and this is the common case - there are different ways to repair the ontology. The easiest way is still to just add the missing is-a relations to the ontology. For instance, Figure 1 shows a small part of a pizza ontology based on [26], that is relevant for our discussions. Assume that we have detected that MyPizza $\sqsubseteq$ FishyMeatyPizza and MyFruttiDiMare $\sqsubseteq$ NonVegetarianPizza are missing is-a relations. Obviously, adding MyPizza $\sqsubseteq$ FishyMeatyPizza and MyFruttiDiMare $\sqsubseteq$ NonVegetarianPizza to the ontology will repair the missing is-a structure. However, there are other more interesting possibilities. For instance, adding AnchoviesTopping $\sqsubseteq$ FishTopping and ParmaHamTopping $\sqsubseteq$ MeatTopping will also repair the missing is-a structure.

Pizza $\sqsubseteq$ $\top$
PizzaTopping $\dot{\sqsubseteq}$ $\top$
hasTopping $\sqsubseteq$ $\top \times \top$
AnchoviesTopping $\dot{\sqsubseteq}$ PizzaTopping
MeatTopping $\dot{\sqsubseteq}$ PizzaTopping
HamTopping $\sqsubseteq$ MeatTopping
ParmaHamTopping $\dot{\sqsubseteq}$ PizzaTopping
FishTopping $\dot{\sqsubseteq}$ PizzaTopping $\sqcap$ ¬MeatTopping
TomatoTopping $\dot{\sqsubseteq}$ PizzaTopping $\sqcap$ ¬MeatTopping $\sqcap$ ¬FishTopping
GarlicTopping $\dot{\sqsubseteq}$ PizzaTopping $\sqcap$ ¬MeatTopping $\sqcap$ ¬FishTopping
MyPizza $\doteq$ Pizza $\sqcap$ $\exists$ hasTopping.AnchoviesTopping $\sqcap$ $\exists$ hasTopping.ParmaHamTopping
FishyMeatyPizza $\doteq$ Pizza $\sqcap$ $\exists$ hasTopping.FishTopping $\sqcap$ $\exists$ hasTopping.MeatTopping
MyFruttiDiMare $\doteq$ Pizza $\sqcap$ $\exists$ hasTopping.AnchoviesTopping
  $\sqcap$ $\exists$ hasTopping.GarlicTopping $\sqcap$ $\exists$ hasTopping.TomatoTopping
  $\sqcap$ $\forall$ hasTopping.(AnchoviesTopping $\sqcup$ GarlicTopping $\sqcup$ TomatoTopping)
VegetarianPizza $\doteq$ Pizza $\sqcap$ ¬ $\exists$ hasTopping.FishTopping $\sqcap$ ¬ $\exists$ hasTopping.MeatTopping
NonVegetarianPizza $\doteq$ Pizza $\sqcap$ ¬VegetarianPizza

Fig. 1. A pizza ontology

Another more informative way[1] to repair the missing is-a structure is to add AnchoviesTopping $\dot{\sqsubseteq}$ FishTopping and ParmaHamTopping $\dot{\sqsubseteq}$ HamTopping. Essentially, these other possibilities to repair the ontology include missing is-a relations (e.g. AnchoviesTopping $\dot{\sqsubseteq}$ FishTopping) that were not originally detected by the detection algorithm.[2] We also note that from a logical point of view, adding AnchoviesTopping $\dot{\sqsubseteq}$ MeatTopping and ParmaHamTopping $\dot{\sqsubseteq}$ FishTopping also repairs the missing is-a structure. However, from the point of view of the domain, this solution is not correct. Therefore, as it is the case for all approaches for debugging modeling defects, a domain expert needs to validate the logical solutions.

The contributions of this paper are threefold. First, we show that the problem of finding possible ways to repair the missing is-a structure in an ontology in general can be formalized as a generalized version of the TBox abduction problem (Section 3). Second, we propose an algorithm to generate different ways to repair the missing is-a structure in $\mathcal{ALC}$ ontologies that can be represented using acyclic terminologies (Section 4). Third, we discuss a system that allows a domain expert to repair the missing is-a structure in Section 5. We discuss the functionality and user interface of the repairing system and show an example run. Further, we discuss related work in Section 6 and conclude in Section 7. We continue, however, with some necessary preliminaries in Section 2.

2   Preliminaries

In this paper we deal with ontologies represented in the description logic $\mathcal{ALC}$ with acyclic terminologies (e.g. [1]). Concept descriptions are defined using con-

---

[1] This is more informative in the sense that the former is derivable from the latter. Adding ParmaHamTopping $\dot{\sqsubseteq}$ HamTopping, also allows to derive ParmaHamTopping $\dot{\sqsubseteq}$ MeatTopping as the ontology already includes HamTopping $\dot{\sqsubseteq}$ MeatTopping.
[2] Therefore, the approach discussed in this paper can also be seen as a detection method that takes already found missing is-a relations as input.

structors as well as concept and role names. As constructors $\mathcal{ALC}$ allows concept conjunction (C ⊓ D), disjunction (C ⊔ D), negation (¬ C), universal quantification (∀ r.C) and existential quantification (∃ r.C).[3] In this paper we consider ontologies that can be represented by a TBox that is an acyclic terminology. An acyclic terminology is a finite set of concept definitions (i.e. terminological axioms of the form $C \doteq D$ where C is a concept name) that neither contains multiple definitions nor cyclic definitions.[4] An ABox contains assertional knowledge, i.e. statements about the membership of individuals (interpreted as elements in the domain) to concepts (C(i)) as well as relations between individuals (r(i,j)).[5] A knowledge base contains a TBox and an ABox. A model of the TBox/ABox/knowledge base satisfies all axioms of the TBox/ABox/knowledge base. A knowledge base is consistent if it does not contain contradictions.

An important reasoning service is the checking of (un)satisfibility of concepts (a concept is unsatisfiable if it is necessarily interpreted as the empty set in all models of the TBox/ABox/knowledge base, satisfiable otherwise). A TBox is incoherent if it contains an unsatisfiable concept.

Checking satisfiability of concepts in $\mathcal{ALC}$ can be done using a tableau-based algorithm (e.g. [1]). To test whether a concept C is satisfiable such an algorithm starts with an ABox containing the statement C(x) where x is a new individual and it is usually assumed that C is normalized to negation normal form. It then applies consistency-preserving transformation rules to the ABox (Figure 2). The ⊓-, ∀- and ∃-rules extend the ABox while the ⊔-rule creates multiple ABoxes. The algorithm continues applying these transformation rules to the ABoxes until no more rules apply. This process is called completion and if one of the final ABoxes does not contain a contradiction (we say that it is open), then satisfiability is proven, otherwise unsatisfiability is proven. One way of implementing this approach is through completion graphs which are directed graphs in which every node represents an ABox. Application of the ⊔-rule produces new nodes with one statement each, while the other rules add statements to the node on which the rule is applied. The ABox for a node contains all the statements of the node as well as the statements of the nodes on the path to the root. Satisfiability is proven if at least one of the ABoxes connected to a leaf node does not contain a contradiction, otherwise unsatisfiability is proven.

In this paper we assume that an ontology O is represented by a knowledge base containing a TBox that is an acyclic terminology and an empty ABox. In this case reasoning can be reduced to reasoning without the TBox by unfolding the definitions. However, for efficiency reasons, instead of running the previously described satisfiability checking algorithm on an unfolded concept description,

---

[3] C and D represent concepts, and r represents a role.

[4] We observe that the TBox in Figure 1 is not an acyclic terminology as there are statements of the form $A \sqsubseteq C$. However, it is possible to create an equivalent TBox that is a acyclic terminology by replacing the statements of the form $A \sqsubseteq C$ with $A \doteq C \sqcap \overline{A}$ where $\overline{A}$ is new atomic concept. See [18].

[5] i and j represent individuals. In the completion graph in Section 4 statements of the form C(i) are written as $i : C$, and statements of the form r(i,j) are written as irj.

$\sqcap$-rule: if the ABox contains $(C_1 \sqcap C_2)(x)$, but it does not contain both $C_1(x)$ and $C_2(x)$, then these are added to the ABox.
$\sqcup$-rule: if the ABox contains $(C_1 \sqcup C_2)(x)$, but it contains neither $C_1(x)$ nor $C_2(x)$, then two ABoxes are created representing the two choices of adding $C_1(x)$ or adding $C_2(x)$.
$\forall$-rule: if the ABox contains $(\forall\ r.C)(x)$ and $r(x,y)$, but it does not contain $C(y)$, then this is added to the ABox.
$\exists$-rule: if the ABox contains $(\exists\ r.C)(x)$ but there is no individual z such that $r(x,z)$ and $C(z)$ are in the ABox, then $r(x,y)$ and $C(y)$ with y an individual name not occurring in the ABox, are added.

Fig. 2. Transformation rules (e.g. [1])

the unfolding is usually performed on demand within the satisifiability checking algorithm. It has been proven that satisfiability checking w.r.t. acyclic terminologies is PSPACE-complete in $\mathcal{ALC}$ [23].

3   An Abduction Problem

In our setting, a missing is-a relation in an ontology O represented by a knowledge base KB, is an is-a relation between named concepts that is not derivable from the KB, but that is correct according to the intended domain. We assume that we have a set M of missing is-a relations (but not necessarily all) for O. Then, the is-a structure of O can be repaired by adding is-a relations (or axioms of the form $C \sqsubseteq D$) between named concepts to O such that the missing is-a relations can be derived from the repaired ontology. This repair problem can be formulated as a generalized version of the TBox abduction problem.

Definition 1. *Let KB be a knowledge base in $\mathcal{L}$, and for $1 \le i \le m$: C_i, D_i are concepts that are satisfiable w.r.t. KB, such that $KB \cup \{\ C_i \sqsubseteq D_i \mid 1 \le i \le m\}$ is coherent. A solution to the generalized TBox abduction problem for $(KB, \{(C_i, D_i) \mid 1 \le i \le m\})$ is any finite set $S_{GT} = \{G_j \sqsubseteq H_j \mid j \le n\}$ of TBox assertions in $\mathcal{L}'$ such that $\forall\ i: KB \cup S_{GT} \models C_i \sqsubseteq D_i$. The set of all such solutions is denoted by $S_{GT}(KB, \{(C_i, D_i) \mid 1 \le i \le m\})$.*

In our setting the language $\mathcal{L}$ is $\mathcal{ALC}$ and $\mathcal{L}'$ only allows *named* concepts. We say that any solution in $S_{GT}(KB, \{(C_i, D_i)\}_i)$ is a *repairing action*. A repairing action is thus a set of is-a relations. When $m = 1$, this definition of the generalized TBox abduction problem coincides with the TBox abduction problem as formulated in [8], which therefore deals with repairing one missing is-a relation. Further, we have that $S_{GT}(KB, \{(C_i, D_i)\}_i) = \cap_i\ S_{GT}(KB, \{(C_i, D_i)\})$.[6]

This shows that solving a generalized TBox abduction problem can be done by solving m TBox abduction problems and then taking the intersection of the solutions. In practice, however, this leads to a number of difficulties. First, it

---

[6] A solution for all missing is-a relations is also a solution for each missing is-a relation and therefore, in the intersection of the solutions concerning one missing is-a relation at the time. Further, a solution in the intersection of the sets of solutions for each of the missing is-a relations, allows, when added to the knowledge base, to derive all missing is-a relations, and is therefore, a solution for the generalized TBox abduction problem as well.

Input: The ontology O represented by knowledge base KB and a set of missing is-a relations M.
Output: Set of repairing actions $Rep(M)$.
Algorithm
1. For every missing is-a relation $A_i \sqsubseteq B_i$ in M:
    1.1 G = completion graph after running a tableaux algorithm with $A_i \sqcap \neg B_i$ as input on KB;
    1.2 *Leaf-ABoxes* = get ABoxes of the leaves from the completion graph G;
    1.3 For every open ABox $\mathcal{A} \in$ *Leaf-ABoxes*:
        1.3.1 $R_{\mathcal{A}} = \emptyset$;
        1.3.2 For every individual x_j in $\mathcal{A}$;
            1.3.2.1 $Pos_{x_j} = \{P \mid x_j : P \in \mathcal{A} \wedge P \text{ is a named concept}\}$;
            1.3.2.2 $Neg_{x_j} = \{N \mid x_j : \neg N \in \mathcal{A} \wedge N \text{ is a named concept}\}$;
            1.3.2.3 $R_{\mathcal{A}} = R_{\mathcal{A}} \cup \{P \sqsubseteq N \mid P \in Pos_{x_j} \wedge N \in Neg_{x_j}\}$;
    1.4 $Rep(A_i, B_i) = \emptyset$;
    1.5 As long as there are different choices:
        1.5.1 Create a repairing action ra by choosing one element from each set $R_{\mathcal{A}}$;
        1.5.2 $Rep(A_i, B_i) = Rep(A_i, B_i) \cup \{ra\}$;
        1.5.3 Remove reduncancy in $Rep(A_i, B_i)$;
        1.5.4 Remove incoherent solutions from $Rep(A_i, B_i)$;
2. $Rep(M) = \{M\}$;
3. As long as there are different choices:
    3.1 Create a repairing action rp by choosing one element from each $Rep(A_i, B_i)$
        and taking the union of these elements;
    3.2 $Rep(M) = Rep(M) \cup \{rp\}$;
4. Remove reduncancy in $Rep(M)$;
5. Remove incoherent solutions from $Rep(M)$;

Fig. 3. Basic algorithm for generating repairing actions using completion graph

would mean that a domain expert will need to choose between large sets of repairing actions for all the missing is-a relations at once, and this may be a very hard task. Further, due to the size of the solution space, even generating all solutions for one TBox abduction problem is, in general, infeasible. Also, many of the solutions will not be interesting for a domain expert (e.g. [20]). For instance, there may be solutions containing is-a relations that do not contribute to the actual repairing. Some solutions may introduce unintended equivalence relations. A common way to limit the number of possible solutions is to introduce constraints, e.g. minimality.

4 Algorithm for Generating Repairing Actions

Basic Algorithm. The basic algorithm for generating repairing actions for a set of given missing is-a relations for an ontology is shown in Figure 3. In this first study we assume that the existing structure in the ontology is correct. Also, as stated in the definition in Section 3, adding the missing is-a relations to the ontology does not lead to incoherence.

In step 1 a set of repairing actions is generated for each missing is-a relation (thereby solving a TBox abduction problem for each missing is-a relation). In step 1.1 we run the satisfiability checking algorithm with unfolding on demand as described in Section 2, on KB with input $A_i \sqcap \neg B_i$, and we collect the ABoxes of the leaves in step 1.2. As $A_i \sqsubseteq B_i$ is a missing is-a relation, it cannot be derived from KB and thus the completion graph will have open leaf ABoxes. We then generate different ways to close these ABoxes in step 1.3. For each individual x in an open leaf ABox we collect the concepts in the statements of the form $x{:}A$ in Pos_x and the concepts in the statements of the form $x{:}\neg B$ in Neg_x where A and B are named concepts. The ABox can be closed if $A \sqsubseteq B$ is added to the ontology for any $A \in Pos_x$ and $B \in Neg_x$. Indeed, with this extra information $x{:}A$ could be unfolded and $x{:}B$ would be added to the ABox, and this gives a contradiction with $x{:}\neg B$ which was already in the ABox. A repairing action for the missing is-a relation is then a set of $A \sqsubseteq B$ that closes each open leaf ABox. In step 1.5 we generate such sets by selecting one such axiom per open leaf ABox and remove redundancy based on the sub-set relation. If a repairing action is a super-set of another repairing action, it is removed. Further, we remove solutions that introduce incoherence.

In step 2 the repairing actions set for all missing is-a relations is initialized with the set of missing is-a relations. Therefore, there will always be at least one repairing action. In step 3 additional repairing actions are generated by combining repairing actions for the individual missing is-a relations. As repairing actions are sets, there are no duplicates in a repairing action. In step 4 we remove redundancy based on the sub-set relation. In step 5 we remove solutions that introduce incoherence. We note that there always will be at least one solution that does not introduce incoherence (i.e. M or a sub-set of M).

A repairing action is a set of statements of the form $A \sqsubseteq B$ with A and B named concepts. We note that for acyclic terminologies, after adding such a statement, the resulting TBox is not an acyclic terminology anymore. If this is needed, then instead of adding $A \sqsubseteq B$ the following should be done (see [18]). If there is no definition for A yet in the TBox, then add $A \doteq B \sqcap \overline{A}$ with $\overline{A}$ a new atomic concept. If there is already a definition for A in the TBox, say $A \doteq C$ then change this definition to $A \doteq B \sqcap C$.

Our proposed algorithm is sound and it generates solutions to a TBox abduction problem that are minimal in the sense that repairing actions only contain necessary information for repairing the missing is-a relations. Further, we check the generated solutions for the introduction of incoherence.[7]

As an example, consider the acyclic terminology equivalent to the ontology in Figure 1 and M = {MyPizza $\sqsubseteq$ FishyMeatyPizza, MyFruttiDiMare $\sqsubseteq$ Non-VegetarianPizza}. For the missing is-a relation MyPizza $\sqsubseteq$ FishyMeatyPizza the completion graph obtained after running the satisfiability check on $MyPizza \sqcap \neg FishyMeatyPizza$ is shown in Figure 4. The completion graph contains 17 nodes of which 11 are leaf nodes. Of these leaf nodes 6 are closed and the repairing actions will be based on the 5 open leaf nodes. Closing all open leaf ABoxes will lead to 11 non-redundant repairing actions of which 8 lead to incoherence. The remaining repairing actions are {MyPizza $\sqsubseteq$ FishyMeaty-Pizza}, {AnchoviesTopping $\sqsubseteq$ FishTopping, ParmaHamTopping $\sqsubseteq$ MeatTopping } and {ParmaHamTopping $\sqsubseteq$ FishTopping, AnchoviesTopping $\sqsubseteq$ MeatTopping}.

---

[7] For the proof of the properties of the algorithm we refer to [18].

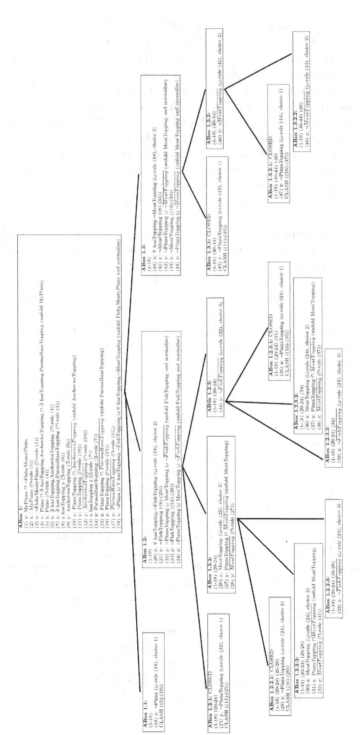

Fig. 4. Completion graph for $MyPizza \sqcap \neg FishyMeatyPizza$

In a similar way, for missing is-a relation MyFruttiDiMare $\dot{\sqsubseteq}$ NonVegetarian-Pizza, we find the following non-redundant coherence-preserving repairing actions: {MyFruttiDiMare $\dot{\sqsubseteq}$ NonVegetarianPizza}, {AnchoviesTopping $\dot{\sqsubseteq}$ Fish-Topping} and {AnchoviesTopping $\dot{\sqsubseteq}$ MeatTopping}. After combining the repairing actions for the individual missing is-a relations, and removing redundancy and incoherence-introducing repairing actions, we obtain the following 5 solutions for the repairing of the missing is-a structure: {MyPizza $\dot{\sqsubseteq}$ FishyMeatyPizza, MyFruttiDiMare $\dot{\sqsubseteq}$ NonVegetarianPizza}, {AnchoviesTopping $\dot{\sqsubseteq}$ FishTopping, ParmaHamTopping $\dot{\sqsubseteq}$ MeatTopping}, {ParmaHamTopping $\dot{\sqsubseteq}$ FishTopping, AnchoviesTopping $\dot{\sqsubseteq}$ MeatTopping}, {MyPizza $\dot{\sqsubseteq}$ FishyMeatyPizza, AnchoviesTopping $\dot{\sqsubseteq}$ FishTopping}, and {MyPizza $\dot{\sqsubseteq}$ FishyMeatyPizza, AnchoviesTopping $\dot{\sqsubseteq}$ MeatTopping}.

Optimization. The basic algorithm may produce many redundant solutions already in step 1. For instance, if the root node in the completion graph contains $x{:}A$ and $x{:}\neg B$ with A and B named concepts, then $A \dot{\sqsubseteq} B$ appears in every open leaf ABox as a possible way of closing that ABox. To deal with this issue we modify the algorithm such that at each node with an open ABox it generates Pos_{x_j} and Neg_{x_j}. However, Pos_{x_j} and Neg_{x_j} only contain the concepts that are not already in a Pos_{x_j} and Neg_{x_j} related to an ancestor node. For instance, if $A \in Pos_x$ related to the root node, then A will not appear in any Pos_x related to another node. For each ABox we then generate a set $R_{\mathcal{A}} = \bigcup_x \{ P \dot{\sqsubseteq} N \}$ where $P \in Pos_x$ for the current node or any ancestor node, $N \in Neg_x$ for the current node or any ancestor node, and at least one of P and N occurs in the current node. This allows us to reduce the redundancy in step 1. An open leaf ABox can now be closed by using an element from $R_{\mathcal{A}}$ from the leaf node or from any ancestor node. When generating repairing actions in step 1.5 we then make sure that when an element related to a non-leaf node is chosen, that no additional element from any descendant of that node is selected to close any leaf ABoxes that are descendants of that non-leaf node. For instance, if any element from the root's $R_{\mathcal{A}}$ is chosen, then no other elements should be chosen. For an example see [18].

Extension. The algorithm can be extended to generate additional repairing actions for every individual missing is-a relation. In step 1.5 if $A \dot{\sqsubseteq} B$ is used as one of the is-a relations in a repairing action then also $S \dot{\sqsubseteq} T$ where S is a super-concept of A and T is a sub-concept of B could be used. Therefore, the extended algorithm generates two sets of concepts for every is-a relation $A \dot{\sqsubseteq} B$ in a repairing action, Source set containing named super-concepts of A and Target set containing named sub-concepts of B. Further, to not introduce non-validated equivalence relations where in the original ontology there are only is-a relations, we remove the super-concepts of B from Source, and the sub-concepts of A from Target. Alternative repairing actions for a repairing action {$A_1 \dot{\sqsubseteq} B_1$, ..., $A_n \dot{\sqsubseteq} B_n$} are then repairing actions {$S_1 \dot{\sqsubseteq} T_1$, ..., $S_n \dot{\sqsubseteq} T_n$} such that $(S_i, T_i) \in Source(A_i, B_i) \times Target(A_i, B_i)$.

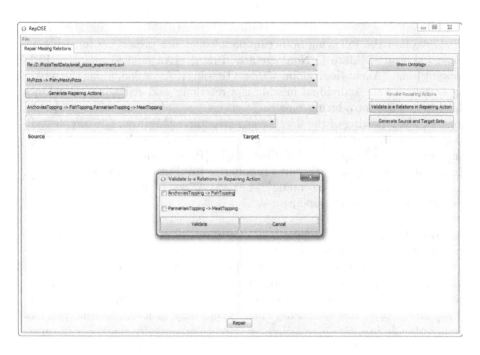

Fig. 5. Screenshot - Validating is-a relations in a repairing action

5   Implementation

We have implemented a system that supports the user to repair missing is-a relations. In our system the user loads the ontology and the missing is-a relations
from the 'File' menu. The missing is-a relations are then shown in a drop-down
list (e.g. MyPizza $\sqsubseteq$ FishyMeatyPizza[8] in the second drop-down list in Figure
5). The user then chooses a missing is-a relation to repair.[9] At any time the user
can switch between different missing is-a relations. Once a missing is-a relation
is chosen for repairing the user generates repairing actions for it by clicking the
'Generate Repairing Actions' button. This covers step 1 in Figure 3 and was
implemented in Java using Pellet (version 2.3.0) [31]. The satisfiability checker
in Pellet was modified in order to extract full completion graphs. Furthermore,
to increase performance and account for higher level of non-determinism, ontologies are first passed through Pellint [22] before running the algorithm. The
computed repairing actions are shown in the drop-down list under the button.
Each repairing action consists of one or more is-a relations. In Figure 5 the
user has chosen to repair MyPizza $\sqsubseteq$ FishyMeatyPizza and the system has generated three repairing actions that do not introduce incoherence ({MyPizza $\sqsubseteq$
FishyMeatyPizza}, {AnchoviesTopping $\sqsubseteq$ FishTopping, ParmaHamTopping $\sqsubseteq$

---

[8] In the system C $\sqsubseteq$ D is shown as C $\rightarrow$ D.
[9] As we repair one is-a relation at a time, there may be some redundancy in the
solutions.

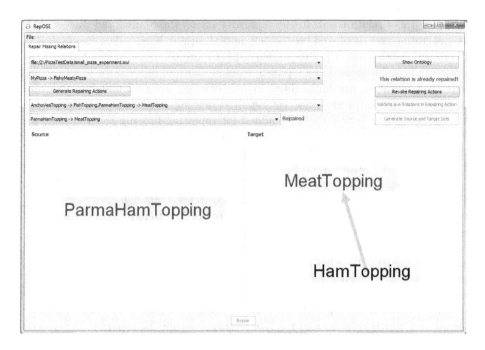

Fig. 6. Screenshot - Repairing using Source and Target sets

MeatTopping}, and {AnchoviesTopping $\sqsubseteq$ MeatTopping, ParmaHamTopping $\sqsubseteq$ FishTopping}). To repair this missing is-a relation the user has to succesfully deal with at least one of the repairing actions, i.e. add all is-a relations (or more informative is-a relations from Source and Target) in at least one of the repairing actions to the ontology. In Figure 5 the user has chosen the second repairing action. When repairing actions are added to the ontology, they will make the missing is-a relation derivable. However, it is not guaranteed that all the is-a relations in the repairing actions are also valid with respect to the domain. Therefore, a domain expert needs to validate the is-a relations in the repairing actions. When the user clicks the 'Validate is-a Relations in Repairing Action' button, a pop-up window (Figure 5) appears where the user can mark all the is-a relations that are correct with respect to the domain model. The repairing actions for all missing is-a relations and the ontology are updated according to the results of the validation. If an is-a relation is validated as incorrect according to the domain, all repairing actions that contain this incorrect is-a relation, for this and for all other missing is-a relations, are removed from the lists of the repairing actions. When an is-a relation is validated as correct it is added to the ontology and it is marked as correct in all repairing actions for all missing is-a relations. When all is-a relations in the current repairing action are validated as correct, they are shown in the last drop-down list (Figure 6). Now the user can repair them one by one.

For each is-a relation within the repairing action the Source and Target sets are generated and displayed on the left and the right hand sides, respectively, within the panel under the drop-down lists (Figure 6). Both panels have zoom control and can be opened in a separate window. The concepts in the is-a relation under consideration are highlighted in red, existing asserted and inferred is-a relations are shown in grey, not yet repaired missing is-a relations in blue and is-a relations that were previously added for repairing missing is-a relations in black. In order to repair the is-a relation the user has to choose one concept from each of the sets and click the 'Repair' button. In Figure 6 the user has chosen to repair ParmaHamTopping $\sqsubseteq$ MeatTopping with ParmaHamTopping $\sqsubseteq$ HamTopping. Upon repair the ontology is updated, i.e. the chosen is-a relation (ParmaHamTopping $\sqsubseteq$ HamTopping) is added to the ontology. A red label next to the drop-down list shows the status (Repaired or Not Repaired) of the selected is-a relation. When all is-a relations within a repairing action are repaired the missing is-a relation is marked as repaired ('This relation is already repaired' label in Figure 6). The other repairing actions are still available for review by the user. These may give information about other possible missing is-a relations. The user can also revoke repairing actions (through the 'Revoke Repairing Actions' button). If the user revokes the repairing action, the missing is-a relation may become not repaired again and the is-a relations within the repairing action are marked as not repaired. All changes in the ontology are revoked and the user can start repairing this missing is-a relation again in the way just described.

Example Run. As an example run, consider the ontology in Figure 1 and missing is-a relations MyPizza $\sqsubseteq$ FishyMeatyPizza and MyFruttiDiMare $\sqsubseteq$ NonVegetarianPizza. After loading the ontology and the missing is-a relations, we can choose a missing is-a relation to repair. Assume we choose MyFruttiDiMare $\sqsubseteq$ NonVegetarianPizza and click the 'Generate Repairing Actions' button. The system will show three repairing actions in the drop-down list: {MyFruttiDiMare $\sqsubseteq$ NonVegetarianPizza}, {AnchoviesTopping $\sqsubseteq$ FishTopping}, and {AnchoviesTopping $\sqsubseteq$ MeatTopping}. We can choose to deal with AnchoviesTopping $\sqsubseteq$ MeatTopping and validate this to be incorrect with respect to the domain. In this case all repairing actions containing this is-a relation will be removed. We could then choose AnchoviesTopping $\sqsubseteq$ FishTopping and validate it to be correct. All is-a relations in this repairing action (i.e. AnchoviesTopping $\sqsubseteq$ FishTopping) are validated to be correct and thus we can continue with the repair of AnchoviesTopping $\sqsubseteq$ FishTopping. In this small example the Source set only contains AnchoviesTopping and the Target set only contains FishTopping. Therefore, we click on these concepts and the 'Repair' button. AnchoviesTopping $\sqsubseteq$ FishTopping will be marked as repaired and also the missing is-a relation MyFruttiDiMare $\sqsubseteq$ NonVegetarianPizza will be marked as repaired.

We can then start repairing MyPizza $\sqsubseteq$ FishyMeatyPizza. The system would have generated as repairing actions that do not introduce incoherence {MyPizza $\sqsubseteq$ FishyMeatyPizza}, {AnchoviesTopping $\sqsubseteq$ FishTopping, ParmaHamTopping $\sqsubseteq$ MeatTopping} and {AnchoviesTopping $\sqsubseteq$ MeatTopping, ParmaHamTopping

$\stackrel{.}{\sqsubseteq}$ FishTopping}. However, as we earlier already validated AnchoviesTopping $\stackrel{.}{\sqsubseteq}$ MeatTopping to be incorrect with respect to the domain, the third repairing action has already been removed. When we choose the second repairing action, AnchoviesTopping $\stackrel{.}{\sqsubseteq}$ FishTopping is already marked as correct (because of earlier validation) and only ParmaHamTopping $\stackrel{.}{\sqsubseteq}$ MeatTopping needs to be validated. We validate this as correct and then choose to repair it. The Source set in this small example contains ParmaHamTopping and the Target set contains HamTopping and MeatTopping. Although we can add ParmaHamTopping $\stackrel{.}{\sqsubseteq}$ MeatTopping, it is more informative (and correct with respect to the domain) to add ParmaHamTopping $\stackrel{.}{\sqsubseteq}$ HamTopping. We therefore choose the latter. All is-a relations in this repairing action are then repaired and thus also MyPizza $\stackrel{.}{\sqsubseteq}$ FishyMeatyPizza.

6 Related Work

Debugging Ontologies. *Detecting* missing is-a relations can be done in a number of ways (see Section 1). There is, however, not much work on the *repairing* of missing is-a structure. In [20] we addressed this in the setting of taxonomies.

There is more work on the debugging of semantic defects. Most of it aims at identifying and removing logical contradictions from an ontology. Standard reasoners are used to identify the existence of a contradiction, and provide support for resolving and eliminating it [9]. In [29] minimal sets of axioms are identified which need to be removed to render an ontology coherent. In [16,15] strategies are described for repairing unsatisfiable concepts detected by reasoners, explanation of errors, ranking erroneous axioms, and generating repair plans. In [10] the focus is on maintaining the consistency as the ontology evolves through a formalization of the semantics of change for ontologies. In [24] and [13] the setting is extended to repairing ontologies connected by mappings. In this case, semantic defects may be introduced by integrating ontologies. Both works assume that ontologies are more reliable than the mappings and try to remove some of the mappings to restore consistency. The solutions are often based on the computation of minimal unsatisfiability-preserving sets or minimal conflict sets. The work in [28] further characterizes the problem as mapping revision. Using belief revision theory, the authors give an analysis for the logical properties of the revision algorithms. Another approach for debugging mappings is proposed in [32] where the authors focus on the detection of certain kinds of defects and redundancy. The approach in [14] deals with the inconsistencies introduced by the integration of ontologies, and unintended entailments validated by the user.

Work that deals with both modeling and semantic defects includes [5] where the authors propose an approach for detecting modeling and semantic defects within an ontology based on patterns and antipatterns. Some suggestions for repairing are also given. In [19] we provided a method to detect and repair wrong and missing is-a structure in taxonomies connected by mappings.

A different setting is the area of modular ontologies where the ontologies are connected by directional mappings and where knowledge propagation only occurs in one direction. Regarding the detection of semantic defects, within a framework based on distributed description logics, it is possible to restrict the propagation of local inconsistency to the whole set of ontologies (e.g. [30]).

Abductive Reasoning in Description Logics. In [8] four different abductive reasoning tasks are defined - (conditionalized) concept abduction, ABox abduction, TBox abduction and knowledge base abduction. Concept abduction deals with finding sub-concepts. Abox abduction deals with retrieving abductively instances of concepts or roles that, when added to the knowledge base, allow the entailment of a desired ABox assertion. Knowledge base abduction includes both ABox and TBox abduction.

Most existing approaches for DL abduction focus on ABox and concept abduction and are mostly based on existing proof techniques such as semantic tableaux and resolution. Since the number of possible solutions is very large, the approaches introduce constraints. The work in [17] proposes a goal-oriented approach where only actions which contribute to the solution are chosen in the proof procedures. The method is both complete and sound for consistent and semantically minimal solutions. Since the set of solutions can contain some inconsistent and non-minimal solutions additional checks are required. A practical approach for ABox abduction, based on abductive logic programming, was proposed in [7]. In order to use existing abductive logic programming systems it is necessary to do a transformation to a plain Datalog program. The solutions are consistent and minimal given a set of abduciles. However, the approach does not guarantee completeness since the translation to a Datalog program is approximate and in some cases a solution would not be found. An approach for conditionalised concept abduction that uses a variation of the semantic tableaux and two labeling functions was proposed in [4]. The two labeling functions $T()$ and $F()$ represent true and false formulas in a tableaux where the solutions are formed from concepts which have at least one constraint in $F()$ of every open branch. This choice is non-deterministic and can be used to select solutions based on some criteria. The algorithm also contains a consistency check which implies that produced solutions are always consistent.

There has not been much work related to TBox abduction, which is the most relevant abduction problem for this paper. The work in [12] proposes an automata-based approach to TBox abduction using abduciles representing axioms that can appear in solutions. It is based on a reduction to the axiom pinpointing problem which is then solved with automata-based methods. A PTIME algorithm is proposed for the language $\mathcal{EL}$.

All of the presented approaches to description logic abduction work with relatively inexpressive ontologies, such as $\mathcal{EL}$ and $\mathcal{ALC}$. However, some recent work [6] describes a type of conditionalised concept abduction called structural abduction which is applicable to $\mathcal{SH}$.

7   Conclusion

This paper formalized repairing the is-a structure in $\mathcal{ALC}$ acyclic terminologies as a generalized TBox abduction problem, provided a solution based on semantic tableaux, and discussed a system.

There are a number of interesting aspects for future work. First, we intend to extend the algorithm to deal with more expressive ontologies. Further, it may be useful to consider also solutions introducing incoherence as they may lead to the detection of other kinds of modeling defects such as wrong is-a relations. In this case we do not assume anymore that the existing structure is correct. As a domain expert may need to deal with many possible solutions, other useful extensions are mechanisms for ranking the generated repairing actions and among those recommending repairing actions, e.g. based on domain knowledge.

Acknowledgements. We acknowledge the financial support of the Swedish Research Council (Vetenskapsrådet), the Swedish National Graduate School in Computer Science (CUGS) and the Swedish e-Science Research Centre (SeRC)

References

1. Baader, F., Sattler, U.: An overview of tableau algorithms for description logics. Studia Logica 69, 5–40 (2001)
2. Bada, M., Hunter, L.: Identification of OBO nonalignments and its implication for OBO enrichment. Bioinformatics 24(12), 1448–1455 (2008)
3. Cimiano, P., Buitelaar, P., Magnini, B.: Ontology Learning from Text: Methods, Evaluation and Applications. IOS Press (2005)
4. Colucci, S., Di Noia, T., Di Sciascio, E., Donini, F., Mongiello, M.: A uniform tableaux-based approach to concept abduction and contraction in ALN. In: International Workshop on Description Logics, pp. 158–167 (2004)
5. Corcho, O., Roussey, C., Vilches, L.M., Pérez, I.: Pattern-based owl ontology debugging guidelines. In: Workshop on Ontology Patterns, pp. 68–82 (2009)
6. Donini, F., Colucci, S., Di Noia, T., Di Sciasco, E.: A tableaux-based method for computing least common subsumers for expressive description logics. In: 21st International Joint Conference on Artificial Intelligence, pp. 739–745 (2009)
7. Du, J., Qi, G., Shen, Y.-D., Pan, J.: Towards practical abox abduction in large owl dl ontologies. In: 25th AAAI Conference on Artificial Intelligence, pp. 1160–1165 (2011)
8. Elsenbroich, C., Kutz, O., Sattler, U.: A case for abductive reasoning over ontologies. In: OWL: Experiences and Directions (2006)
9. Flouris, G., Manakanatas, D., Kondylakis, H., Plexousakis, D., Antoniou, G.: Ontology Change: Classification and Survey. Knowledge Engineering Review 23(2), 117–152 (2008)
10. Haase, P., Stojanovic, L.: Consistent Evolution of OWL Ontologies. In: Gómez-Pérez, A., Euzenat, J. (eds.) ESWC 2005. LNCS, vol. 3532, pp. 182–197. Springer, Heidelberg (2005)
11. Hearst, M.: Automatic acquisition of hyponyms from large text corpora. In: 14th International Conference on Computational Linguistics, pp. 539–545 (1992)
12. Hubauer, T., Lamparter, S., Pirker, M.: Automata-based abduction for tractable diagnosis. In: International Workshop on Description Logics, pp. 360–371 (2010)

13. Ji, Q., Haase, P., Qi, G., Hitzler, P., Stadtmüller, S.: RaDON — repair and diagnosis in ontology networks. In: Aroyo, L., Traverso, P., Ciravegna, F., Cimiano, P., Heath, T., Hyvönen, E., Mizoguchi, R., Oren, E., Sabou, M., Simperl, E. (eds.) ESWC 2009. LNCS, vol. 5554, pp. 863–867. Springer, Heidelberg (2009)
14. Jiménez-Ruiz, E., Cuenca Grau, B., Horrocks, I., Berlanga, R.: Ontology Integration Using Mappings: Towards Getting the Right Logical Consequences. In: Aroyo, L., Traverso, P., Ciravegna, F., Cimiano, P., Heath, T., Hyvönen, E., Mizoguchi, R., Oren, E., Sabou, M., Simperl, E. (eds.) ESWC 2009. LNCS, vol. 5554, pp. 173–187. Springer, Heidelberg (2009)
15. Kalyanpur, A., Parsia, B., Sirin, E., Cuenca-Grau, B.: Repairing Unsatisfiable Concepts in OWL Ontologies. In: Sure, Y., Domingue, J. (eds.) ESWC 2006. LNCS, vol. 4011, pp. 170–184. Springer, Heidelberg (2006)
16. Kalyanpur, A., Parsia, B., Sirin, E., Hendler, J.: Debugging Unsatisfiable Classes in OWL Ontologies. Journal of Web Semantics 3(4), 268–293 (2006)
17. Klarman, S., Endriss, U., Schlobach, S.: Abox abduction in the description logic ALC. Journal of Automated Reasoning 46, 43–80 (2011)
18. Lambrix, P., Dragisic, Z., Ivanova, V.: Get my pizza right: Repairing missing is-a relations in ALC ontologies, extended version (2012), http://arxiv.org/abs/1210.7154
19. Lambrix, P., Liu, Q.: Debugging is-a structure in networked taxonomies. In: 4th International Workshop on Semantic Web Applications and Tools for Life Sciences, pp. 58–65 (2011)
20. Lambrix, P., Liu, Q., Tan, H.: Repairing the Missing is-a Structure of Ontologies. In: Gómez-Pérez, A., Yu, Y., Ding, Y. (eds.) ASWC 2009. LNCS, vol. 5926, pp. 76–90. Springer, Heidelberg (2009)
21. Lambrix, P., Strömbäck, L., Tan, H.: Information Integration in Bioinformatics with Ontologies and Standards. In: Bry, F., Maluszynski, J. (eds.) Semantic Techniques for the Web. LNCS, vol. 5500, pp. 343–376. Springer, Heidelberg (2009)
22. Lin, H., Sirin, E.: Pellint - a performance lint tool for pellet. In: OWL: Experiences and Directions (2008)
23. Lutz, C.: Complexity of terminiological reasoning revisited. In: Ganzinger, H., McAllester, D., Voronkov, A. (eds.) LPAR 1999. LNCS, vol. 1705, pp. 181–200. Springer, Heidelberg (1999)
24. Meilicke, C., Stuckenschmidt, H., Tamilin, A.: Repairing Ontology Mappings. In: 22nd Conference on Artificial Intelligence, pp. 1408–1413 (2007)
25. MeSH. Medical subject headings, http://www.nlm.nih.gov/mesh/
26. Pizza Ontology v1.5, http://www.co-ode.org/ontologies/pizza/2007/02/12/
27. PubMed, http://www.ncbi.nlm.nih.gov/pubmed/
28. Qi, G., Ji, Q., Haase, P.: A Conflict-Based Operator for Mapping Revision. In: Bernstein, A., Karger, D.R., Heath, T., Feigenbaum, L., Maynard, D., Motta, E., Thirunarayan, K. (eds.) ISWC 2009. LNCS, vol. 5823, pp. 521–536. Springer, Heidelberg (2009)
29. Schlobach, S.: Debugging and Semantic Clarification by Pinpointing. In: Gómez-Pérez, A., Euzenat, J. (eds.) ESWC 2005. LNCS, vol. 3532, pp. 226–240. Springer, Heidelberg (2005)
30. Serafini, L., Borgida, A., Tamilin, A.: Aspects of distributed and modular ontology reasoning. In: 19th International Joint Conference on Artificial Intelligence, pp. 570–575 (2005)
31. Sirin, E., Parsia, B., Cuenca Grau, B., Kalyanpur, A., Katz, Y.: Pellet: A practical owl-dl reasoner. Journal of Web Semantics 5(2), 51–53 (2007)
32. Wang, P., Xu, B.: Debugging ontology mappings: a static approach. Computing and Informatics 27, 21–36 (2008)

Ontological Modeling of Interoperable Abnormal States

Yuki Yamagata[1], Hiroko Kou[1], Kouji Kozaki[1], Riichiro Mizoguchi[2],
Takeshi Imai[3], and Kazuhiko Ohe[3]

[1] The Institute of Scientific and Industrial Research, Osaka University
8-1 Mihogaoka, Ibaraki, Osaka, 567-0047 Japan
{yamagata,kou,kozaki,miz}@ei.sanken.osaka-u.ac.jp
[2] Research Center for Service Science School of Knowledge Science Japan Advanced Institute
of Science and Technology
1-1 Asahidai, Nomi, Ishikawa, 923-1292 Japan
mizo@jaist.ac.jp
[3] Department of Medical Informatics, Graduate School of Medicine, The University of Tokyo
7-3-1 Hongo, Bunkyo-ku, Tokyo, 113-0033 Japan
{ken,ohe}@hcc.h.u-tokyo.ac.jp

Abstract. Exchanging huge volumes of data is a common concern in various
fields. One issue has been the difficulty of cross-domain sharing of knowledge because of its highly heterogeneous nature. We constructed an ontological model of
abnormal states from the generic to domain-specific level. We propose a unified
form to describe an abnormal state as a "property", and then divide it into an
"attribute" and a "value" in a qualitative form. This approach promotes interoperability and flexibility of quantitative raw data, qualitative information, and gener-
ic/abstract knowledge. By developing an *is-a* hierarchal tree and combining causal
chains of diseases, 17,000 abnormal states from 6000 diseases can be captured as
generic causal relations and are reusable across 12 medical departments.

Keywords: ontology, abnormal state, property, disease, interoperability.

1   Introduction

Recent advances in science and technology, together with the increasing diversity and
volume of data, have led to recognition of the importance of data exchange. Scientists
observe objects through experimentation to identify the attributes they possess and the
values of those attributes; in other words, scientific data are represented by attributes
and values. Many ontologies have been developed based on this approach, such as
BFO [1], DOLCE [2], and OGMS [3], and they have contributed to understanding the
semantics of heterogeneous data. However, one problem with these ontologies is the
difficulty of exchanging their concepts because of their individual formulations.
YAMATO has focused on integrating these differing formulations [4].

In this study, we developed an ontology of properties which allows interoperable
representation in a consistent manner.

Understanding and analyzing the cause of failures and anomalies is a common issue in various domains, such as machinery, aviation, materials, and medicine.

H. Takeda et al. (Eds.): JIST 2012, LNCS 7774, pp. 33–48, 2013.

Handling of appropriate knowledge representations and management of anomalies is of common concern to engineers, and knowledge sharing will contribute to the development of infrastructures for reliability/security engineering, mass-production technology, and quality assurance.

The aim of our work is to systematically develop an anomaly ontology that can capture the intrinsic properties of anomalies from content-oriented view.

In this paper, we focus on cases in the medical domain. We define abnormal states and introduce our representation model to guide the modeling of how to understand and capture the target world in the next section. Then, in Section 3, we describe our ontology of anomalies. The top level of the anomaly ontology defines very basic and generic concepts, which are "interoperable" across different domains, for example, "small in size" is interoperable between machinery and medicine. The middle-level concepts are common i.e., interoperable between diseases, e.g., "blood stenosis" in the medical domain and lower-level concepts are designed to specific-context dependent (e.g., disease-dependent) knowledge. In Section 4, we show an application in which we developed a causal chains of abnormal states covering over 6,000 diseases, capable of being used across domains by 12 medical departments, and we describe the contributions of our model to disease ontology. In Section 5, we discuss related work. Finally, we present concluding remarks and outline plans for future work.

2   Definition of Abnormal State

2.1   Basic Policy for Definition of Abnormal State

In order to understand knowledge about anomalies systematically, it is important to clearly capture essential characteristics of the abnormal states and to conceptualize them from a consistent viewpoint.

There are two different views of representations: one is content-oriented view which focuses on how to understand and capture the target world; and the other is form-oriented view, like F-logic, DL-Lite, and OWL, which focuses on dealing with how to formalize the representation of the content with syntax or axioms in logic. Our study deals with the topic of the former, i.e., content-oriented representation.

In this section, we focus on cases in the medical domain. We define abnormal states and introduce our representation model with a content-oriented viewpoint to guide the modeling of how to understand and capture *anomalies used in definition of disease*s. A property is an abstraction of a characteristic possessed by an individual, having the value of an attribute, and is inhered in any particular entity, whereas a state is a time-indexed property that changes with time [4]. For example, when the state is "hungry" at time T1, it is represented by "being hungry" or not. Note that properties and attribute values are intrinsically different. A property such as "tall", as in "He is tall", is different from an attribute value such as "large", as in "His height is large". A property is further divided into an attribute and an attribute value.

Here, when discussing a disease in the medical domain or a failure in the engineering domain, "being in an abnormal state or not" is an important and common issue. In the medical domain, various types of representations for anomalies are used, such as "blood glucose level 250 mg/dL, "glucose concentration is high", and "hyperglycemia".

Based on YAMATO [4], which is an upper ontology that was carefully designed to cover both quality and quantity ontologies, we classified abnormal states into the three categories shown in Table 1.

Table 1. Representation of anomaly

Representation	Example	Usage
quantitative representation	blood glucose level 250mg/dl blood pressure 200 mmHg	Diagnostics
qualitative representation	glucose concentration is high blood pressure is high	
property representation	hyperglycemia hypertension	Definition of disease

In many diagnostic situations, a quantitative representation is indispensable, because identifying a precise value for each patient by clinical tests is the main task. However, the definition of a disease is different. In the above three instances, "being hyperglycemic" might be a common and intrinsic nature. Actually, as explained in many textbooks or guidelines, to illustrate disease, most abnormal states are described as property representations; for example, to explain diabetes, "Diabetes is characterized by hyperglycemia caused by impaired insulin action..." [5].

In this way, to capture the essentials of abnormal states, a property representation seems to be the most natural and intrinsic. Therefore, first, we captured abnormal states as properties, represented by a tuple like <Property (P), Property Value (Vp)>. Basically, Property Value takes a Boolean value, i.e., <existence / non-existence>. For example, "constriction (stenosis)" is described as <constriction, existence>. In addition, when necessary, a degree value[1] can also be used for describing the degree of the Property Value.

An ontology is a theory of concepts which explain the target world. When explaining abnormal states, it says "having abnormal values". Abnormal states also can be explained by some kind of disturbance of the homeostasis in human body. While the former does not necessarily have connotation of bad for health, the latter does. Therefore, making a decision about the latter is not the job of ontology researchers but rather is the one of medical experts based on the judgment of the medical knowledge. For example, answering a question whether the high HDL cholesterol[2] level is "abnormal state" or not in the latter sense is not a task of ontologists but that of medical experts, which is not discussed in this article.

---

[1] The degree value of Property Value is, for example, "mild/moderate/severe". For example, <diarrhea (P), severe (Vp)>.

[2] HDL cholesterol is known as good cholesterol in medicine.

In practice, however, a property is too abstract to represent the precise meaning of data. In the case of the above "constriction", it would have a more concrete meaning, say, <cross-sectional area, small>.

Therefore, we expanded the specification of a property by decomposing it into a tuple: <Attribute (A), Attribute Value (V)>. We adopt Attribute Value in a qualitative form, i.e., Qualitative Value (Vql). This approach contributes to promoting consistency in representation, as well as interoperability between quantitative raw data and generic/abstract anomaly knowledge (see Section 2.3).

In the engineering domain, most properties can be decomposed into a set of <A, V>, whereas in clinical medicine, some properties cannot be decomposed, because the precise mechanisms in the human body remain elusive. In such cases, for example, deformation and pain, the property representation could be an undecomposed one: <Property (P), Property Value (Vp)>. Whether such abnormal states represented in terms of properties defined above makes sense or not is dependent on the advance of medicine.

One of the advantages of representation in terms of property is that it is substantially unaffected by small parameter fluctuations. Another advantage is that representation in terms of property in an abstract form makes it easier to capture the essentials of each disease. Although a diagnostic task requires a quantitative representation rather than a qualitative representation in terms of property, this is clearly a different task from defining disease.

2.2   Anomaly Representation

2.2.1   Standard Representation

In the human body, abnormal states are highly diverse. They have a variety of grain sizes, from cell and organs to action level. In addition, related agents involved with these states, such as chemicals, viruses, bacteria, and so on, are also extremely diverse. In this section, we apply our representation model to medical abnormal states and examine whether we can represent them appropriately and consistently.

Because an attribute cannot exist alone but always exists in association with an independent object, it is necessary to identify what it inheres in (hereinafter referred to as "target object"). For example, in the case of "intestinal stenosis", the target object of its attribute "cross-sectional area" is neither the blood vessels nor other organs but the intestine (Fig. 1, upper left).

In the case of "hypertension", blood is the target object of "pressure". In order to represent the target object of an attribute, we introduce "Object" and decompose a property into a triple: <Object (O), Attribute (A), Attribute Value (V)>. For example, "gastric dilatation" and "hypertension" are decomposed into <stomach, volume, large> and <blood, pressure, high>, respectively. This is a standard representation model of anomalies.

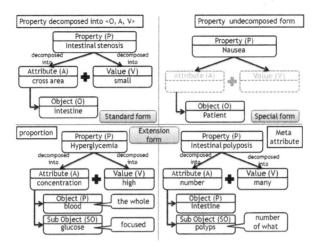

Fig. 1. Framework for abnormal states in medical domain

2.2.2   Advanced Representation

Although many properties are represented in terms of the standard triple, some are not. These are properties defined by a ratio and a meta-attribute and need a more advanced representation.

A)   Ratio

In the case of a representation involving a ratio, the target object having the ratio must be identified. In addition, an   advanced representation for what will be focused on ("focused object") is needed. For example, in the case of a representation for "hyperglycemia", concentration is a known attribute of blood, so blood is identified as Object. However, since concentration means the ratio of the focused object relative to the whole mixture, glucose is needed for representation as the focused object. Therefore, we introduce Sub-Object (SO) to represent a focused object.

There are different kinds of ratio, depending on what is focused on. Therefore, Object and Sub-Object vary according to the kind of ratio as follows:

m/n ratio (no unit): Represented by the ratio of the focused object (m) relative to the whole (mixture) (n), which has no unit:

In this kind of ratio, the whole mixture (n) is identified as Object having a ratio as its attribute, and focused object (m) is defined as Sub-Object, which represents "the ratio of what". For instance, since blood glucose concentration is the amount of glucose (sugar) present in the blood, it is a property of blood as the whole mixture, and the focused object is glucose. Accordingly, the representation of "hyperglycemia" is a quadruple, <blood (O), glucose (SO), concentration (A), high (V)> (Fig. 1, bottom left), where blood is defined as Object, and glucose is Sub-Object (Table 2, No. 1).

Table 2. Representations used with ratio

No.	Proportion Classification	Property (P)	Value	Attribute (A)	value (V)	Object (O)	Sub Object (SO)	Ratio
1	m/n (focused on m/ the whole)	high m ratio	existence	ratio	high/ low	the whole	focused	m/n
	example	hyperglycemia	existence	Glucose concentration	high	blood	Glucose	Glucose/ Blood
2-1	m/n (focused on m)	high m ratio	existence	ratio	high/ low	object	m	m/n
	example	high Albumin ratio	existence	Albumin ratio	high	urine	Albumin	Albumin/ Creatinine
2-2	m/n (focused on the ratio)	high/low m/n ratio	existence	ratio	high/ low	object	blank	m/n
	example	increased A/G ratio	existence	A/G ratio	high	blood	blank	Albumin/ Globulin

2- m/n ratio (of same object): Represented by the ratio of attribute m to attribute n of the same object:

2-1 m/n (focused on m): Represented by the ratio of the focused attribute (m) to another attribute (n) of the same object:

The thing having both attributes m and n is identified as Object, and the focused attribute m is defined as Sub-Object. For example, Urine Albumin-to-Creatinine Ratio (UACR) is the ratio of albumin level to creatinine level in the urine. The albumin is Sub-Object, which is focused on, and urine is defined as Object. Therefore, "high UACR" is represented as a quadruple <Urine (O), Albumin (SO), Albumin ratio (A), high (V)> (Table 2, No. 2-1).

2-2 m/n ratio (focused on the ratio): Defined by the ratio of attribute m to attribute n of the same object:

In this kind of ratio, what is focused on is simply the ratio itself rather than either attribute. For instance, in the Albumin to Globulin Ratio (A/G ratio), which is the ratio of the albumin level to globulin level in the blood, blood is defined as Object, though Sub-Object is null since what is focused on is neither the albumin nor globulin level but the A/G ratio itself. Thus, "decreased A/G ratio" is described as a triple: <blood (O), A/G ratio (A), low (V)> (Table 2, No. 2-2). Accordingly, in this kind of ratio, the thing having both attributes m and n is identified as Object, whereas Sub-Object is null.

B) Meta Attribute

In the property "intestinal polyposis", color and size are attributes of polyps. However, "many polyps" is not an attribute of "polyps" because it is not an inherent nature.

Following the meta-attribute in YAMATO, in which "number of curves" is identified as a meta-attribute of a road which has many curves, we regard "the number of polyps" as a meta-attribute of the intestine. By introducing "Sub-object", the property "intestinal polyposis" is decomposed into a quadruple <intestine (O), polyps (SO), number (A), many (V)>, where intestine is identified as Object, and polyps are described as Sub-object, which collectively represent "number of polyps" (Fig. 1, bottom right).

Applying the concept of Sub-Object provides a flexible representation in the case of complicated concepts like ratio and meta-attribute and allows us to determine Objects as distinct targets with consistency. As shown in Fig. 1, various descriptions for abnormal states are supported by our framework.

In summary, the key point of our model is, first, to capture an abnormal state as a property and introduce a property representation framework, and second, to decompose the representation into a standard triple <Object (O), Attribute (A), Attribute Value (V)>. This framework ensures a consistent representation of various simple abnormal states. Furthermore, introducing the concept of "Sub-Object" is effective in increasing the flexibility and consistency of representations in complicated cases, such as those involving ratios and meta-attributes, described as a quadruple <Object (O), Sub-Object (SO), Attribute (A), Attribute Value (V)>.

2.3   Interoperability between Property and Attribute

In medical institutions or hospitals, huge volumes of diagnostic/clinical test data have been accumulated and stored, most of which are quantitative data: e.g., blood pressure (systolic) 180 mm Hg. By using our representation model, quantitative data can be described as <Object (O), Attribute (A), Quantitative Value (Vqt)>. For example, in the case of "blood pressure (systolic) 180 mm Hg", it is decomposed into <blood (O), pressure (A), 180 mmHg (Vqt)>.

Here, a threshold[3] based on a generic value used in the domain is introduced, and a qualitative value can be obtained by comparing the threshold value with quantitative value. For example, above the threshold (e.g. 130mmHg), the Quantitative Value (180mmHg) is converted to the qualitative value "high".

Therefore, the quantitative representation <blood (O), pressure (A), 180 mmHg (Vqt)> can be converted into a qualitative representation <blood (O), pressure (A), high (Qualitative value (Vql))>, which is a decomposition of the property representation for "hypertension". Accordingly, to deal with raw data, a quantitative representation as <Object (O), Attribute (A), Quantitative Value (Vqt)> can be used.

In Section 2.2, we introduced the property representation framework <Object (O), Property (P), Property Value (Vp)>, which is decomposed into its qualitative representation <Object (O), Attribute (A), Qualitative Value (Vql)>. This enables interchangeability from a quantitative representation to a property representation. For example, <blood (O), pressure (A), 180 mmHg (Vqt)> can be exchanged with <blood

---

[3] How to set the threshold and identifying the precise value is a diagnostic task at the instance level. We do not discuss them further in this paper.

(O), hypertension (P), severe (Vp)>, which is sufficient to be judged as "being an anomaly" or not. Therefore, our approach contributes to promote interoperability between quantitative raw data, such as clinical examination data, and generic/abstract anomaly knowledge.

That is to say, our anomaly representation model is classified into three interoperable groups as follows:

1.1. Raw data representation :<O, A, Vqt>
1.2. Abnormal state representation: <O, P, Vp>
1.3. Specific abnormal representation: <O, P, Vp, So>.

3   *Is-a* hierarchy of Anomaly Ontology

In this section, we present our approach for building a framework for knowledge systematization using ontological engineering.

In general, domain experts work with strongly domain-specific knowledge, consequently making it more difficult to share common and generic knowledge. To cope with this problem, in developing our anomaly ontology design, we have to cover both specific concepts (i.e., context-dependent concepts) and basic/generic concepts across multiple domains, namely, context-independent concepts.

To build a framework for knowledge systematization using ontological engineering, we propose the following three-layer ontological model of abnormal states with an *is-a* hierarchical tree (Fig. 2):

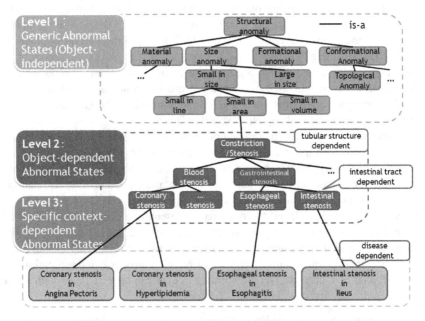

Fig. 2. Three-layer ontological model of abnormal states

1- Top layer (Level 1): Generic abnormal states
2- Middle layer (Level 2): Object-dependent abnormal states
3- Bottom layer (Level 3): Specific context-dependent abnormal states

The details are discussed in the following subsections.

3.1   Level 1: Generic Abnormal States

The top level of the anomaly ontology defines and provides very basic and generic concepts commonly found in several objects, such as cracking, deformation, color changes, dysfunction, and so on. They do not depend on any structures; in other words, they are primitive states[4] that are very general and usable in a wide range of domains, such as machinery, materials, aviation, and medicine.

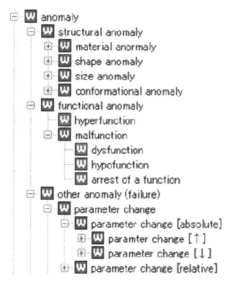

Fig. 3. Top-level categories related to an abnormal state

Consider any abnormal states in multiple domains. Generally, they can be classified into two main groups: structure-related and function-related. Therefore, in this study, the top-level category of generic abnormal states has three subclasses: "structural anomaly", "functional anomaly", and "other anomaly" (Fig. 3). A structural anomaly is defined as an abnormal state associated with structure. It is sub-categorized into material anomaly (e.g., hardening), formation anomaly (e.g., deformation), size anomaly, and conformational anomaly, such as topological anomaly (e.g., dislocation), structural defects (e.g., adduct/loss of parts) etc., while still retaining the identity of the structural body.

Generic abnormal states are context-independent states found in several objects, i.e., object-independent. They are represented in a recursive structure, with reference to Level 2 abnormal states. For example, "small in area" is defined as a lower-level size anomaly at Level 1, which is referred to in various domains and is specified as Object-dependent concepts at Level 2.

A functional anomaly is defined as an abnormal state related to failure to function properly. It is classified into hyperfunction and malfunction, the latter of which is subcategorized into dysfunction, hypofunction, and function arrest, respectively.

---

[4] In this paper, we deal with a single primitive state. A composite state, composed of more than one single state, e.g., "torsional vibration", which is composed of torsion and vibration, is omitted due to space limitation.

Other abnormal states include parameter anomalies, which are classified into increased or decreased parameter, depending on whether or not the attribute has a higher or lower value than a threshold level. Various generic abnormal states related to parameter change, e.g., increased/decreased velocity, increased/decreased pressure, etc., are defined by using general physical parameters, such as velocity, pressure, and concentration.

3.2   Level 2: Object-Dependent Abnormal States

Top level concepts at Level 2 are dependent on generic structures, such as "tubular structure", "bursiform structure", etc., which are common and are used in many domains. By identifying the target object and specializing generic abnormal states of Level 1, an *is-a* hierarchal tree at Level 2 was developed with consistency. For instance, by specializing "small in area" at Level 1, "constriction of tubular structure", where the cross-sectional area has become narrowed, is defined at Level 2, and this is further specialized in the definitions "water pipe narrowing", "intestinal stenosis", and so on.

In the lower level of the tree, abnormal states dependent on domain-specific objects are defined. In the medical domain, for example, "constriction of tubular structure" at the upper-left in Fig. 4 is specified as "vascular stenosis" dependent on "blood vessels" (Fig. 4, middle left.), and is further specified into "coronary stenosis" dependent on coronary artery (Fig. 4, lower left.). Thus, we can represent concepts at all required granularities in the expert domain.

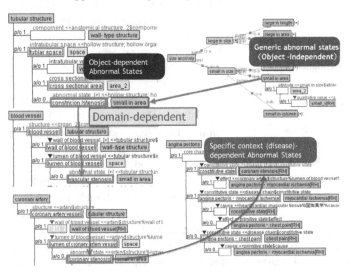

Fig. 4. Specialization of abnormal states from generic level to specific level

Here, in general, the difficulty arises in how fine a granularity needs to be supported in our ontology. In the medical domain, for example, as compared to "polyp formation", "gastric polyp formation" in a specific organ (the stomach) might be redundant and dispensable. However, it is widely known that abnormal states on one component subsequent influence adjacent components, which causes other abnormal states. Actually, a medical concept like gastric polyp is different from nasal polyp; the former causes gastric cancer, whereas the latter causes dysosmia. Therefore, there is a need for distinct abnormal states at specific organ levels.

From an ontological engineering point of view, our framework for modeling abnormal states is intended to capture abnormal states from the generic level to the specific domain-level, so as to provide abnormal states in a necessary number of specific organ/tissue/cell layers in the medical domain.

Note that, although such abnormal states of specific cells are defined at Level 2, they are distinct from context-dependent concepts at Level 3.

For example, hypertension, which means high blood pressure, can be defined in a context-independent manner at Level 2 with reference to level 3 concepts used in various diseases. For instance, how hypertension causes renal artery sclerosis should be defined in a context-dependent manner, i.e., a disease-dependent manner (e.g., renal artery sclerosis-dependent) at Level 3.

3.3   Level 3: Specific Context-dependent Abnormal States

Level 3 concepts are captured as context-dependent abnormal states, which are specialized from abnormal states at Level 2 in specific diseases or machine failure. For example, "coronary stenosis" dependent on coronary artery at Level 2 is defined as a constituent of the disease angina pectoris at Level 3, which causes myocardial ischemia (Fig. 2, lower right), which is also defined as a constituent of hyperlipidemia.

In summary, our ontology can represent various anomalies with consistency. In our ontological approach, common concepts can be kept distinct from specific ones and can be defined as appropriate according to their context. By building an anomaly ontology with an *is-a* hierarchy tree, higher level concepts are more generic and can be shared as cross-domain common knowledge, and lower-level concepts are designed to represent domain-specific knowledge for the required granularity, which is suitable for practical expertise. Consequently, our approach can provide backbone concepts with a machine understandable description, which supports the development of an infrastructure for anomaly knowledge of failures/diseases and will be useful for a wide range of applications.

4   Application of Anomaly Ontology

We have been developing a disease ontology in our Japanese Medical Ontology project. We define a disease as a dependent continuant constituted of one or more causal chains of clinical disorders appearing in a human body and initiated by at least one disorder [6].

Each disease is structured as a directed acyclic graph (DAG), which is composed of nodes (clinical disorders) and relations. We can introduce an *is-a* relation between diseases using chain-inclusion relationship between causal chains as follow:

Disease A is a super type of disease B if the core causal chain of disease A is included in that of disease B. The inclusion of nodes is judged by taking an *is-a* relation between the nodes into account, as well as sameness of the nodes. For example, that diabetes and type-1 diabetes are respectively defined as *<deficiency of insulin →* *elevated level of glucose in the blood>* and <destruction of pancreatic beta cells → lack of insulin in the blood → deficiency of insulin → elevated level of glucose in the blood>. Then, we get *<type-1 diabetes is-a diabetes>*.

Currently, clinicians from 12 medical departments have described causal chains consisting of over 6000 diseases and approximately 17,000 abnormal states (Table 3).

We defined upper level concepts (classes) such as clinical disorders (abnormal states), causal chains, causal relationships (cause and effect), etc. based on YAMATO. In addition, by using these abnormal states, we developed an anomaly ontology with a three-layer model, as described in the previous section, and have also applied our representation model as described in Section 2.

Abnormal states are defined as causes or results of each disease, i.e., context-dependent concepts at Level 3 described in one medical department. For instance, in the Department of Gastroenterology, one clinician defines "intestinal stenosis" as the cause of ileus. Each Level 3 abnormal state is stored as an organ-dependent abnormal state in the Level 2 leaf. Furthermore, at upper levels, abnormal states are more sharable. For example, "intestinal stenosis" and "esophagostenosis" are able to share "stenosis of digestive tract" at Level 2. Moreover, "coronary stenosis" can be shared with a corresponding upper level 2 concept, i.e. "tubular stenosis", via the *is-a* hierarchical tree of the anomaly ontology.

Table 3. Statistics

Medical Department	Number of Abnormal state	Number of Disease
Allergy and Rheumatoid	806	101
Cardiovascular Medicine	2,289	550
Diabetes and Metabolic Diseases	1,989	445
Orthopedic Surgery	1,121	208
Respiratory Medicine	1,739	788
Neurology	1,893	397
Ophthalmology	1,306	561
Nephrology and Endocrinology	868	196
Hematology and Oncology	354	415
Dermatology	908	1,086
Pediatrics	2,334	879
Otorhinolaryngology	1,118	470
Total	16,725	6,096

Level 2 leaf abnormal states are able to be reused as reference information for other diseases at all 12 medical departments. For example, level 3 "coronary stenosis" described in the Department of Cardiovascular Medicine is first stored as a Level 2 leaf abnormal state, and later it can be reused as a result of Level 3 "accumulation of cholesterol" in hyperlipidemia at another department, namely, the Department of Diabetes and Metabolic Diseases (Fig. 5).

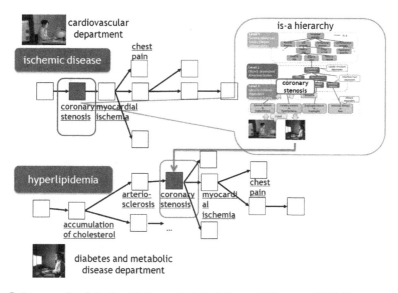

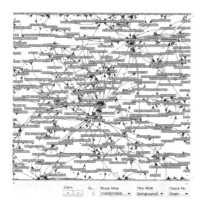

Fig. 5. An example of sharing of abnormal states between different medical departments

Note that, once a causal chain related to an abnormal state is described, another clinician can use it as a generic causal chain. Consequently, a clinician at the Department of Diabetes and Metabolic Diseases can annotate the generic progression of "coronary stenosis".

Fig. 6. An example of the dynamic generation of generic causal chain of a disease

Though the causal chains of each disease are described at particular medical departments, generic causal chains can be generated by combining causal chains that include the same abnormal states, which allows all causal relationships, including 17,000 abnormal states from 12 medical departments, to be visualized (Fig. 6).

Since our approach can manage both the causal relations of diseases and the related *is-a* hierarchical structure of the anomaly ontology, it might be a good infrastructure for managing various kinds of anomaly knowledge.

In this sense, using our infrastructure, which connects abnormal states to various granularities of abnormal states via (1) causal chains of disease i.e., horizontal relations among the same level, and (2) an *is-a* hierarchy in the ontological structure, i.e., vertical relations among the different level, will facilitate cross-domain usage among heterogeneous concepts of abnormal states and diseases/failures. This method should aid in understanding the mechanisms that cause diseases and lead to more-complete understanding of related abnormal states. Furthermore, knowledge regarding canonical causal chains of upper-level abnormal states can be annotated

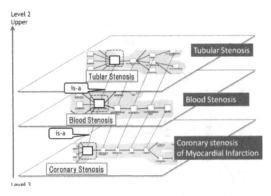

Fig. 7. Causal chain layers

from the relations among real causal chains by using the anomaly ontology and causal chains (Fig. 7).

For example, from a causal chain including "coronary stenosis" at Level 2, we could obtain an upper-level causal chain including "blood stenosis" and the even higher level "tube stenosis".

If we trace the vertical relation of each abnormal state from some causal chains of diseases that are understood, a common scenario might be generated as a canonical causal chain, which will provide a clue to revealing the mechanisms of other diseases that remain poorly understood.

5   Related Work

There are at least three methods of describing qualities in BFO [1], DOLCE [2] and Galen [7]. BFO recommends the <Entity, Property> (e.g., <John, tall>) formalism, whereas DOLCE recommends <Entity, Attribute, Value> (e.g., <John, height, 180cm>) and Galen recommends <Entity, Property, Value> (e.g., <John, tallness, large>). However, all three descriptions have the same meaning, namely, John's height.

A property description such as <bone, deformed> seems not to be covered by DOLCE whereas it is covered by BFO, which does not cover <Entity, Attribute, Value>. The YAMATO ontology covers all these three kinds of descriptions and supports interoperability among them.

Since it is based on the YAMATO ontology, our representation model is not only able to manage the three kinds of description but can also be used as a reference ontology for representing properties (qualities) among these upper ontologies.

In the biotechnology community, Phenotypic Quality (PATO) [8] is a famous ontology of phenotypic qualities, defining composite phenotypes and phenotype annotations, and the current version of PATO adopts property descriptions, e.g., <eye, red>. Our ontology is compatible with the latest PATO description, and also with older versions of EAV (Entity + Attribute + Value) .

In the medical domain, many medical ontologies have been developed for realizing sophisticated medical information systems, such as Ontology for General Medical Science (OGMS) [3], DO [9], IDO [10]. However, they do not include sufficient information regarding the relations between abnormal states in one disease. In future, we plan to link to an external ontology such as OGMS and provide useful information about causal relationships in diseases so that the concepts complement each other.

6 Conclusions

In this paper, first, we discussed various complicated issues of abnormal state concepts and introduced a representation model that handles 17,000 abnormal states from approximately 6000 diseases by capturing properties, which are consistently decomposed into attributes. Next, we proposed a systematic method for developing an anomaly ontology from the generic level to the specific domain level. We have also developed a hierarchical tree for capturing both commonality and specificity discretely. Our model demonstrated that common knowledge about abnormal states can be shared at the upper level of abnormal states and is reusable for describing other diseases in different medical departments.

From preliminary studies on abnormal states from our ontological tree, the number of top level (Level 1) generic abnormal states, which are not dependent on the human body and are common concepts in multiple domains, is about 100.

Furthermore, with a combination of our anomaly ontology and causal chains of diseases, we can capture all causal relations of 17,000 abnormal states in approximately 6,000 diseases from 12 medical departments.

In the medical domain, e-health needs data exchange, such as Electronic health records (EHR), in order to exchange data appropriately, and it is necessary not only to manage data as quantitative values but also to capture the intrinsic essentials as semantics. Hence, our representation model has interoperability between qualitative data and the properties of anomalies, as shown in Section 2.3, which might contribute to organizing an integrated system in which various anomalies from raw data to abstract knowledge are accessible and manageable by computers. Our approach provides various useful information for a better understanding of the essentials of abnormal states in disease, and in addition, provides practical usability from the raw data level to the semantics level.

As shown in Fig. 8, from the raw data level to the anomaly knowledge level, a three-level architecture that consists of (1) a database for raw clinical data, (2) a

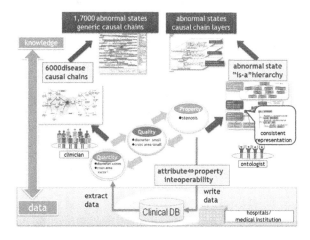

Fig. 8. Integrated system architecture for anomaly knowledge

processor of interoperating quantitative, qualitative and property information, and (3) a knowledge space for abnormal states, where an ontology and causal chains are developed as a backbone, might be suitable.

Acknowledgements. A part of this research was supported by the Ministry of Health, Labor and Welfare, Japan, through its "Research and development of medical knowledge base databases for medical information system" and by the Japan Society for the Promotion of Science (JSPS) through its "Funding Program for World-Leading Innovative R&D on Science and Technology (FIRST Program)".

References

1. Grenon, P., et al.: Biodynamic Ontology: Applying BFO in the Biomedical Domain. In: Pisanelli, D.M. (ed.) Ontologies in Medicine, pp. 20–38. IOS, Amsterdam (2004)
2. Guarino, N.: Some Ontological Principles for Designing Upper Level Lexical Resources. In: Proc. of International Conference on Lexical Resources and Evaluation (1998)
3. Scheuermann, R.H., Ceusters, W., Smith, B.: Toward an Ontological Treatment of Disease and Diagnosis. In: Proc. of the 2009 AMIA Summit on Translational Bioinformatics, San Francisco, pp. 116–120 (2009)
4. Mizoguchi, R.: YAMATO: Yet Another More Advanced Top-level Ontology. In: Proceedings of the Sixth Australasian Ontology Workshop, pp. 1–16 (2010)
5. Kuzuya, T., et al.: Report of the Committee on the Classification and Diagnostic Criteria of Diabetes Mellitus. Journal of the Japan Diabetes Society 53(6), 450–467 (2010)
6. Mizoguchi, R., et al.: River Flow Model of Diseases. In: Proc. of ICBO 2011, pp. 63–70 (2011)
7. OpenGALEN, http://www.opengalen.org/
8. Gkoutos, G.V., et al.: Ontologies for the description of mouse phenotypes. Comp. Funct. Genomics 5, 545–551 (2004)
9. Osborne, J.D., et al.: Annotating the human genome with Disease Ontology. BMC Genomics 10(1), S6 (2009)
10. Cowell, L.G., Smith, B.: Infectious Disease Ontology. In: Sintchenko, V. (ed.) Infectious Disease Informatics, ch. 19, pp. 373–395 (2010)

SkyPackage: From Finding Items to Finding a Skyline of Packages on the Semantic Web

Matthew Sessoms and Kemafor Anyanwu

Semantic Computing Research Lab, Department of Computer Science
North Carolina State University,
Raleigh, North Carolina, USA
{mwsessom,kogan}@ncsu.edu
http://www.ncsu.edu

Abstract. Enabling complex querying paradigms over the wealth of available Semantic Web data will significantly impact the relevance and adoption of Semantic Web technologies in a broad range of domains. While the current predominant paradigm is to retrieve a list of items, in many cases the actual intent is satisfied by reviewing the lists and assembling compatible items into lists or packages of resources such that each package collectively satisfies the need, such as assembling different collections of places to visit during a vacation. Users may place constraints on individual items, and the compatibility of items within a package is based on global constraints placed on packages, like total distance or time to travel between locations in a package. Finding such packages using the traditional item-querying model requires users to review lists of possible multiple queries and assemble and compare packages manually.

In this paper, we propose three algorithms for supporting such a package query model as a first class paradigm. Since package constraints may involve multiple criteria, several competing packages are possible. Therefore, we propose the idea of computing a skyline of package results as an extension to a popular query model for multi-criteria decision-making called skyline queries, which to date has only focused on computing item skylines. We formalize the semantics of the logical query operator, *SkyPackage*, and propose three algorithms for the physical operator implementation. A comparative evaluation of the algorithms over real world and synthetic-benchmark RDF datasets is provided.

Keywords: RDF, Package Skyline Queries, Performance.

1 Introduction

The Linking Open Data movement and broadening adoption of Semantic Web technologies has created a surge in amount of available data on the Semantic Web. The rich structure and semantics associated with such data provides an opportunity to enable more advanced querying paradigms that are useful for different kinds of tasks. An important class of such advanced querying models is the *skyline query* model for supporting multi-criteria retrieval tasks. While

H. Takeda et al. (Eds.): JIST 2012, LNCS 7774, pp. 49–64, 2013.
© Springer-Verlag Berlin Heidelberg 2013

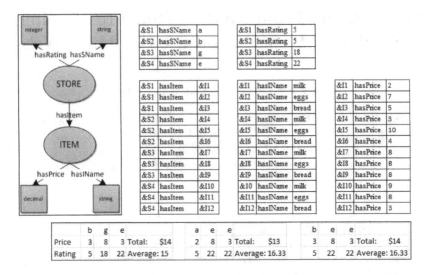

Fig. 1. Data For E-Commerce Example

this class of queries have been considered as an extension to traditional pattern matching queries in relational [2][7][10] and more recently Semantic Web data models [3], it is important to consider more complex underlying query models. While pattern matching paradigms are predicated on finding individual single items, often times a user is interested in finding item combinations or packages that satisfy certain constraints. For example, a student may be interested in finding a combination of e-learning resources that cover a set of topics, with constraints like overall rating of resources should be high, and if being purchased, total cost should be low.

As a more concrete example, assume stores expose Semantic Web enabled catalogs, and a customer wishes to make a decision about what combination of stores would best meet his or her purchase needs. The customer wishes to purchase milk, eggs, and bread and is willing to make the purchase from multiple stores as long as the total price spent is minimized, and the overall average rating across stores is good or maximized, i.e., the stores are known to have good quality products. Additional constraints could include minimizing total distance traveled between the stores and some constraints about open shopping hours.

The example shows that the target of the query is in fact a set of items (a set of stores), where the query should return multiple combinations or "packages". Since the preferences are specified over aggregated values for elements in a package, the process of producing combinations will need to precede the preference checking. One possible package for the e-commerce example, given the sample data in Fig 1, is *aee* (i.e., buying milk from store *a*, eggs from store *e* and bread also from store *e*). Another possibility is *bee* (i.e., milk from store *b*, eggs and bread from store *e* as in the previous case). The bottom right of Fig 1 shows the total price and average rating for packages *aee*, *bee* and *bge*. We see that *aee* is a better package than *bee* because it has a smaller total price and the same average rating. On the other hand, *bge* and *aee* are incomparable because although *bge*'s total price is worse than *aee*'s, its average rating is better.

In [5][11][12], the idea of package recommendation has been explored. This problem assumes a single criterion such that an optimal package can be selected or a top-k of packages based on techniques similar to collaborative filtering for recommendation. With respect to skyline queries, different indexing strategies for optimizing skyline computation over a single relation or restricted join structures e.g. just two relations i.e. single join [10], star-like schemas [14] have been investigated. The challenges of skylining over RDF data models which do not conform to the requirement of being contained in two relations or having restricted schema structures, were explored in [3]. The proposed approach interleaved the join and skyline phases in such a way that the information about the already explored values was used to prune the non-valuable candidates from entering the skyline phase. However, these existing techniques all focus on computing item-based skylines. The main challenge in terms of package skylines is dealing with the increase in search space due to the combinatorial explosion.

In this paper, we introduce the concept of *package skyline queries* and propose query evaluation techniques for such queries. Specifically, we contribute the following: 1) a formalization of the logical query operator, *SkyPackage*, for package skyline queries over an RDF data model, 2) two families of query processing algorithms for physical operator implementation based on the traditional vertical partitioned storage model and a novel storage model here called *target descriptive, target qualifying* (TDTQ), and 3) an evaluation of the three algorithms proposed over synthetic and real world data. The rest of the paper is organized as follows. The background and formalization of our problem is given in Section 2. Section 3 introduces the two algorithms based on the vertical partitioned storage model, and Section 4 presents an algorithm based on our TDTQ storage model. An empirical evaluation study is reported in Section 5, and related work is described in Section 6. The paper is concluded in Section 7.

2   Preliminaries

Specifying the e-commerce example as a query requires the use of a graph pattern structure for describing the *target* of the query (*stores*), and the *qualification* for the desired targets e.g. stores *should sell* at least one of *milk* or *bread* or *eggs*. We call these *target qualifiers*. The second component of the query specifications concerns the *preferences*, e.g., *minimizing the total cost*. In our example, preferences are specified over the aggregates of target attributes (datatype properties), e.g., maximizing the average over store ratings, as well as the attributes of target qualifiers, e.g., minimizing total price. Note that the resulting graph pattern structure (we ignore the details of preferences specification at this time) involves a union query where each branch computes a set of results for one of the target qualifiers, shown in Fig 2.

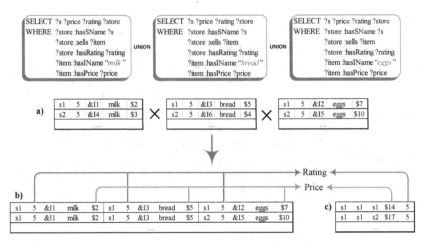

Fig. 2. Dataflow for the Skyline Package Problem in Terms of Traditional Query Operators

2.1   Problem Definition

Let D be a dataset with property relations P_1, P_2, .., P_m and GP be a graph pattern with triple patterns TP_i, TP_j, .., TP_k (TP_x denotes triple pattern with property P_x). $[[GP]]_D$ denotes the answer relation for GP over D, i.e., $[[GP]]_D = P_i \bowtie P_j \bowtie ... \bowtie P_k$. Let $var(TP_x)$ and $var(GP)$ denote the variables in the triple and graph pattern respectively. We designate the return variable $r \in var(GP)$, e.g., (?store), representing the target of the query (stores) as the *target* variable. We call the triple patterns such as (*?item hasName "milk"*) *target qualifying constraints* since those items determine the stores that are selected.

We review the formalization for preferences given in [7]. Let $Dom(a_1)$, ..., $Dom(a_d)$ be the domains for the columns $a_1, .., a_d$ in a d-dimensional tuple $t \in [[GP]]_D$. Given a set of attributes B, then a preference PF over the set of tuples $[[GP]]_D$ is defined as $PF := (B; \succ_{PF})$ where $\succ_{PF}$ is a strict partial order on the domain of B. Given a set of preferences $PF_1, ..., PF_m$, their combined Pareto preference PF is defined as a set of equally important preferences.

For a set of d-dimensional tuples R and preference $PF = (B; \succ_{PF})$ over R, a tuple $r_i \in R$ dominates tuple $r_j \in R$ based on the preference (denoted as $r_i \succ_{PF} r_j$), iff $(\forall(a_k \in B)(r_i[a_k] \succeq r_j[a_k]) \wedge \exists(a_l \in B)(r_i[a_l] \succ r_j[a_l]))$

Definition 1 (Skyline Query). *When adapting preferences to graph patterns we associate a preference with the property (assumed to be a datatype property) on whose object the preference is defined. Let PF_i denote a preference on the column representing the object of property P_i). Then, for a graph pattern $GP = TP_1, .., TP_m$ and a set of preferences $PF = PF_i, PF_j, .., PF_k$, a skyline query $SKYLINE[[[GP]]_D, PF]$ returns the set of tuples from the answer of GP such that no other tuples dominate them with respect to PF and they do not dominate each other.*

The extension of the skyline operator to packages is based on two functions *Map* and *Generalized Projection*.

Definition 2 (Map and Generalized Projection). . *Let* $\mathcal{F} = \{f_1, f_2, ...f_k\}$ *be a set of k mapping functions such that each function* $f_j(B)$ *takes a subset of attributes* $B \subseteq A$ *of a tuple t, and returns a value x.* **Map** $\hat{\mu}_{[\mathcal{F},\mathcal{X}]}$ *(adapted from [7]) applies a set of k mapping functions* $\mathcal{F}$ *and transforms each d-dimensional tuple t into a k-dimensional output tuple* t' *defined by the set of attributes* $\mathcal{X} = \{x_1, x_2, ...x_k\}$ *with* x_i *generated by the function* f_i *in* $\mathcal{F}$.

Generalized Projection $\prod_{colr_x, colr_y, colr_z, \hat{\mu}_{[\mathcal{F},\mathcal{X}]}}(R)$ *returns the relation* $R'(colr_x, colr_y, colr_z, .., x_1, x_2, ...x_k)$. *In other words, the generalized projection outputs a relation that appends the columns produced by the map function to the projection of R on the attributes listed, i.e.* $colr_x, colr_y, colr_z$.

Definition 3 (SkyPackage Graph Pattern Query). . *A* SkyPackage *graph pattern query is graph pattern* $GP_{[\{c_1, c_2, ..., c_N\}, \mathcal{F} = \{f_1, f_2, ..., f_k\} \{PF_{P_i}, PF_{P_j}, ...PF_{P_k}\}, r]}$ *such that :*

1. c_i *is a qualifying constraint.*
2. $r \in \text{var}(GP)$ *is called the target of the query e.g., stores.*
3. PF_{P_i} *is the preference specified on the property* P_i, *i.e., actually the object of* p_i. f_i *is the mapping function of* P_i.

Definition 4 (SkyPackage Query Answer). *The answer to a skypackage graph pattern query* R_{SKY} *can be seen the result of the following steps:*

1. $R_{product} = [[GP_{c_1}]] \times [[GP_{c_2}]] \times ... \times [[GP_{c_N}]]$ *such that* $[[GP_{c_x}]]$ *is the result of evaluating the branch of the union query with constraint* c_x. *Fig 2.(b) shows the partial result of the crossproduct of the three subqueries in (a) based on the 3 constraints on milk, bread and eggs.*
2. $R_{project} = \prod_{r_1, r_2, ...r_N, \hat{\mu}_{[\mathcal{F}=\{f_1, f_2, ...f_k\}, \mathcal{X}=\{x_1, x_2, ...x_k\}]}}(R_{product})$ *where* r_i *is the column for the return variable in subquery i's result,* $f_1 : (domc1(o_1) \times domc2(o_1) \times ... \times domcN(o_1)) \rightarrow \mathbb{R}$ *where* $domc1(o_1)$ *is the domain of values for the column representing the object of* P_1 *e.g. column for object of hasPrice, in* $[[GP_{c_1}]]$. *The functions in our example would be* $total_{hasPrice}$, $average_{hasRating}$. *The output of this step is shown in Fig 2.(c)*
3. $SKYLINE[R_{project}, \{PF_{P'_i}, PF_{P'_j}, ...PF_{P'_k}\}]$ *such that* $\{PF_{P'_i}$ *is the preference defined on the aggregated columns produced by the map function (denoted by* P'_i *e.g. minimizing totalprice.*

In other words, the result of a skypackage query is the result of skylining over the extended generalized projection (with map functions consisting of desired aggregates) of the combination (product) relation.

3   Algorithms for Package Skyline Queries over Vertical Partitioned Tables

We present in this section two approaches: *Join, Cartesian Product, Skyline* (*JCPS*) *and* $RDFSkyJoinWithFullHeader - Cartesian$ $Product, Skyline$

$(RSJFH - CPS)$, for solving the package skyline problem. These approaches assume data is stored in vertically partitioned tables (VPTs) [1].

3.1 *JCPS* Algorithm

The formulation of the package skyline problem suggests a relatively straight-forward algorithm involving multiple *joins*, followed by a *Cartesian product* to produce all combinations, followed by a single-table *skyline* algorithm (e.g., block-nested loop), called *JCPS*.

Consider the VPTs *hasIName*, *hasSName*, *hasItem*, *hasPrice*, and *hasRating* obtained from Fig 1. Solving the skyline package problem using *JCPS* involves the following steps:

1. $I \leftarrow hasIName \bowtie hasSName \bowtie hasItem \bowtie hasPrice \bowtie hasRating$
2. Perform Cartesian product on I twice (e.g., $I \times I \times I$)
3. Aggregate *price* and *rating* attributes
4. Perform a single-table skyline algorithm

Steps 1 and 2 can be seen in Fig 3a, which depicts the Cartesian product being performed twice on I to obtain all store packages of size 3. As the product is being computed, the *price* and *rating* attributes are aggregated, as shown in Fig 3b. Afterwards, a single-table skyline algorithm is performed to discard all dominated packages with respect to total price and average rating.

item	price	store	rating
milk	2	A	5
eggs	7	A	5
⋮			

item	price	store	rating
milk	2	A	5
eggs	7	A	5
⋮			

item	price	store	rating
milk	2	A	5
eggs	7	A	5
⋮			

(a) Cartesian Product of Join Result

item1	item2	item3	store1	store2	store3	total price	average rating
milk	eggs	bread	A	B	C	10	5
milk	eggs	bread	B	B	A	7	5
⋮							

(b) Cartesian Product Result

Fig. 3. Resulting tables of *JCPS*

Algorithm 1 contains the pseudocode for such an algorithm. Solving the skyline package problem using *JCPS* requires all VPTs to be joined together (line 2), denoted as I. To obtain all possible combinations (i.e., packages) of targets, multiple Cartesian products are performed on I (lines 3-5). Afterwards, *equivalent* skyline attributes are aggregated (lines 6-8). Equivalent skyline attributes, for example, of the e-commerce motivating example would be price and rating attributes. Aggregation for the price of milk, eggs, and bread would be performed to obtain a total price. Finally, line 9 applies a single-table skyline algorithm to remove all dominated packages.

Algorithm 1. JCPS

Input: $VPT_1, VPT_2, \ldots VPT_x$ containing skyline attributes $s_1, s_2, \ldots, s_y$, and corresponding aggregation functions $\mathcal{A}_{s_1}(T), \mathcal{A}_{s_2}(T), \ldots, \mathcal{A}_{s_y}(T)$ on table T

Output: Package Skyline $\mathcal{P}$

1: $n \leftarrow$ package size
2: $I \leftarrow VPT_1 \bowtie VPT_2 \bowtie \cdots \bowtie VPT_x$
3: **for all** $i \in [1, n-1]$ **do**
4:    $I \leftarrow I \times I$
5: **end for**
6: **for all** $i \in [1, y]$ **do**
7:    $I \leftarrow \mathcal{A}_{s_i}(I)$
8: **end for**
9: $\mathcal{P} \leftarrow$ skyline(I)
10: **return** $\mathcal{P}$

The limitations of such an algorithm are fairly obvious. First, many unnecessary joins are performed. Furthermore, if the result of joins is large, the input to the Cartesian product operation will be very large even though it is likely that only a small fraction of the combinations produced will be relevant to the skyline. The exponential increase of tuples after the Cartesian product phase will result in a large number of tuple-pair comparisons while performing a skyline algorithm. To gain better performance, it is crucial that some tuples be pruned before entering into the Cartesian product phase, which is discussed next.

3.2 $RSJFH - CPS$ Algorithm

A pruning strategy that prunes the input size of the Cartesian product operation is crucial to achieving efficiency. One possibility is to exploit the following observation: *skyline packages can be made up of only target resources that are in the skyline result when only one constraint (e.g., milk) is considered* (note that a query with only one constraint is equivalent to an item skyline query).

Lemma 1. *Let* $\rho = \{p_1 p_2 \ldots p_n\}$ *and* $\rho' = \{p_1 p_2 \ldots p'_n\}$ *be packages of size* n *and* ρ *be a skyline package. If* $p_n \preceq_{C_n} p'_n$*, where* C_n *is a qualifying constraint, then* ρ' *is not a skyline package.*

Proof. Let $x_1, x_2, \ldots, x_m$ be the preference attributes for p_n and p'_n. Since $p_n \preceq_{C_n} p'_n$, $p_n[x_j] \preceq p'_n[x_j]$ for some $1 \leq j \leq n$. Therefore, $\mathcal{A}_{1 \leq i \leq n}(p_i[x_j]) \preceq \mathcal{A}_{1 \leq i \leq n}(p'_i[x_j])$, where $\mathcal{A}$ is an aggregation function and $p'_i \in \rho'$. Since for any $1 \leq k \leq n$, where $k \neq j$, $\mathcal{A}_{1 \leq i \leq n}(p_i[x_k]) = \mathcal{A}_{1 \leq i \leq n}(p'_i[x_k])$. This implies that $\rho \preceq \rho'$. Thus, ρ' is not a skyline package. $\square$

As an example, let $\rho = \{p_1 p_2\}$ and $\rho' = \{p_1 p_2'\}$ and x_1, x_2 be the preference attributes for p_1, p_2, p_2'. We define the attribute values as follows: $p_1 = (3, 4), p_2 = (3, 5)$, and $p_2' = (4, 5)$. Assuming the lowest values are preferred, $p_2 \preceq p_2'$ and $p_2[x_1] \preceq p_2'[x_1]$. Therefore, $\mathcal{A}_{1 \leq i \leq 2}(p_i[x_1]) \preceq \mathcal{A}_{1 \leq i \leq 2}(p_i'[x_1])$. In other words, $(p_1[x_1] + p_2[x_1]) \preceq (p_1[x_1] + p_2'[x_1])$, i.e., $(3 + 3 = 6 \preceq 7 = 3 + 4)$. Since all attribute values except $p_2'[x_1]$ remained unchanged, by definition of skyline we conclude $\rho \preceq \rho'$.

This lemma suggests that the skyline phase can be pushed ahead of the Cartesian product step as a way to prune the input of the $JCPS$. Even greater performance can be obtained by using a skyline-over-join algorithm, $RSJFH$ [3], that combines the skyline and join phase together. $RSJFH$ takes as input two VPTs sorted on the skyline attributes. We call this algorithm $RSJFH - CPS$. This lemma suggests that skylining can be done in a divide-and-conquer manner where a skyline phase is introduced for each constraint, e.g., milk, (requiring 3 phases for our example) to find all potential members of skyline packages which may then be fed to the Cartesian product operation.

Given the VPTs $hasIName, hasSName, hasItem, hasPrice$, and $hasRating$ obtained from Fig 1, solving the skyline package problem using $RSJFH - CPS$ involves the following steps:

1. $I^2 \leftarrow hasSName \bowtie hasItem \bowtie hasRating$
2. For each target t (e.g., milk)
   (a) $I_t^1 \leftarrow \sigma_t(hasIName) \bowtie hasPrice$
   (b) $S_t \leftarrow RSJFH(I_t^1, I^2)$
3. Perform a Cartesian product on all tables resulting from step 2b
4. Aggregate the necessary attributes (e.g., price and rating)
5. Perform a single-table skyline algorithm

Fig 4a shows two tables, where the left one, for example, depicts step (a) for milk, and the right table represents I^2 from step 1. These two tables are sent as input to $RSJFH$, which outputs the table in Fig 4b. These steps are done for each target, and so in our example, we have to repeat the steps for *eggs* and *bread*. After steps 1 and 2 are completed (yielding three tables, e.g., *milk*, *eggs*, and *bread*), a Cartesian product is performed on these tables, as shown in Fig 4c, which produces a table similar to the one in Fig 3b. Finally, a single-table skyline algorithm is performed to discard all dominated packages.

Algorithm 2 shows the pseudocode for $RSJFH - CPS$ The main difference between $JCPS$ and $RSJFH - CPS$ appears in line 5-8. For each target, a *select* operation is done to obtain all like targets, which is then joined with another VPT containing a skyline attribute of the targets. This step produces a table for *each* target. After the remaining tables are joined, denoted as I^2 (line 4), each target table I_i^1 along with I^2 is sent as input to $RSJFH$ for a skyline-over-join operation. The resulting target tables undergo a Cartesian product phase (line 9) to produce all possible combinations, and then all equivalent attributes are

hasIName	⋈	hasPrice		⋈	hasSName	⋈	hasItem	⋈	hasRating	
&I1	milk	&I1	2		&S1	A	&S1	&I1	&S1	5
&I4	milk	&I4	3		&S1	A	&S1	&I2	&S1	5

(a) $RSJFH$ (skyline-over-join) for milk

item	price	store	rating
milk	2	A	5
milk	3	B	5
⋮			

(b) $RSJFH$'s result for milk

item	price	store	rating		item	price	store	rating		item	price	store	rating
milk	2	A	5	✖	eggs	3	B	5	✖	bread	5	A	4
⋮					⋮					⋮			

(c) Cartesian product on all targets (e.g., milk, eggs, and bread)

Fig. 4. Resulting tables of $RSJFH$

Algorithm 2. RSJFH-CPS

Input: $VPT_1, VPT_2, \ldots VPT_x$ containing skyline attributes $s_1, s_2, \ldots, s_y$, and
corresponding aggregation functions $\mathcal{A}_{s_1}(T), \mathcal{A}_{s_2}(T), \ldots, \mathcal{A}_{s_y}(T)$ on table T
Output: Package Skyline $\mathcal{P}$
1: $n \leftarrow$ package size
2: $t_1, t_2, \ldots, t_n \leftarrow$ targets of the package
3: VPT_1 contains targets and VPT_2 contains a skyline attribute of the targets
4: $I^2 \leftarrow VPT_3 \bowtie \cdots \bowtie VPT_x$
5: **for all** $i \in [1, n]$ **do**
6:     $I_i^1 \leftarrow \sigma_{t_i}(VPT_1) \bowtie VPT_2$
7:     $S_i \leftarrow RSJFH(I_i^1, I^2)$
8: **end for**
9: $T \leftarrow S_1 \times S_2 \times \cdots \times S_n$
10: **for all** $i \in [1, n]$ **do**
11:     $I \leftarrow \mathcal{A}_{s_y}(T)$
12: **end for**
13: $\mathcal{P} \leftarrow \text{skyline}(I)$
14: **return** $\mathcal{P}$

---

aggregated (lines 10-12). Lastly, a single-table skyline algorithm is performed to discard non-skyline packages (line 13). Since a skyline phase is introduced early in the algorithm, the input size of the Cartesian product phase is decreased, which significantly improves execution time compared to $JCPS$.

4   Algorithms for Package Skyline Queries over the TDTQ Storage Model

In this section, we present a more efficient and feasible method to solve the skyline package problem. We discuss a devised storage model, *Target Descriptive, Target Qualifying* (TDTQ), and an overview of an algorithm, *SkyJCPS*, that exploits this storage model.

4.1   The TDTQ Storage Model

While the previous two approaches, $JCPS$ and $RSJFH - CPS$, rely on VPTs, the next approach is a multistage approach in which the first phase is analogous to the build phase of a hash-join. In our approach, we construct two types of tables: *target qualifying* tables and *target descriptive* tables, called $TDTQ$. Target qualifying tables are constructed from the target qualifying triple patterns (*?item hasIName "milk"*) and the triple patterns that associate them with the targets (*?store sells ?item*). In addition to these two types of triple patterns, a triple pattern that describes the target qualifier that is associated with a preference is also used to derive the target qualifying table. In summary, these three types of triple patterns are joined per given constraint and a table with the target and preference attribute columns are produced. The left three tables in Fig 5 show the the target qualifying tables for our example (one for each constraint). The target descriptive tables are constructed per target attribute that is associated with a preference, in our example *rating* for stores. These tables are constructed by joining the triple patterns linking the required attributes and produce a combination of attributes and preference attributes (store name and store rating produced by joining hasRating and hasSName). The rightmost table in Fig 5 shows the target descriptive table for our example.

MILK		EGGS		BREAD		RATING	
store	price	store	price	store	price	store	value
a	2	g	5	e	3	e	5
b	3	a	7	b	4	i	5
f	3	c	8	a	5	c	12
e	4	e	8	i	8	f	13
g	6	h	9	g	5	h	14
e	9	b	10	f	6	g	18
h	9	d	10	h	6	a	20
d	10			d	10	d	21
						b	22

Fig. 5. Target Qualifying (*milk, eggs, bread*) and Target Descriptive (*rating*) Tables for E-commerce Example

We begin by giving some notations that will aid in understanding of the TDTQ storage model. In general, the build phase produces a set of partitioned tables $T_1, \ldots, T_n, T_{n+1}, \ldots, T_m$, where each table T_i consists of two attributes, denoted by T_i^1 and T_i^2. We omit the subscript if the context is understood or if the identification of the table is irrelevant. $T_1, \ldots, T_n$ are the target qualifying tables where n is the number of qualifying constraints. $T_{n+1}, \ldots, T_m$ are the target descriptive tables, where $m - (n + 1) + 1 = m - n$ is the number of target attributes involved in the preference conditions.

4.2 *CPJS* and *SkyJCPS* Algorithms

Given the TDTQ storage model presented previously, one option for computing the package skyline would be to perform a Cartesian product on the target qualifying tables, and then joining the result with the target descriptive tables. We call this approach *CPJS* (*Cartesian product, Join, Skyline*), which results in time and space complexity. Given n targets and m target qualifiers, n^m possible combinations exist as an intermediate result prior to performing a skyline algorithm. Each of these combinations is needed since we are looking for a set of packages rather than a set of points. Depending on the preferences given, additional computations such as aggregations are required to be computed at query time. Our objective is to find all package skylines *efficiently* by eliminating unwanted tuples before we perform a Cartesian product. Algorithm 3 shows the *CPJS* algorithm for determining package skylines.

Algorithm 3. CPJS

Input: $T_1, T_2, \ldots T_n, T_{n+1}, \ldots, T_m$
Output: Package Skyline $\mathcal{P}$
1: $I \leftarrow T_1 \times T_2 \times \cdots \times T_n$
2: **for all** $i \in [n + 1, m]$ **do**
3:    $I \leftarrow I \bowtie T_i$
4: **end for**
5: $\mathcal{P} \leftarrow \text{skyline}(I)$
6: **return** $\mathcal{P}$

CPJS begins by finding all combinations of targets by performing a Cartesian product on the target qualifying tables (line 1). This resulting table is then joined with each target descriptive table, yielding a single table (line 3). Finally, a single-table skyline algorithm is performed to eliminate dominated packages (line 5).

Applying this approach to the data in Fig 5, one would have to compute all 448 possible combinations before performing a skyline algorithm. The number of combinations produced from the Cartesian product phase can be reduced by initially introducing a skyline phase on each target, e.g., milk, as we did in for $RSJFH - CPS$. We call this algorithm $SkyJCPS$. Although similarities to $RSJFH - CPS$ can be observed, $SkyJCPS$ yields better performance due to the reduced number of joins. Fig 4a clearly illustrates that $RSJFH - CPS$ requires four joins before an initial skyline algorithm can be performed. All but one of these joins can be eliminated by using the TDTQ storage model. To illustrate $SkyJCPS$, given the TDTQ tables in Fig 5, solving the skyline package problem involves the following steps:

1. For each target qualifying table TQ_i (e.g., milk)
   (a) $I_{TQ_i} \leftarrow (TQ_i) \bowtie rating$
   (b) $I'_{TQ_i} \leftarrow \text{skyline}(I_{TQ_i})$
2. $CPJS(I'_{TQ_1}, \ldots, I'_{TQ_i}, rating)$

Since the dominating cost of answering skyline package queries is the Cartesian product phase, the input size of the Cartesian product can be reduced by performing a single-table skyline algorithm over each target.

5   Evaluation

The main goal of this evaluation was to compare the performance of the proposed algorithms using both synthetic and real datasets with respect to package size scalability. In addition we compared the the feasibility of answering the skyline package problem using the VPT storage model and the TDTQ storage model.

5.1   Setup

All experiments were conducted on a Linux machine with a 2.33GHz Intel Xeon processor and 40GB memory, and all algorithms were implemented in Java SE 6. All data used was converted to RDF format using the Jena API and stored in Oracle Berkeley DB.

We compared three algorithms, $JCPS$, $SkyJCPS$, and $RSJFH-CPS$. During the skyline phase of each of these algorithms, we used the *block-nested-loops* (BNL) [2] algorithm. The package size metric was used for the scalability study of the algorithms. Since the Cartesian product phase is likely to be the dominant cost in skyline package queries, it is important to analyze how the algorithms perform when the package size grows, which increases the input size of the Cartesian product phase.

5.2   Package-Size Scalability - Synthetic Data

Since we are unaware of any RDF data generators that allow generation of different data distributions, the data used in the evaluations were generated using a synthetic data generator [2]. The data generator produces relational data in different distributions, which was converted to RDF using the Jena API. We generated three types of data distributions: correlated, anti-correlated, and normal distributions. For each type of data distribution, we generated datasets of different sizes and dimensions.

In Fig 6 we show how the algorithms perform across packages of size 2 to 5 for the a triple size of approximately 635 triples. Due to the exponential increase of the Cartesian product phase, this triple size is the largest possible in order to evaluate all three algorithms. For all package sizes, $SkyJCPS$ performs better than $JCPS$ because of the initial skyline algorithm performed to reduce the input size of the Cartesian product phase. $RSJFH$ outperformed $SkyJCPS$

for packages of size 5. We argue that $SkyJCPS$ may perform slightly slower than $RSJFH$ on small datasets distributed among many tables. In this scenario, $SkyJCPS$ has six tables to examine, while $RSJFH$ has only two tables. Evaluation results from the real datasets, which is discussed next, ensure us that $SkyJCPS$ significantly outperforms $RSJFH$ when the dataset is large and distributed among many tables. Due to the logarithmic scale used, it may seem that some of the algorithms have the same execution time for equal sized packages. This is not the case, and since BNL was the single-table skyline algorithm used, the algorithms performed best using correlated data and worst using anti-correlated data.

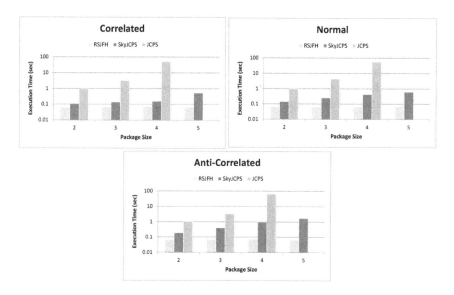

Fig. 6. Package Size Scalability for Synthetic Data

5.3 Package-Size Scalability - Real Dataset

In order to test the algorithms' performance with real data, we used the Movie-Lens[1] dataset, which consists of 10 million ratings and 10,000 movies, and the Book-Crossing[2], which consists of 271,000 books rated by 278,000 users.

We randomly chose five users from each of the datasets, with partiality to those who have rated a large number of movies or books, from the datasets for use in our package-size evaluations. For the MovieLens dataset, the users consisted of those with IDs 8, 34, 36, 65, and 215, who rated 800, 639, 479, 569, and 1,242 movies, respectively. Similarly, for the Book-Crossing dataset, the users consisted of those with IDs 11601, 11676, 16795, 23768, and 23902, who rated 1,571, 13,602, 2,948, 1,708, and 1,503 books, respectively. The target

---

[1] http://www.grouplens.org/node/73/

[2] http://www.informatik.uni-freiburg.de/~cziegler/BX/dataset

descriptive table for MovieLens contained 10,000 tuples, i.e., all the movies, and Book-Crossing contained 271,000 tuples, i.e., all the books. We used the following queries, respectively, for evaluating MovieLens and Book-Crossing datasets: *find packages of n movies such that the average rating of all the movies is high, the release date is high, and each movie-rater has rated at least one of the movie* and *find packages of n books such that the average rating of all the books is high, the publication date is high, and each book-rater has rated at least one of the books,* where $n = 3, 4, 5$.

The results of this experiment can be seen in Fig 10. It is easily observed that *SkyJCPS* performed better in all cases. We were unable to obtain any results from *JCPS* as it ran for hours. Due to the number of joins required to construct the tables in the format required by *RSJFH*, most of its time was spent during the initial phase, i.e., before the Cartesian product phase, and performed the worst. On average, *SkyJCPS* outperformed *RSJFH* by a factor of 1000. Although the overall execution time of the Book-Crossing dataset was longer than the MovieLens dataset, the data we used from the Book-Crossing dataset consisted of more tuples.

 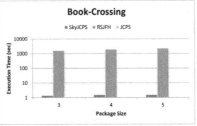

Fig. 7. Package Size Scalability for MovieLens and Book-Crossing Datasets

5.4 Storage Model Evaluation

For each of the above experiments, we evaluated our storage model by comparing the time taken to load the RDF file into the database using our storage model versus using VPTs. All data was indexed using a B-trees. Fig 8 shows the time of inserting the MovieLens and the Book-Crossing datasets, respectively, for packages of size 3, 4, and 5.

In general, the data loading phase using the TDTQ storage model was longer than that of VPTs. The number of tables created using both approaches are not always equal, and either approach could have more tables than the other. Since the time to load the data is roughly the same for each package size, the number of tables created does not necessarily have that much effect on the total time. Our approach imposes additional time because of the triple patterns that must be matched. Since the time difference between the two is small and the database only has to be built once, it is more efficient to use our storage model with *SkyJCPS* than using VPTs.

Fig. 8. Database build

6   Related Work

Much research has been devoted to the subject of preference queries. One kind of preference query is the skyline query, which [2] originally introduced and provided a block nested loops, divide-and-conquer, and B-tree-based algorithms. Later, [4] introduced a sort-filter-skyline algorithm that is based on the same intuition as BNL, but uses a monotone sorting function to all for early termination. Unlike [2] [4] [9], which have to read the whole database at least once, index-based algorithms [6] [8] allow one to access only a part. However, all of these are designed to work on a single relation. As the Semantic Web matures and RDF data is populated, more research needs to be done on preference queries that involve multiple relations. When queries involve aggregations, multiple relations must be joined before any of the above techniques can be used.

Recently, an interest in multi-relation skyline queries has been growing. [3] introduced three skyline algorithms that are based on the concept of a *header point*, which allows some nonskyline tuples to be discarded before succeeding to the skyline processing phase. [7] introduced a sky-join operator that gives the join phase a small knowledge of a skyline. Others have used approximation and top-k algorithms with regards to recommendation. [5] proposes a framework for collaborative filtering using a variation of top-k. However, their set of results do not contain packages but single items. [12] went a step further and used top-k techniques to provide a *composite* recommendation for travel planning. Also, [11] used similar techniques to provide a package recommendation by approximating. Top-k is useful when ranking objects is desired. However, top-k is prone to discard tuples that have a 'bad' value in one of the dimensions, whereas a skyline algorithm will include this object if it is not dominated. [13] explains the concept of top-k dominating queries. They combine the properties of skyline and top-k to yield an algorithm that has the best of both worlds. [10] proposed a novel algorithm called SFSJ (sort-first skyline-join) that computes the complete skyline. Given two relations, access to the two relations is alternated using a pulling strategy, known as adaptive pulling, that prioritizes each relation based on the number of mismatched join values. Although the algorithm has no limitations on the number of skyline attributes, it is limited by two relations.

7 Conclusion and Future Work

This paper addressed the problem of answering package skyline queries. We have formalized and described what constitutes a "package" and have defined the term *skyline packages*. Package querying is especially useful for cases where a user requires multiple objects to satisfy certain constraints. We introduced three algorithms for solving the package skyline problem. Future work will consider the use of additional optimization techniques such as prefetching to achieve additional performance benefits as well as the integration of top-k techniques to provide ranking of the results when the size of query result is large.

Acknowledgments. The work presented in this paper is partially funded by NSF grant IIS-0915865. Thanks to Ling Chen for the meaningful discussions and suggestions. Thanks to Alicia Bieringer, Juliane Foster, and Prachi Nandgaonkar for their fruitful discussions and participation in the initial stage of this work.

References

1. Abadi, D., Marcus, A., Madded, S., Hollenbach, K.: Scalable Semantic Web Data Management Using Vertical Partitioning. In: VLDB 2007 (2007)
2. Borzsonyi, S., Kossmann, D., Stocker, K.: The skyline operator. In: ICDE 2001, pp. 421–430 (2001)
3. Chen, L., Gao, S., Anyanwu, K.: Efficiently evaluating skyline queries on RDF databases. In: Antoniou, G., Grobelnik, M., Simperl, E., Parsia, B., Plexousakis, D., De Leenheer, P., Pan, J. (eds.) ESWC 2011, Part II. LNCS, vol. 6644, pp. 123–138. Springer, Heidelberg (2011)
4. Chomicki, J., Godfrey, P., Gryz, J., Liang, D.: Skyline with presorting. In: ICDE 2003, pp. 717–719 (2003)
5. Khabbaz, M., Lakshmanan, L.V.S.: TopRecs: Top-k algorithms for item-based collaborative filtering. In: EDBT 2011, pp. 213–224 (2011)
6. Kossmann, D., Ramsak, F., Rost, S.: Shooting Stars in the Sky: An Online Algorithm for Skyline Queries. In: VLDB 2002, pp. 275–286 (2002)
7. Raghavan, V., Rundensteiner, E.: SkyDB: Skyline Aware Query Evaluation Framework. In: IDAR 2009 (2009)
8. Papadias, D., Tao, Y., Fu, G., Seeger, B.: Progressive Skyline Computation in Database Systems. In: TODS 2005, pp. 41–82 (2005)
9. Tan, K.L., Eng, P.K., Ooi, B.C.: Efficient Progressive Skyline Computation. In: VLDB 2001, pp. 301–310 (2001)
10. Vlachou, A., Doulkeridis, C., Polyzotis, N.: Skyline query processing over joins. In: SIGMOD 2011, pp. 73–84 (2011)
11. Xie, M., Lakshmanan, L.V.S., Wood, P.T.: Breaking out of the Box of Recommendations: From Items to Packages. In: RecSys 2010, pp. 151–158 (2010)
12. Xie, M., Lakshmanan, L.V.S., Wood, P.T.: CompRec-Trip: a Composite Recommendation System for Travel Planning. In: ICDE 2011, pp. 1352–1355 (2011)
13. Yiu, M.L., Mamoulis, N.: Efficient Processing of Top-k Dominating Queries on Multi-Dimensional Data. In: VLDB 2007, pp. 483–494 (2007)
14. Jin, W., Ester, M., Hu, Z., Han, J.: The Multi-Relational Skyline Operator. In: ICDE 2007, pp. 1276–1280 (2007)

Accessing Relational Data
on the Web with SparqlMap

Jörg Unbehauen, Claus Stadler, and Sören Auer

Universität Leipzig, Postfach 100920, 04009 Leipzig, Germany
{unbehauen,cstadler,auer}@informatik.uni-leipzig.de
http://aksw.org

Abstract. The vast majority of the structured data of our age is stored in relational databases. In order to link and integrate this data on the Web, it is of paramount importance to make relational data available according to the RDF data model and associated serializations. In this article we present *SparqlMap*, a SPARQL-to-SQL rewriter based on the specifications of the W3C R2RML working group. The rationale is to enable SPARQL querying on existing relational databases by rewriting a SPARQL query to exactly one corresponding SQL query based on mapping definitions expressed in R2RML. The SparqlMap process of rewriting a query on a mapping comprises the three steps (1) mapping candidate selection, (2) query translation, and (3) query execution. We showcase our SparqlMap implementation and benchmark data that demonstrates that SparqlMap outperforms the current state-of-the-art.

Keywords: Triplification, SPARQL, RDB2RDF, R2RML.

1 Introduction

The vast majority of the structured data of our age is stored in relational databases. In order to link and integrate this data on the Web, it is of paramount importance to make relational data available according to the RDF data model and associated serializations.

Also, for several reasons, a complete transition from relational data to RDF may not be feasible: Relational databases commonly provide rich functionality, such as integrity constraints; data management according to the relational data model often requires less space and may be simpler than with RDF, such as in cases which would require reification or n-ary relations; the cost of porting existing applications or maintaining actual datasets in both formats may be prohibitive; and RDBs are still a magnitude faster than RDF stores. As it can not be expected that these gaps will close soon, relational data management will be prevalent in the next years. Hence, for facilitating data exchange and integration it is crucial to provide RDF and SPARQL interfaces to RDBMS.

In this article we present *SparqlMap*[1], a SPARQL-to-SQL rewriter based on the specifications of the *W3C R2RML working group*[2]. The rationale is to

---

[1] http://aksw.org/Projects/SparqlMap
[2] http://www.w3.org/TR/r2rml/

H. Takeda et al. (Eds.): JIST 2012, LNCS 7774, pp. 65–80, 2013.

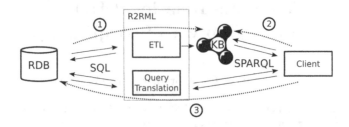

Fig. 1. Two models for mapping relational data to RDF: query rewriting and RDF extraction

enable SPARQL querying on existing relational databases by rewriting a SPARQL query to *exactly one* corresponding SQL query based on mapping definitions expressed in R2RML. The R2RML standard defines a language for expressing how a relational database can be transformed into RDF data by means of term maps and triple maps. In essence, implementations of the standard may use two process models, which are depicted in Figure 1: Either, the resulting RDF knowledge base is materialized in a triple store (1) and subsequently queried using SPARQL (2), or the materialization step is avoided by dynamically mapping an input SPAQRL query into a corresponding SQL query, which renders exactly the same results as the SPARQL query being executed against the materialized RDF dump (3).

SparqlMap is in terms of functionality comparable with *D2R* [3] or *Virtuoso RDF Views* [1] with the focus on performing all query operators in the relational database in a single unified query. This strategy ensures scalability since expensive round trips between the RDBMS and the mapper are reduced and the query optimization and execution of the RDBMS are leveraged. SparqlMap is designed as a standalone application facilitating light-weight integration into existing enterprise data landscapes.

In the following we give a short overview over our approach. The prequisite for rewriting SPARQL queries to SQL is a set of mapping definitions over which queries should be answered. In Section 3 we formalize the mapping and query syntax. The process of rewriting a query on a mapping is performed in the following three steps:

Mapping candidate selection. The initial step, described in Section 4.1, identifies candidate mappings. These are mappings that potentially contribute to the query's result set: Informally, this is the set of mappings that yield triples that could match the triple patterns of the query, as shown in Figure 4. The relation between the candidate mappings and the triple patterns is called a binding.

Query translation. The identified candidate mappings and the obtained bindings enable us to rewrite a SPARQL query to an SQL query. This process is described in Section 4.2.

Query execution. Finally, we show in Section 4.3 how the SPARQL result set is constructed from the SQL result set of the executed SQL query.

We finally evaluate our approach using the BSBM benchmark in Section 5 and show that SparqlMap is on average an order of magnitude faster than state-of-the-art techniques.

The contributions of our work described in this article include in particular (1) the formal description of mapping candidate selection, (2) an approach for efficient query translation and (3) showcasing the implementation and demonstrate its efficiency.

2 Related Work

We identified two related areas of research. First, as many native triple stores are based on relational databases there is considerable research on efficiently storing RDF data in relational schema. Exemplary are both [7] and [5], discussing the translation of a SPARQL into a single SQL query. The translations presented there are however targeted towards database backed triple stores and need to be extended and adopted for usage in a database mapping scenario. Also notable is [8], describing SQL structures to represent facets of RDF data in relational databases. Second, the mapping of relational databases into RDF is a way of leveraging existing data into the Semantic Web. We can differentiate between tools that either expose their data as RDF, Linked Data or expose a SPARQL endpoint for interactive querying of the data. An example for RDF and Linked Data exposition is *Triplify* [2]. Exposing data via a SPARQL endpoints either requires loading transformed data into a SPARQL-enabled triple store or rewriting SPARQL queries into SQL. The answering of SPARQL queries over relational data is the goal of several concepts and implementations. *D2R Server* [3] is a standalone web application, answering SPARQL queries by querying the mapped database. D2R mixes in-database and out-of-database operations. Operators of an incoming SPARQL queries like joins and some filters are performed in the mapped database directly. Other operators are then later executed on the intermediate results directly by D2R. OpenLink's *Virtuoso RDF Views* [1] allows the mapping of relational data into RDF. RDF Views are integrated into the Virtuoso query execution engine, consequently allowing SPARQL query over native RDF and relational data. A further SPARQL-to-SQL tool is *Ultrawrap* [11], which integrates closely with the database and utilizes views for answering SPARQL queries over relational data. However, to the best of our knowledge, there is no approach describing in detail the mapping and translation process for generating a single, unified SQL query.

3 Definitions

In this section we define the syntax of RDF, SPARQL and the mapping. The RDF and SPARQL formalization is closely following [9].

Definition 1 (RDF definition). *Assume there are pairwise disjoint infinite sets I, B, and L (IRIs, blank nodes, and RDF literals, respectively). A triple* $(v_s, v_p, v_o) \in (I \cup B) \times I \times (I \cup B \cup L)$ *is called an RDF triple. In this tuple,* v_s *is the subject,* v_p *the predicate and* v_p *the object. We denote the union* $I \cup B \cup L$ *as by T called RDF terms.*

Using the notion $t.i$ for $i \in \{s, p, o\}$ we refer to the RDF term in the respective position. In the following, the same notion is applied to *triple patterns* and *triple maps*. An RDF *graph* is a set of RDF triples (also called RDF dataset, or simply a dataset). Additionally, we assume the existence of an infinite set V of variables which is disjoint from the above sets. The W3C recommendation SPARQL[3] is a query language for RDF. By using *graph patterns*, information can be retrieved from SPARQL-enabled RDF stores. This retrieved information can be further modified by a query's solution modifiers, such as sorting or ordering of the query result. Finally the presentation of the query result is determined by the *query type*, return either a set of triples, a table or a boolean value. The graph pattern of a query is the base concept of SPARQL and as it defines the part of the RDF graph used for generating the query result, therefore *graph patterns* are the focus of this discussion. We use the same graph pattern syntax definition as [9].

Definition 2 (SPARQL graph pattern syntax). *The syntax of a SPARQL graph pattern expression is defined recursively as follows:*
1. *A tuple from* $(I \cup L \cup V) \times (I \cup V) \times (I \cup L \cup V)$ *is a graph pattern (a triple pattern).*
2. *The expressions* $(P_1$ *AND* $P_2)$, $(P_1$ *OPT* $P_2)$ *and* $(P_1$ *UNION* $P_2)$ *are graph patterns, if* P_1 *and* P_2 *are graph patterns.*
3. *The expression* $(P$ *FILTER* $R)$ *is a graph pattern, if* P *is a graph pattern and* R *is a SPARQL constraint.*

Further the function $var(P)$ returns the set of variables used in the graph pattern P. SPARQL constraints are composed of functions and logical expressions, and are supposed to evaluate to boolean values. Additionally, we assume that the query pattern is well-defined according to [9].

We now define the terms and concepts used to describe the SPARQL-to-SQL rewriting process. The basic concepts are the relational database schema and a mapping for this schema m. The schema ishas a set of relations R and each relation is composed of attributes, denoted as $A_r = (r.a_0, r.a_1, ..., r.a_l)$. A mapping m defines how the data contained in tables or views in the relational database schema s is mapped into an RDF graph g. Our mapping definitions are loosely based on R2RML. An example of such a mapping is depicted in Figure 2 and used further in this section to illustrate the translation process.

Definition 3 (Term map). *A term map is a tuple* $tm = (A, ve)$ *consisting of a set of relational attributes A from a single relation r and a value expression ve that describes the translation of A into RDF terms (e.g. R2RML templates for generating IRIs). We denote by the range* $range(tm)$ *the set of all possible RDF terms that can be generated using this term map.*

---

[3] http://www.w3.org/TR/rdf-sparql-query/

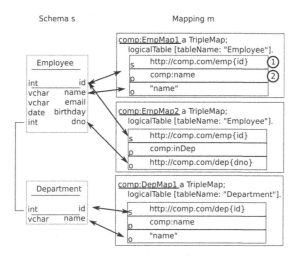

Schema s   Mapping m

Fig. 2. Exemplary mapping of parts of two relations using three triple maps. R2RML's construct `logicalTable` specifies the source relation of a triple map.

Term maps are the base element of a mapping. In Figure 2 an example for such a *term map* is (1). With *ve* being the template `http://comp.com/emp{id}` and $A = \{Employee.id\}$ it is possible to produce resource IRIs for employees. The RDF term (2) in Figure 2 creates a constant value, in this case a property. Consequently, for this RDF term $A = \emptyset$ holds.

Definition 4 (Triple map). *A triple map trm is the triple (tm_S, tm_P, tm_O) of three term maps for generating the subject (position s), predicate (position p) and object (position o) of a triple. All attributes of the three term maps must originate from the same relation r.*

A triple map defines how triples are actually generated from the attributes of a relation (i.e. rows of a table). This definition differs slightly from the R2RML specification, as R2RML allows multiple predicate-object pairs for a subject. These two notions, however, are convertible into each other without loss of generality. In Figure 2 the triple map `comp:EmpMap1` defines how triples describing the name of an employee resource can be created for the relation `Employee`.

A mapping definition $m = (R, TRM)$ is a tuple consisting of a set of relations R and a set of triple maps TRM. It holds all information necessary for the translation of a SPARQL query into a corresponding SQL query. We assume in this context that all data is stored according to the schema s is mapped into a single RDF graph and likewise that all queries and operations are performed on this graph[4].

---

[4] Note, that support for named graphs can be easily added by slightly extending the notion of triple map with an additional term map denoting the named graph.

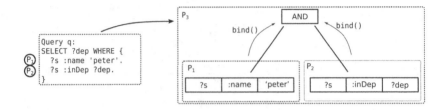

Fig. 3. Mapping candidate selection overview. The patterns of a query parsed into a tree. The bind function recurses over that tree.

4   The SparqlMap Approach

The SparqlMap approach is based on the three steps of mapping candidate selection, query translation and query execution that are discussed in this section.

4.1   Mapping Candidate Selection

Mapping selection is the process of identifying the parts of a mapped graph that can contribute to the solution of a query q. This selection process is executed for every query and forms the basis for the following step – the translation into SQL. The parts of a query that are used for matching the parts of a graph examined are the graph patterns. The graph of a mapped database is the set of triples defined by the *triple maps*. Consequently, we propose the selection of candidate *triple maps* to match the *graph pattern*. The general approach described here aims at first binding each *triple pattern* of q to a set of candidate triple maps, and then to reduce the amount of bindings by determining the unsatisfiability of constraints (e.g join conditions) based on the structure of the SPARQL query.

Before we formally introduce the operators used, we give a brief overview of the process in Figure 3. The simple query q depicted here represents a tree-like structure according to Definition 2. In a bottom-up traversal we first search for mapping candidates for the triple patterns P_1 and P_2. In the next step, indicated by the $bind()$ function, these mapping candidates are examined on the next higher level of the tree. Based on the semantics of P_3 the $bind()$ function reduces the mapping candidates for P_1 and P_2. Before we formally describe this process we define the notion of a binding.

Definition 5 (Triple Pattern Binding). *Let q be a query, with TP_q being its set of triple patterns. Let m be the mapping of a database, with TRM_m being its set of all triple maps. A triple pattern binding tpb is the tuple (tp, TRM), where $tp \in TP_q$ and $TRM \subseteq TRM_m$. We further denote by the set QPB_q for a query q the set of triple pattern bindings TPB, such that there exists for every $tp \in TP_q$ exactly one tpb.*

In this context we assume that in case q contains a triple pattern more than once, for each occurrence there exists in QPB_q a separate tpb. The set TRM for a triple pattern tp is also called the set of mapping candidates for tp. We now define successively the basic terms and functions on the triple pattern bindings and illustrate them using the sample query introduced in Figure 3. In Figure 4 the result of the process is depicted. The dotted squares indicate the triple pattern bindings tpb with their patterns and triple maps.

Definition 6 (Term map compatibility). *We consider two term maps tm_1 and tm_2 to be compatible, if $range(tm_1) \cap range(tm_2) \neq \emptyset$. We further consider a term map tm compatible with an RDF term t, if the term $t \in range(tm)$. A variable v is always considered compatible with a term map.*

With the boolean function $compatible(t_1, t_2)$ we denote the check for compatibility. This function allows us to check, if two term maps can potentially produce the same RDF term and to pre-check constraints of the SPARQL query. Mapping candidates that cannot fulfill these constraints are removed from the binding. Further it allows to check, if a triple map is compatible with a triple pattern of a query. In the example given in Figure 4 term map compatibility is used to bind *triple maps* to *term maps*. At position (1) in Figure 4 the triple pattern P_2 is bound to the term map `:EmpMap2` because the resource IRI at the predicate position of P_2 is compatible with the constant value term map of `:EmpMap2` in the same position. The notion of term compatibility can be extended towards checking bindings for compatibility by the functions *join* and *reduce*.

Definition 7 (Join of triple pattern bindings). *Let $tpb_1 = (tp_1, TRM_1)$ and $tpb_2 = (tp_2, TRM_2)$ be triple pattern bindings. Further, let $V = var(tp1) \cap var(tp2)$ be the set of shared variables.*
We define $join(tpb_1, tpb_2) : \{(trm_a, trm_b) \in TRM_1 \times TRM_2|$ for each variable $v \in V$ the union of the corresponding sets of term maps of trm_a and trm_b is either empty or its elements are pairwise compatible.[5] $\}$

Definition 8 (Reduction of triple pattern bindings). *The function $reduce(tpb_1, tpb_2)$ is defined as $proj(join(tpb_1, tpb_2), 1)$, i.e. the projection of the first component of the tuples obtained by the join operation.*

Reduction is the base operation for minimizing the set of *triple maps* associated with every *triple pattern*. It rules out all candidate tuples that would eventually yield unsatisfiable SQL join conditions. In Figure 4 the reduction process follows the dotted line indicated by (2). The triple patterns P_1 and P_2 share the variable $?s$ which is in both cases in the subject position of the triple pattern. Consequently, each triple map in TPB_1 is compared at the subject position with all subject term maps of TPB_2 for compatibility. If no compatible triple map is found, the triple map is removed from the candidate set. The term map `:DepMap1` in Figure 4 is therefore not included in TPB_3, as the subject of `:DepMap1` is not compatible with the subject of `:EmpMap2`. The reduction function now allows

---

[5] Note, that the same variable may occur multiple times in a triple pattern and therefore map to multiple term maps.

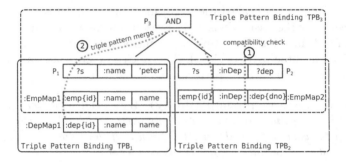

Fig. 4. Binding operations on a sample query. In (1) a simple check for compatibility at the predicate position is performed, in (2) the triples maps are merged between two triple pattern bindings, checking compatibility at the subject position.

(a) $merge(TPB_1, TPB_2)$

$TPB \leftarrow TPB_1 \cup TPB_2$
$TPB' \leftarrow \emptyset$
for $tpb_1 in TPB$ **do**
  for $tpb_2 in TPB$ **do**
    $tpb_1 \leftarrow reduce(tpb_1, tpb_2)$
  end for
  $TPB' \leftarrow TPB' \cup \{tpb_1\}$
end for
return TPB'

(b) $optmerge(TPB_1, TPB_2)$

$TPB' \leftarrow TPB_1$
for $tpb_1 in TPB_1$ **do**
  for $tpb_2 in TPB_2$ **do**
    $tpb_2 \leftarrow reduce(tpb_2, tpb_1)$
  end for
  $TPB' \leftarrow TPB' \cup \{tpb_2\}$
end for
return TPB'

Fig. 5. The merge and optmerge algorithms

the definition of two operators that perform a reduction of mapping candidates along the syntax of a SPARQL query.

For the two sets of *triple pattern bindings* TPB_1 and TPB_2 we define two merge operations for the triple pattern bindings as follows:

Binding merge $merge(TPB_1, TPB_2)$ reduces all triple pattern bindings with each other, as illustrated in Figure 5a.

Binding opt merge $optmerge(TPB_1, TPB_2)$ reduces all triple pattern bindings of TPB_2 with the all triple pattern bindings of TPB_1, as illustrated in Figure 5b.

Both merge operations preevaluate the join conditions of the later SQL execution. The compatibility check for the shared variables of two triple patterns rule out unfulfillable join or respectively left join conditions.

We can use these operators to define the recursive function $bind_m(P)$, which computes for a mapping m and the graph pattern P the set of triple pattern bindings TPB_p, similar to the recursive evaluation function defined in [9].

Definition 9. *Let TRM_m be the set of all triple maps in m, P_1 and P_2 be graph patterns and tp be a triple pattern of a query. The function $bind_m(P)$*

is the recursive binding of the TRM_m to the triple patterns of a query for the following cases:

1. *If P is a triple pattern tp, $bind(P) = \{(tp, TRM_{tp}) \mid TRM_{tp} = \{trm \mid trm \in TRM_m \wedge compatible(trm.s, tp.s) \wedge compatible(trm.p, tp.p) \wedge compatible(trm.o, tp.o)\}\}$.*
2. *If P is $(P_1 \ AND \ P_2)$, $bind(P) = merge(bind_m(P_1), bind_m(P_2))$*
3. *If P is $(P_1 \ OPT \ P_2)$, $bind(P) = optmerge(bind_m(P_1), bind_m(P_2))$*
4. *If P is $(P_1 \ UNION \ P_2)$, $bind(P) = (bind_m(P_1) \cup bind_m(P_2))$*
5. *If P is $(P_1 \ FILTER \ R)$, $bind(P) = \{tpb \mid tpb \in bind_m(P_1) \wedge$ if tpb is sharing variables with R, the constraint is pre-evaluated. If the filter is always false, the term map is not included.$\}$*

The complete binding process can now be illustrated using the example in Figure 4. Starting from the bottom, $bind(P_1)$ evaluates to $TPB_1 = \{(P_1, \{$:empMap1,:depMap1$\})\}$ and $bind(P_2)$ to $TPB_2 = \{(P2, \{$:empMap2$\})\}$. For P_1 the triple map :empMap2 is not bound, because $compatible(P1.p,$:empMap2.p$)$ = *false*. In the next step of the recursion, the pattern binding merge is evaluated between the two sets, creating TPB_3. The sets of triple maps of TPB_1 and of TPB_2 are reduced on the shared variable s. The term map at the subject position of :depMap1 is not compatible with the subject from another triple map of TPB_1 and is not included in TPB_3. Here the recursion halts, the set obtained in the last step represents the full mapping of the query $QPB = TPB_3$. QBP is a set of two triple pattern bindings, each with one triple map, and is ready to be used for creating the SQL query in the next step of the process.

The approach described in Definition 9 has some limitations. Variables used in both sub-patterns of UNIONs are not exploited for reducing the candidate triple maps. This for example could be overcome by using algebraic equivalence transformations which are described in [10]. Another limitation is that not all candidate reductions are applied to triple patterns that are not directly connected by shared variables. The modifications to the query binding strategy dealing with this limitation are part of the future work on SparqlMap.

4.2   Query Translation

Query translation is the process of generating a SQL query from the SPARQL query using the bindings determined in the previous step. We devise a recursive approach to query generation, similar to the one presented for mapping selection. The result of this translation is a nested SQL query reflecting the structure of the SPARQL query. We first describe the function $toCG(tm)$ that maps a term map tm into a set of column expressions CG, called a *column group*. The utilization of multiple columns is necessary for efficient filter and join processing and data type compatibility of the columns in a SQL union. In a CG the columns can be classified as:

Type columns. The RDF term type, i.e. resource, literal or blank node is encoded into these columns using constant value expressions. The column

```
Select  cast(1 as numeric) s_type,          ⎫
        cast(1 as numeric) s_datatype,      ⎬ Type columns
        cast(null as text) s_text,          ⎫
        cast(null as numeric) s_num,        ⎪
        cast(null as bool) s_bool,          ⎬ Literal columns
        cast(null as time) s_time,          ⎭
        cast(2 as numeric) s_reslength,     ⎫
        cast('http://comp.com/emp' as text) s_res_1,  ⎬ Resource
        cast("Employee"."id"  as text) s_res_2        ⎭ columns
```

Fig. 6. Column group for variable ?s of graph pattern P_1

expression `cast(1 as numeric) s_type` declares the RDF terms produced
in this column group to be resources.

Resource columns. The IRI value expression VE is embedded into multiple
columns. This allows the execution of relational operators directly on the
columns and indexes.

Literal columns. Literals are cast into compatible types to allow SQL UNION
over these columns.

In Figure 6 the column group created for the variable ?s of triple pattern P_1
is depicted. The following aspects of query translation require the definition of
additional functions.

Align. The alignment of of two select statements is a prerequisite for performing
a SQL union as the column count equality and data type compatibility of
the columns are mandatory. The function $align(s_1, s_2)$ for two SQL select
statements s_1 and s_2 returns s_1' by adding adding columns to s_1 such that
s_1' contains all columns defined in s_2. The columns added do not produce
any RDF terms.

Join. Filter conditions are performed on column groups, not directly on columns.
As already outlined in [6] this requires embedding the filter statements into
a conditional check using **case** statements. This check allows the database
to check for data types and consequently to select the correct columns of
a column group for comparison. For two SQL queries s_1, s_2 the function
$joinCond(s_1, s_2)$ calculates the join condition as an expression using **case**
statements, checking the column groups bound to of shared variables.

Filter. For R being a filter expression according to Definition 2 (5), the function
$filter_f(R)$ translates a filter expression into an equivalent SQL expression on
column groups.

Triple Pattern. The RDF literals and resources used in a triple pattern tp are
implicit predicates and need to be made explicit. The function $filter_p(tp)$
maps these triple patterns into a set of SQL predicates, similar to the definition of $filter_f(R)$.

Alias. The renaming of a *column group* CG is performed by the function $alias($
$CG, a)$ that aliases all column expressions of CG, by adding the prefix a. For
the scope of this paper, we assume that proper alias handling is performed
and is not further explicitly mentioned.

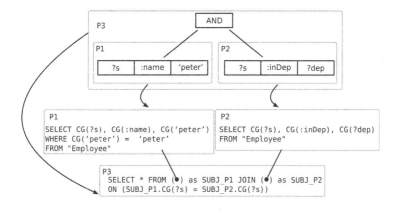

Fig. 7. Query nesting for a sample query

Using the previously defined function $toCG(tm)$ and the usage of the SQL operators *JOIN, LEFT JOIN and UNION* we can now devise a simple recursive translation function.

Definition 10 (Query translation). *Let QPB be a query pattern binding, P_1 and P_2 be graph patterns and tp be a triple pattern and $tpb = (tp, TRM)$ be triple pattern binding for tp with the set of term maps TRM. The relation for each $trm \in TRM$ is denoted r. The translation of a graph pattern P into a SQL query $Q_{sql} = t_{QPB}(P)$ is performed as follows.*

1. *If P is a triple pattern:* $t_{QPB}(P) = $ UNION ALL $\{\forall trm \in TRM :$ SELECT $toCG(trm.s), toCG(trm.p), toCG(trm.o)$ FROM r WHERE $filter_p(P)\}$
2. *If P is (P_1 AND P_2):* $t_{QPB}(P) = $ SELECT * FROM ($t_{QPB}(P_1)$) p1 JOIN ($t_{QPB}(P_2)$) p2 ON($joinCond(p1, p2)$)
3. *If P is (P_1 OPT P_2):* $t_{QPB}(P) = $ SELECT * FROM ($t_{QPB}(P_1)$) p1 LEFT JOIN ($t_{QPB}(P_2)$) p2 ON($joinCond(p1, p2)$)
4. *If P is (P_1 UNION P_2):* $t_{QPB}(P) = ($ $align(t_{QPB}(P_1), t_{QPB}(P_2)))$ UNION ($align(t_{QPB}(P_2), t_{QPB}(P_1))$)
5. *If P is (P_1 FILTER R):* $t_{QPB}(P) = $ SELECT * FROM $t_{QPB}(P)$ WHERE $filter_f(R)$

The translation of the example of Figure 4 is depicted in Figure 7. The column groups are indicated here by the notion (CG(t)), where t is the RDF term or variable the column group was created for.

4.3   Query Execution

The SQL query created as outlined in the previous section can now be executed on the mapped database. The result set of this SQL query is then mapped into a SPARQL result set depending on the query type of the SPARQL query. Each row of a SQL result set produces for every column group an RDF term which then can be used to create the SPARQL result set. In the case of an SPARQL SELECT query, for each projected variable the corresponding column group is used to generate the RDF terms. We use the result set for the query initially described in Figure 3 to illustrate the result set translation process. The following listing presents a result set snippet for the column group of variable ?*dep*. For brevity, literal columns are collapsed into to `dep_lits`.

```
dep_type|dep_datatype|dep_lits|dep_reslength|    dep_res_1    |dep_res_2
--------|------------|--------|-------------|-----------------|---------
       1|           1|    null|            2|http://comp.com/dep|       1
--------|------------|--------|-------------|-----------------|---------
```

According to `dep_type` the RDF term is a resource. The IRI of the resource is generated from 2 columns, indicated by `dep_reslength`. The IRI is constructed by concatenating the prefix from `s_res_1` with the percent-encoded[6] value from `dep_res_2`. The SPARQL result set corresponding to the sample query is consequently:

```
<sparql xmlns="http://www.w3.org/2005/sparql-results#">
  <head><variable name="dep"/></head>
  <results> <result>
      <binding name="dep"><uri>http://comp.com/dep1</uri></binding>
  </result> </results>
</sparql>
```

4.4   Implementation and Optimizations

We implemented the mapping candidate selection, query translation and query execution in SparqlMap, a standalone SPARQL-to-SQL rewriter. It utilizes ARQ[7] for SPARQL parsing and Java servlets for exposing a SPARQL endpoint. Our implementation is available under an open-source license on the project page[8]. We implemented several optimizations for achieving both better SQL query runtime and efficient mapping candidate selection. After parsing the query filter expressions are pushed into the graph patterns as describe in [10]. Further, when searching for mapping candidates, the patterns are grouped according to their subject for reducing search time. As described in [7] we flatten out the nested SQL structure described in Section 4.2 with the goal of minimizing self-joins.

5   Evaluation

We evaluated SparqlMap using the *Berlin Sparql Benchmark* (BSBM) [4]. We selected BSBM because of its widespread use and the provision of both RDF and relational data with corresponding queries in SPARQL and SQL. As a baseline

---

[6] as defined in http://tools.ietf.org/html/rfc3986
[7] http://jena.apache.org/documentation/query/
[8] http://aksw.org/projects/SparqlMap

tool *D2R* [3] was chosen since D2R is the most mature implementation and also implemented as standalone application, interacting with a remote database. As outlined in Section 2, D2R performs only a part of the operations of a SPARQL query in the database. Therefore multiple SQL queries will be issued for a single SPARQL query.

Setup. We used a virtualized machine with three AMD Opteron 4184 cores and 4 GB RAM allocated. Data resided in all cases in a *PostgreSQL* 9.1.3 database[9], with 1 GB of RAM. Additional indices to the BSBM relational schema were added, as described in the BSBM results document[10]. Both D2R and SparqlMap were allocated 2 GB of RAM. We utilized D2R version 0.8 with the mapping provided by BSBM[11]. All benchmarks employed the explore use case and were performed on a single thread with 50 warm-up and 500 measurement runs. In our evaluation we performed two different runs of BSBM. In the first run the SPARQL endpoints of SparqlMap, D2R and D2R with fast mode enabled (D2R-fast) were tested. Additionally, the PostgreSQL database was directly benchmarked. Due to D2R running out of memory when executing query 7 and 8[12] with 100M triples we additionally run the benchmark without Q7 and Q8. In a second run, we compare the efficiency of the generated SQL by recording the SQL generated by SparqlMap during a benchmark session and executing the SQL directly.

Overall runtimes. The results of the first run are presented for each query in Figure 8, the total runtime of this run is depicted in Figure 9 (b). A first observation is the performance advantage of SQL over SPARQL. Both D2R and SparqlMap are Java applications and therefore add overhead to the queries. Especially simple select queries like Q2 or Q11 with short execution times for BSBM-SQL show extreme differences compared to the SPARQL-to-SQL rewriters. We therefore focus on the discussion of the first benchmark run on a comparison between D2R and SparqlMap. For the 1M and 25M triple dataset SparqlMap outperforms D2R-fast in the vast majority of the queries. Comparing the total runtime, as presented in Figure 9 (b), for the 1M dataset SparqlMap is overall 5 times faster than D2R-fast and for the 25M dataset SparqlMap is overall 90 times faster than D2R-fast. It can be clearly seen, that SparqlMap outperforms D2R-fast by at least an order of magnitude and has superior scaling characteristics. For larger data sets (i.e. above 1M triples) SparqlMap's performance is also relatively close to the one of handcrafted SQL. It is noteworthy, that the huge difference in overall runtime can be attributed mainly to the slow performance of D2R-fast in Q7 and Q8.

---

[9] http://www.postgresql.org/

[10] http://www4.wiwiss.fu-berlin.de/bizer/BerlinSPARQLBenchmark/results/index.html#d2r

[11] http://www4.wiwiss.fu-berlin.de/bizer/BerlinSPARQLBenchmark/V2/results/store_config_files/d2r-mapping.n3

[12] Cf.: http://sourceforge.net/mailarchive/message.php?msg_id=28051074

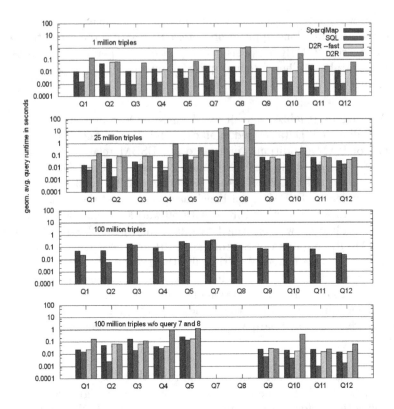

Fig. 8. Berlin SPARQL Benchmark evaluation with 1M, 25M and 100M triples, comparing SparqlMap, D2R with and without fast mode and native SQL. Values are average runtimes of queries in seconds for each query on a log scale.

Query specific performance. In general, for queries 1, 2, 3, 4, 5, 9, 10, 11 and 12 SparqlMap, D2R and D2R-fast show performance figures in the same order of magnitude. However, when examining the SQL generated by D2R and SparqlMap and relate it to their performance as presented in Figure 8 we can identify three categories.

1. Queries, which both mappers translate into a single SQL query with a similar runtime (Q1, Q5, Q10, Q12). These queries scale well for both mappers and the differences towards SQL decreases as the constant overhead imposed by the mapping contributes less to query execution time.
2. Queries that D2R maps into multiple SQL queries, that still yield a performance similar to SparqlMap (Q2, Q3, Q4, Q9, Q11). These queries however retrieve result sets of moderate sizes and may therefore in general scale worse than queries executed in a single SQL query. An example for this is Q3, where the ratio of the average execution time for SparqlMap and D2R-fast increases from 0.84 for the 1m dataset to a factor of 2.84 for the 25m dataset.
3. Queries mapped by D2R into multiple queries with low selectivity yielding huge intermediate result sets (Q7, Q8). In comparison with SparqlMap the

queries show poor scalability and performance. The average execution time ratio for Q7 between SparqlMap and D2R increases from 20 for the 1m dataset to 60 for the 25m dataset.

SQL performance comparison. The second benchmark run compares the SQL generated by SparqlMap (SparqlMap-SQL) with the SQL queries defined for BSBM (BSBM-SQL). This second run allows a direct comparison of efficiency between the handcrafted BSBM-SQL and the SQL generated by SparqlMap. The benchmark compares pure SQL efficiency, cutting out the overhead imposed by finding and translating the queries and is depicted in Figure 9 (a). In general we can observe that SparqlMap adds some overhead compared to the BSBM-SQL. SparqlMap generates for Q2 structurally similar SQL compared to the one of BSBM-SQL. The queries of SparqlMap-SQL, however, contain additional information required to generate RDF terms, such as the datatype, and therefore result in a twice as high average query runtime when compared to BSBM-SQL. A more drastic example is Q12, where BSBM-SQL is approximately 3.6 times faster than SparqlMap-SQL, or Q3 which takes on average 3.8 times longer in SparqlMap. Due to the large amount of tables used in this query, the overhead is more significant. In Q1 we observe that SparqlMap-SQL is slightly faster than BSBM-SQL (factor 1.1). Also in Q5 SparqlMap-SQL slightly outperforms BSBM-SQL by a factor of 1.2. We analyzed this difference and attribute it to the nested queries in BSBM-SQL. Especially the multiple nesting of subqueries in BSBM-SQL Q5 presents a challenge for the PostgreSQL query optimizer. The flat SQL structure generated by SparqlMap performs better here. Q11 is an example where SparqlMap-SQL performs significantly worse than BSBM-SQL (factor 34.0). The unbound predicate of Q11 leads SparqlMap to create a SQL query containing multiple unions of subselects. The lean structure of BSBM-SQL explains this gap.

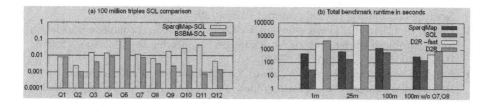

Fig. 9. (a) Comparison of BSBM-SQL with the SQL generated by SparqlMap for the 100M dataset. Average runtime in seconds on a log scale. (b) Total benchmark runtime in seconds on a log scale.

6   Conclusion and Outlook

In this paper we presented a formalization of the mapping from relational databases to RDF, which allows a translation of SPARQL queries into single unified

SQL queries. The benchmarking of our implementation shows clearly, that such an efficient mapping into a single query leveraging the SQL query optimizer and processor during query execution and results in significantly improved scalability when compared to non-single-SQL RDB2RDF implementations. In future work, we aim to improve the support for R2RML. We will also add further optimizations, such as support for the SPARQL REDUCED construct, which can boost the execution of certain queries. The query generation overhead can be substantially reduced by enabling prepared SPARQL queries, where a SPARQL query template is already precompiled into the corresponding SQL query template and subsequently reoccurring queries using the template do not have to be translated anymore. During our evaluation, we gained the impression, that BSBM is not best suited for benchmarking RDB2RDF mapping approaches. Certain features that are particularly challenging for an RDB2RDF tool (such as queries over the schema) are not part of BSBM. We plan to perform additional benchmarks and also to evaluate SparqlMap with large-scale real-world data.

References

1. Mapping relational data to rdf with virtuoso's rdf views, `http://virtuoso.openlinksw.com/whitepapers/relational%20rdf%20views%20mapping.html`
2. Auer, S., Dietzold, S., Lehmann, J., Hellmann, S., Aumueller, D.: Triplify - lightweight linked data publication from relational databases. In: 18th International World Wide Web Conference, p. 621 (April 2009)
3. Bizer, C., Cyganiak, R.: D2r server – publishing relational databases on the semantic web. Poster at the 5th Int. Semantic Web Conf., ISWC 2006 (2006)
4. Bizer, C., Schultz, A.: The berlin SPARQL benchmark. Int. J. Semantic Web Inf. Syst. 5(2), 1–24 (2009)
5. Chebotko, A., Lu, S., Fotouhi, F.: Semantics preserving sparql-to-sql translation. Data and Knowledge Engineering 68(10), 973–1000 (2009)
6. Cyganiak, R.: A relational algebra for SPARQL. Technical report, Digital Media Systems Laboratory, HP Laboratories Bristol (2005)
7. Elliott, B., Cheng, E., Thomas-Ogbuji, C., Ozsoyoglu, Z.M.: A complete translation from sparql into efficient sql. In: Int. Database Engineering & Applications Symp., IDEAS 2009, pp. 31–42. ACM (2009)
8. Lu, J., Cao, F., Ma, L., Yu, Y., Pan, Y.: An Effective SPARQL Support over Relational Databases. In: Christophides, V., Collard, M., Gutierrez, C. (eds.) SWDB-ODBIS 2007. LNCS, vol. 5005, pp. 57–76. Springer, Heidelberg (2008)
9. Pérez, J., Arenas, M., Gutierrez, C.: Semantics and complexity of sparql. ACM Trans. Database Syst. 34(3), 16:1–16:45 (2009)
10. Schmidt, M., Meier, M., Lausen, G.: Foundations of sparql query optimization. In: 13th Int. Conf. on Database Theory, ICDT 2010, pp. 4–33. ACM (2010)
11. Sequeda, J.F., Miranker, D.P.: Ultrawrap: SPARQL Execution on Relational Data. Poster at the 10th Int. Semantic Web Conf., ISWC 2011 (2011)

Protect Your RDF Data!

Sabrina Kirrane[1,2], Nuno Lopes[1], Alessandra Mileo[1], and Stefan Decker[1]

[1] Digital Enterprise Research Institute
National University of Ireland, Galway
{firstname.lastname}@deri.ie
http://www.deri.ie
[2] Storm Technology, Ireland
http://www.storm.ie

Abstract. The explosion of digital content and the heterogeneity of enterprise content sources have pushed existing data integration solutions to their boundaries. Although RDF can be used as a representation format for integrated data, enterprises have been slow to adopt this technology. One of the primary inhibitors to its widespread adoption in industry is the lack of fine grained access control enforcement mechanisms available for RDF. In this paper, we provide a summary of access control requirements based on our analysis of existing access control models and enforcement mechanisms. We subsequently: (i) propose a set of access control rules that can be used to provide support for these models over RDF data; (ii) detail a framework that enforces access control restrictions over RDF data; and (iii) evaluate our implementation of the framework over real-world enterprise data.

1 Introduction

Data on the web and within the enterprise is continuously increasing and despite advances in Information Technology (IT), it is still difficult for employees to find relevant information in a timely manner. This problem is further magnified when related information is segregated in different software systems. Bridging the information contained in such systems is necessary to support employees in their day-to-day activities. For instance, a high-level view of a customer would enable IT support staff to quickly respond to issues raised by that customer or additional information on drug compounds would allow a pharmaceutical company to identify potential issues with a new drug early in the drug development process.

The Resource Description Framework (RDF) is a flexible format used to represent data on the web which can also be used for data integration [21]. An additional benefit is that RDF data can easily be supplemented with complementary information from Linked Open Data (LOD) sources. It is possible to use Relational Database to RDF (RDB2RDF) techniques to translate existing relational data into RDF. However, given security is extremely important to enterprises we cannot simply extract the data and ignore the access control policies that have been placed on that data. Although we could extend RDB2RDF to

H. Takeda et al. (Eds.): JIST 2012, LNCS 7774, pp. 81–96, 2013.

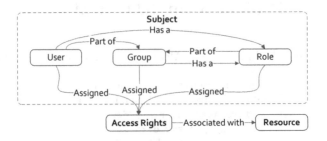

Fig. 1. Components of an access control statement

extract both the data and the access control information, RDF still does not have a mechanism to enforce existing access control policies over RDF.

Javanmardi et al. [9], Ryutov et al. [16], Amini and Jalili [1], Costabello et al. [4], Sacco et al. [17] all propose mechanisms to supplement RDF with access control. However, each of the authors adopt a top-down approach modelling access control based on RDF data structures. Our approach is rather to extract, model and enforce existing access control mechanisms over RDF data. Therefore, we adopt a bottom up approach examining the access control requirements based on existing software engineering and database access control approaches. Based on this analysis, we make the following contributions (i) identify the core access control models that need to be supported; (ii) propose a set of rules that are necessary for the enforcement of these models over RDF; and (iii) present an access control enforcement framework for RDF. In addition, we detail our implementation of the proposed enforcement framework and examine the overall performance of our prototype over real enterprise data.

The remainder of the paper is structured as follows: in Section 2, we provide an overview of commonly used access control models. Section 3 describes how these models can be used in conjunction with RDF and details the rules necessary to propagate access control policies, based on existing access control models, over RDF data. The framework and our specific implementation are described and evaluated in Section 4. Finally, in Section 5 we discuss related work and we conclude and outline directions for future work in Section 6.

2   Analysis of Enterprise Access Control Models

An *Access Control Model* provides guidelines on how access to system resources and data should be restricted. Whereas an *Access Control Policy* (ACP) details the actual authorizations and access restrictions to be enforced. Such permissions and prohibitions are specified as individual *Access Control Statements* (ACS), represented as a tuple: $\langle S, R, AR \rangle$ where S denotes the subject, R denotes the resource and AR represents the Access Rights. Fig. 1 depicts the relationship between the access control terms that are used in this paper. A *Resource* is used to denote the information to be protected (e.g. database records, application

Table 1. Categorisation of access control models

Enterprise	Data Models	Open Systems
MAC	VBAC	ABAC
DAC	OBAC	
RBAC		

objects or website Uniform Resource Identifiers (URIs)). *Users* represent individuals requesting access to resources. We note that such individuals may or may not be known in advance of the information request. *Groups* are collections of users or roles with common features (e.g. contributors, supervisors or management). *Roles* are used to assign a set of access rights to a set of individuals and groups, for example by department (human resources, sales or marketing) or task (insurance claim processing, reporting or invoicing). The term *Subject* is an umbrella term used to collectively refer to users, roles and groups. *Access Rights* are generally defined as permissions and prohibitions applied to resources and granted to subjects.

A high level categorisation of the models discussed in this section is presented in Table. 1. In keeping with our enterprise data integration use case we focus on access control models commonly used in enterprises; models applicable to other data representation formats; and those that are relevant for distributed open systems. An overview of the models is presented below. As the models will be discussed in the next section we label them M_x where x is the name of the access control model.

M_{MAC}: *Mandatory Access Control (MAC) [18].* In this model access to resources is restricted through mandated policies determined by a central authority (e.g. mainframes and lightweight directory services).

M_{DAC}: *Discretionary Access Control (DAC) [18].* Similar to MAC, access to resource is constrained by a central access control policy, however in contrast to MAC users are allowed to override the central policy (e.g. employees may be allowed to grant others access to their own resources).

M_{RBAC}: *Role Based Access Control (RBAC) [19].* Generally speaking, RBAC involves grouping a set of access rights that describe the responsibilities or tasks that can be performed by a role (e.g. manager, sales person and clerk). RBAC is the most commonly used access control model in enterprise applications.

M_{VBAC}: *View-based Access Control (VBAC) [8].* In relational database systems VBAC is used to simultaneously grant access to one or more tables, tuples or fields. A similar approach is used in Object Oriented Access Control (OBAC) [7] where access rights are granted to application objects.

M_{ABAC}: *Attribute Based Access Control (ABAC) [14].* In distributed open environments where the requester is unknown prior to the submission of the request, an alternative means of granting access is required. In ABAC [14], access control decisions are based on attributes (e.g. *employer=storm, policyNumber=565656*), in the form of digitally signed documents, that are sent by the requester to the server.

3 Extending RDF to Support Existing Access Control Models

In this section we examine how the access control models identified in the previous section can be used to protect RDF data. We propose a set of access control rules required to propagate and enforce access control policies, based on these models, over RDF data.

3.1 Annotated RDF Model

Our work relies on an extension of RDF, called Annotated RDF [22], that allows domain specific meta-information to be attached to RDF triples. The access rights are represented as annotations attached to individual RDF triples. The Annotated RDF access control domain model and the annotated RDFS inference rules are based on our previous work presented in [13] and therefore we only provide a short overview of the domain model below. For the definitions of other domains, namely the fuzzy, temporal and provenance domains, and an overview of how such domains can be combined the reader is referred to Zimmermann et al. [23].

In the following examples an NQuads [5] format is used to associate access control meta-information with a triple. However the framework itself does not dictate the use of an NQuads RDF serialisation, therefore alternative approaches such as using reification or mapping annotations to triple hashcodes would work also. Throughout the paper we assume the default prefix `http://urq.deri.org/enterprise#`.

In Annotated RDF, the domain annotations must follow a pre-defined structure that facilitates inferencing and querying. Software applications generally provide four basic functions of persistent storage *Create, Read, Update and Delete* (CRUD). Intuitively data permissions and prohibitions are modelled in a similar fashion, whereby subjects are granted CRUD access to resources. *Read, update,* and *delete* could be represented as a 3-tuple ACL (R, U, D), where R specifies the formula for *read* permission, U for *update* permission and D for *delete*. A *create* permission has a different behaviour as it would not be attached to any specific triple but rather to a named graph and as such enforcement would need to be handled using rules. In this paper we use separate *Access Control Lists* (ACLs) to represent the permissions as follows:

`:TopSecret1 a :Project` "(*readACL, updateACL, deleteACL*)"

However as our current implementation focuses on read access, we omit both the *updateACL* and *deleteACL* in the following examples and present only the *readACL* as "(*readACL*)". In the access control domain, each ACL is composed of one or more *Access Control Statements* (ACSs) which in turn contain

Access Control Elements (ACEs) in the form of *usernames, roles, groups* or *attributes*. Negated elements are used to explicitly deny access to an entity, for example ¬:mary (where :mary is a username) indicates that :mary is *denied* access. If no annotation is present it is assumed that there is no access control information to be applied to the triple.

Example 1 (RDF Annotation). The following annotated triple:

:TopSecret1 a :Project "([[:manager], [¬:mary]])"

states that the users identified with :manager have read access to the triple however :mary is explicitly denied access.   □

To avoid duplicate and conflicting ACLs, the model relies on a normalisation procedure that checks for redundant ACSs and applies a conflict resolution mechanism if necessary. The conflict resolution is applied in scenarios where an annotation statement contains both a positive and a negative access control element for the same entity, e.g [:mary, ¬:mary]. There are two different ways to resolve conflicts in annotation statements, either apply: (i) a *brave conflict resolution* (allow access); or (ii) a *safe conflict resolution* (deny access).

3.2   Support for Existing Access Control Models

In this section we demonstrate how each of the access control models, presented in 2, can be handled either by the domain model or the enforcement framework.

M_{MAC} *and* M_{DAC}. Given both models relate to the entire ACP as opposed to an ACS, a permission management module is necessary to enable system administrators to specify standard access control policies and system users to perform discretionary updates to some policies.

M_{RBAC}. RBAC can be represented using a single value element notation. The following quad demonstrate how a single annotation can simultaneously cater for users, roles and groups.

:WestCars1 a :Project "([[:mary, :manager, :salesDept]])"

M_{ABAC}. A key–value pair element representation is needed to support attributes. The annotation elements outlined below are used to demonstrate how a subjects employer can be represented using key–value pairs.

:WestCars1 a :Project "([[(:employer ,:storm)]])"

M_{VBAC}. VBAC is concerned with granting access to resources based on a logical abstraction as opposed to the physical representation, as such we propose a number of rules that allow us to grant access based on common data abstractions. For example, the ability to grant access to RDF data that share, a common RDF subject (RDB tuples) or the same RDF object (RDB attributes). In addition we support traditional access control mechanisms by allowing access to be granted based on hierarchical data structures.

Table 2. Overview of access control rules

	M_{MAC}	M_{DAC}	M_{RBAC}	M_{VBAC}	M_{ABAC}	$CRUD$
R1				✓		
R2			✓			
R3	✓	✓	✓	✓	✓	
R4	✓	✓	✓	✓	✓	
R5						✓

3.3   Proposed Access Control Rules

This section introduces a set of access control rules that enable the administration and the enforcement of the different models presented in the previous section. Table. 2 provides an overview of the rules required by each of the models. The table also includes a column labelled $CRUD$. Although access rights are not a model per se, we have included them here as a specific rule is required to infer access rights based on permission hierarchies.

These rules can be used to associate permissions with data that may or may not be extracted from existing systems. Propagation chains can be broken by explicitly specifying permissions for a particular triple. However, the rules may need to be extended to consider provenance as it may not be desirable to propagate permissions to related data from different sources.

In each of the rules, we represent variables with the ? prefix. In general $?S$, $?P$, and $?O$ refer to data variables while $?\lambda$ and $?E$ refer to annotation variables and annotation element variables respectively. The premises (represented above the line) correspond to a conjunction of quads, possibly with variables for any position. Whereas the conclusion (below the line) corresponds to a single quad where the variables are instantiated from the premises. We also assume the premises may include functions, for example the *member* function is true, if and only if, a given access control element is included in the access control list provided. The $\oplus_{ac}$ operation is used to combine the access control information associated with two triples. This operation is used to combine complete lists as well as to combine single access control elements with access control lists, which intuitively adds the new element to the list.

Resource Based Access. The most basic access control rule involves granting a subject access rights to a resource. Normally the access is explicit and therefore no inference is required. However in order to provide support for M_{VBAC} we need the ability to associate access rights with several triples. As such we need a rule to propagate access rights to all triples with the same RDF subject. Given the following triples:

```
:Invoice1 a :Document "([[:john]])"
:Invoice1 :located "/dms/projs/docs"
```

we can infer that:

```
:Invoice1 a :Document "([[:john]])"
:Invoice1 :located "/dms/projs/docs" "([[:john]])"
```

Rule 1. Assuming that we have access rights, denoted by λ_1, associated with a triple. We can use this rule to propagate λ_1 to all triples with the same subject.

$$\frac{?S\ ?P_1\ ?O_1\ ?\lambda_1,\ \ ?S\ ?P_2\ ?O_2\ ?\lambda_2}{?S\ ?P_2\ ?O_2\ (?\lambda_2 \oplus_{ac} ?\lambda_1)} \tag{R1}$$

Hierarchical Subjects Inheritance. In M_{RBAC} access control subjects are organised hierarchically and lower-level subjects inherit the access rights of higher-level subjects. For example, in a role hierarchy access rights allocated to the :manager role will be inherited by all individuals in the organisation that have been granted the :manager role. Given the following triples:

```
:Invoice1 a :Document "([[:manager]])"
:john :inheritsFrom :manager
```

we can infer that:

```
:Invoice1 a :Document "([[:manager],[:john]])"
```

Rule 2. If $?E_1$ and $?E_2$ are access rights and $?E_2$ inherits from $?E_1$ then triples with access rights $?E_1$ should also have $?E_2$.

$$\frac{?S\ ?P\ ?O\ ?\lambda_1,\ ?E_2\ \text{:inheritsFrom}\ ?E_1, member(?E_1, ?\lambda_1)}{?S\ ?P\ ?O\ (?\lambda_1 \oplus_{ac} ?E_2)} \tag{R2}$$

Hierarchical Subjects Subsumption. Like $R2$ Access Control subjects can be organised to form a hierarchy. However in this instance we are talking about an organisation structure as opposed to a role hierarchy, in which case higher-level subjects will subsume the access rights of lower-level subjects. For example managers implicitly gain access to resources that their subordinates have been explicitly granted access to. Given the following triples:

```
:Invoice2 a :Document "([[:john]])"
:mary :hasSubordinate :john
```

we can infer that:

```
:Invoice2 a :Document "([[:john], [:mary]])"
```

If an $?E_2$ has subordinate $?E_1$ then $?E_2$ will be granted access to any triples $?E_1$ has access to. Since this rule differs from Rule (R2) only in the given vocabulary, the same rule can be reused if we replace *:inheritsFrom* with *:hasSubordinate*.

Hierarchical Resources Inheritance. Similar principles can also be applied to resources. In several of the access control models resources are organised into a hierarchy and lower-level resources inherit the access rights of higher-level resources. For example document libraries are often organised into a hierarchy. Given the following triples:

```
:dmsProjs a :DocumentLibrary "([[:employee]])"
:dmsProjsRpts a :DocumentLibrary
:dmsProjsRpts :isPartOf :dmsProjs
```

we can infer that:

```
:dmsProjRpts a :DocumentLibrary "([[:employee]])"
```

Rule 3. Here the rule states that if an $?S_2$ is part of $?S_1$ then the access rights of triples with $?S_1$ i.e. λ_1 should be propagated to $?S_2$.

$$\frac{?S_1\ ?P_1\ ?O_1\ ?\lambda_1,\ ?S_2\ ?P_2\ ?O_2\ ?\lambda_2,\ ?S_2\ \text{:isPartOf}\ ?S_1}{?S_2\ ?P_2\ ?O_2\ (?\lambda_2\oplus_{ac}?\lambda_1)} \qquad \text{(R3)}$$

Resources Categorisation. However resources can also be categorised by type, for example views or objects in M_{VBAC}, and all resources of a particular type can inherit access rights assigned to the type. Access rights placed on a :report object can be propagated to all objects of type :report. Given the following triples:

```
:Report a :Document "([[:employee]])"
:Report1 a :Report
```

we can infer that:

```
:Report1 a :Report "([[:employee]])"
```

Rule 4. Below the rule states that if $?O_2$ is a type of $?O_1$ then the access rights of triples with $?O_1$ i.e. λ_1 should be propagated to $?O_2$.

$$\frac{?S_1\ ?P_1\ ?O_1\ ?\lambda_1,\ ?S_2\ ?P_2\ ?O_2\ ?\lambda_2,\ ?O_2\ \text{a}\ ?O_1}{?S_2\ ?P_2\ ?O_2\ (?\lambda_2\oplus_{ac}?\lambda_1)} \qquad \text{(R4)}$$

Hierarchical Access Rights Subsumption. The access rights themselves can form a hierarchy whereby each permission level can include the access rights of the permission level below it. For example, we can assume **update** access from **delete** access and **read** access from **update** access. Given the following triples:

```
:Invoice3 a :Document "([[]], [[:john]], [[]])"
```

we can infer that:

```
:Invoice3 a :Document "([[:john]], [[:john]], [[]])"
```

Rule 5. Assuming that the ACL is a 3-tuple (R, U, D) and the permission hierarchy is stored as RDF. This rule states that if **update** is part of **delete** and **read** is part of **update** then the delete access rights should be propagated to the update annotation and the update access rights to the read annotation.

$$\frac{?S\ ?P\ ?O\ (?\lambda_1, ?\lambda_2, ?\lambda_3)}{?S\ ?P\ ?O\ ((?\lambda_1\oplus_{ac}?\lambda_2\oplus_{ac}?\lambda_3), (?\lambda_2\oplus_{ac}?\lambda_3), ?\lambda_3)} \qquad \text{(R5)}$$

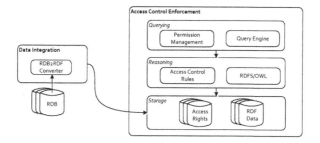

Fig. 2. RDF Data Integration and Access Control Enforcement Framework

4 Framework, Implementation and Evaluation

Based on our analysis of existing access control models this section describes the minimal set of components necessary for a data integration and access control enforcement framework. It provides an overview of our implementation and presents an experimental evaluation of our prototype which focuses on the: (i) the *RDB2RDF* data integration; (ii) the *reasoning engine*; and (iii) the *query engine*. The aim of this evaluation is simply to show the feasibility of our approach and, although we present different dataset sizes, at this point we are not looking at improving scalability and thus do not propose any kind of optimisations.

4.1 Access Control Enforcement Framework

An overview of the proposed framework which is depicted in Fig. 2, is composed of two core modules: *Data Integration* and *Access Control Enforcement*. The Data Integration module is responsible for the conversion of existing RDB data and access control policies to RDF. Whereas the Access Control Enforcement module caters for the management of access rights and enables authenticated users to query their RDF data. We do not claim that this is an exhaustive list of the system components but rather the minimal components any access control framework requires. Noticeably, one missing component is the *authentication* component, which we do not focus on in this paper. Authentication can be achieved by relying on WebId and self-signed certificates that work transparently over HTTPS.

Data Integration. The Data Integration module is responsible for the extraction of data and associated access rights from the underlying relational databases (RDBs). The information extracted is subsequently transformed into RDF and persisted in the *RDF Data* and *Access Rights* stores. Ideally, the data integration step would be carried out in conjunction with a domain expert, for example to assist in defining an R2RML [6] mapping that extracts and converts the relational data into RDF. In a pure Semantic Web scenario, the data integration module is optional, in which case the data and access rights need to be populated either manually or automatically via specialised software systems.

Storage. The integrated data retrieved from the original relational databases is stored in the RDF Data store and the access control policies over the RDF data are stored in the Access Rights store. A link between the two stores is necessary so that access policies can be related to existing graphs, triples, or resources. Any RDF store can be used as a back-end for storing this data. We do not restrict the representation format used for the ACSs which may be stored as quads, associated with reified triples or mapped to triple hash-codes.

Reasoning. For this component we consider two distinct forms of inference: (a) data inference, where new triples are deduced from existing ones; and (b) access rights inference, where new ACSs are deduced from existing ones. For data inferencing we rely on well established forms of inference in the Semantic Web such as RDF Schema (RDFS) or the Web Ontology Language (OWL). However, if access control policies have been assigned to the data we also need a mechanism to propagate these policies when we infer new triples from existing ones. In addition Rule based Access Control reasoning is required to support both permission management and data querying over RDF based on access control policies. Either backward or forward chaining could be used for both the data and the access rights inference.

Querying. SPARQL [20] is the standard query language for RDF however the language itself does not provide any form of access control. It thus needs to be extended to ensure access is only given to RDF triples the user has been granted access to either explicitly or implicitly.

4.2   Implementation

In our prototype the data integration is performed using XSPARQL [12], a transformation and query language for XML, RDB, and RDF, that allows data to be extracted from existing RDBs. Whereas the access control enforcement framework is a Prolog implementation of the Annotated RDF [23] framework which enables reasoning over annotated triples.

RDB2RDF. XSPARQL was chosen over other alternative RDB2RDF approaches as both RDB and XML data formats are used extensively in organisations and also it has built-in support for the NQuad format. In this paper we focus on relational databases but the ability to extract XML, from files or web-services, is desirable.

Annotated RDF. Our Prolog Annotated RDF [23] implementation, allows domain specific meta-information in the form of access rights to be attached to RDF triples. An overview of our annotated RDF access control domain model, presented in [13], was provided in 3.1. The reasoning component is implemented by extension of the RDFS inference rules. It allows the Annotated RDF reasoning engine to automatically determine the annotation values for any inferred triples based on the annotations of the premises. As such, we are actually performing data and annotation reasoning in the same step. We also support reasoning over the access rights alone, by allowing permissions to be propagated according to the rules presented in Section 3.3. With the exception of (R5) our prototype provides for the presented rules, either in the form of inference rules or by explicit rules in the domain modelling. Our prototype currently does not cater for rule (R5) because the existing implementation only handles *read* permissions. However we are in the process of extending this prototype to support the other forms of access rights.

AnQL. Our query engine relies on an extension of SPARQL, AnQL [11], which allows the annotated RDF data to be queried. AnQL extends SPARQL to consider also a fourth element in the query language. However, for the enforcement of access rights, the end user must not be allowed to write AnQL queries as this would pose a security risk, thus we rely on a translation step from SPARQL into AnQL. The Query Engine takes a SPARQL query and transparently expands it into an AnQL query, using the list of credentials provided by the authentication module. As authentication is outside the scope of this paper, we simply extract end user credentials from the RDB, and maintain a mapping between the enterprise employees and their usernames, groups and roles. The AnQL query is subsequently executed against the Annotated RDF graph, which guarantees that the user can only access the triples annotated with their credentials.

4.3   Evaluation

The benchmark system is a virtual machine, running a 64-bit edition of Windows Server 2008 R2 Enterprise, located on a shared server. The virtual machine has an Intel(R) Xeon(R) CPU X5650 @ 2.67GHz, with 4 shared processing cores and 5GB of dedicated memory. For the evaluation we use XSPARQL to extract both the data and the access rights from two separate software application databases: a document management system (DMS) and a timesheet system (TS). We subsequently use the rules specified in 3.3 to propagate the permissions to relevant triples. In the case of the TS we only assigned existing access rights to one triple and let R1 propagate the permissions to all triples with the same subject. As for the DMS we extracted the existing URI hierarchy and the associated permissions and used R3 to propagate the permissions to all data extracted depending on their location in the hierarchy. Existing type information was extracted from both systems and propagated using R4. Finally we input the organisation structure as facts in the Prolog application and used R2 to propagate permissions based on this hierarchy. As our prototype only supports read access we did not consider R5 in our evaluation.

The different datasets (DS_1, DS_2, DS_3, and DS_4) use the same databases, tables, and XSPARQL queries and differ only on the number of records that are retrieved from the databases.

Table 3 provides a summary of each dataset, stating the number of database records queried, the number of triples generated, and the size in MegaBytes (MB) of the NQuads representation of the triples.

Table 3. Dataset description

	DS_1	DS_2	DS_3	DS_4
database records	9990	17692	33098	63909
triples	62296	123920	247160	493648
file size (MB)	7.6	14.9	29.9	59.6

Table 4. Prototype evaluation

	DS_1	DS_2	DS_3	DS_4
RDB2RDF (sec)	39	52	92	179
Import (sec)	3	5	10	19
Inference engine (sec)	3218	11610	32947	58978
Inferred triples	59732	117810	237569	473292

RDB2RDF and Inference. Table 4 includes the time the data extraction process took in seconds (sec), the time it took to import the data into Prolog (sec), the evaluation time of the access control rules detailed in 3.3 (sec), and the number of triples inferred in this process. Fig. 3a provides a high level overview of the times for each of the datasets. Based on this simple experiment we have hints that the extraction process and the loading of triples into Prolog behave linearly but more data intensive tests are still required. As the inferencing times are highly dependent on both the rules and the data further experimentation is required in this area.

Query Engine. For the evaluation of the AnQL engine we created three separate query sets $1TP$, $2TP$ and $3TP$. The query sets were each composed of three queries with: one triple pattern $1TP$; two triple patterns $2TP$; or three triple patterns $3TP$. Each query was run without annotations ($\emptyset$), with a credential that appears in the dataset ($\exists$) and with a credential that does not appear in the dataset ($\nexists$). The results displayed in Table 5 were calculated as an average of 20 response times excluding the two slowest and fastest times. Based on this experiment we can see that there is an overhead for the evaluation of annotations when you query using a single triple pattern Fig. 3b. However queries with a combination of annotations and either two or three triple patterns, Fig. 3c and Fig. 3d respectively, out perform their non annotated counterparts. Such behaviour is achieved in our implementation by pushing the filters into the triple patterns as opposed to querying the data and subsequently filtering the results. The experiments do not show a significant performance increase over the four datasets, between queries with annotations Fig. 3e and those without

Table 5. Query execution time in seconds

		DS_1	DS_2	DS_3	DS_4
$1TP$	$\emptyset$	0.0000	0.0003	0.0013	0.0042
	$\exists$	0.0065	0.0159	0.0300	0.0654
	$\not\exists$	0.0081	0.0189	0.0316	0.0670
$2TP$	$\emptyset$	0.0247	0.0544	0.1497	0.2679
	$\exists$	0.0228	0.0638	0.0898	0.1845
	$\not\exists$	0.0094	0.0198	0.0338	0.0690
$3TP$	$\emptyset$	0.0169	0.0482	0.1322	0.2213
	$\exists$	0.0241	0.0593	0.0943	0.1741
	$\not\exists$	0.0101	0.0192	0.0316	0.0609

annotations Fig. 3f. Furthermore we can see that there is no overhead associated with queries where the annotation is not present in the dataset. In fact such queries are actually more efficient when the query is made up of two or three triple patterns.

5 Related Work

In recent years, there has been an increasing interest access control policy specification and enforcement using Semantic Technology, KAos [3], Rein [10] and PROTUNE [2] are well known works in this area. Policy languages can be categorised as general or specific. In the former the syntax caters for a diverse range of functional requirements (e.g. access control, query answering, service discovery and negotiation), whereas in contrast the latter focuses on just one functional requirement. Two of the most well-known access control languages, KAos [3] and Rei [10], are in fact general policy languages. Although the models and the corresponding frameworks are based on semantic technology the authors do not consider applying their model or framework to RDF data.

The authors of Concept-Level Access Control (CLAC) [15], Semantic-Based Access Control (SBAC) [9] and the semantic network access control models proposed by Ryutov et al. [16] and Amini and Jalili [1] respectively, all propose access control models for RDF graphs and focus on policy propagation and enforcement based on semantic relations. Qin and Atluri [15] propose policy propagation of access control based on the semantic relationships among concepts. Javanmardi et al. [9], Ryutov et al. [16] and Amini and Jalili [1] enhance the semantic relations by allowing policy propagation based on the semantic relationships between the subjects, objects, and permissions.

Costabello et al. [4] propose a lightweight vocabulary which defines fine-grained access control policies for RDF data. They focus specifically on modelling and enforcing access control based on contextual data. Sacco et al. [17] present a privacy preference ontology which allows for the application of fine-grained

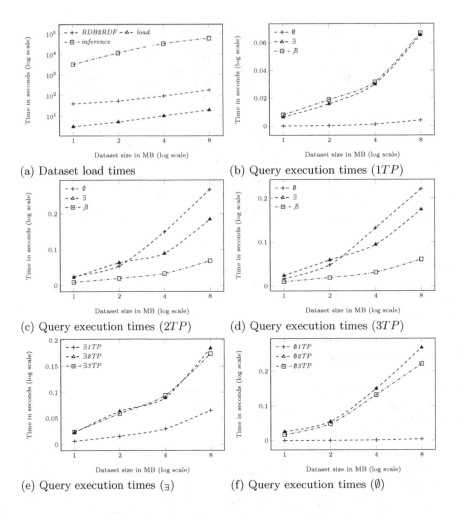

Fig. 3. Load and Query times across 4 datasets

access control policies to an RDF file. Both Costabello et al. [4] and Sacco et al. [17] propose frameworks that rely on SPARQL ask queries to determine access to resources based on the defined vocabularies.

In contrast to existing proposals the solution we present is tightly coupled with the cornerstone Semantic Web technology (RDF, RDFS and SPARQL). We propose an integrated solution which extracts data and access rights from the enterprise software systems into RDF and an enforcement framework which handles reasoning over both the data and the access control statements. Approaches to date can be summarised as top-down approaches where the authors model access control based on RDF data structures and the access control requirements of open systems. We adopt a bottom up approach showing how existing access control models can be applied to RDF data.

6 Conclusions and Future Work

RDF is a flexible data representation format that can greatly benefit an enterprise, not only as a self-describing data model for their existing public data but also as a means of extending their own data with RDF data freely available on the Web. The introduction of access control policies and enforcement over RDF also allows organisations to selectively and securely share information with third parties, for example with other business partners, supplier or clients. In this paper we proposed a set of rules that allow for the enforcement of commonly used access control models and presented the components necessary for an RDF access control enforcement framework, while detailing our own implementation of this framework. We also presented some preliminary performance evaluation of the prototype by using enterprise data supplied by our project partner. Our approach is to transform SPARQL queries and user credentials into AnQL queries. As a side effect from the user perspective queries with two or three patterns appear to run faster. However as we have seen from the benchmark results, it is still not scalable enough for deployment in an enterprise environment.

As for future work, a combination of both the access control and the provenance domains is planned along with a more scalable implementation of the proposed framework. In addition, we propose a study of whether more expressive rules are required and their application over access control annotated data.

Acknowledgements. This work is supported in part by the Science Foundation Ireland under Grant No. SFI/08/CE/I1380 (Lion-2), the Irish Research Council Enterprise Partnership Scheme and Storm Technology Ltd. We would like to thank Aidan Hogan for his valuable comments.

References

1. Amini, M., Jalili, R.: Multi-level authorisation model and framework for distributed semantic-aware environments. IET Information Security 4(4), 301 (2010)
2. Bonatti, P.A., De Coi, J.L., Olmedilla, D., Sauro, L.: Rule-based policy representations and reasoning. In: Bry, F., Małuszyński, J. (eds.) Semantic Techniques for the Web. LNCS, vol. 5500, pp. 201–232. Springer, Heidelberg (2009)
3. Bradshaw, J.M., Dutfield, S., Benoit, P., Woolley, J.D.: KAoS: Toward an industrial-strength open agent architecture. In: Software Agents (1997)
4. Costabello, L., Villata, S., Delaforge, N.: Linked data access goes mobile: Context-aware authorization for graph stores. In: LDOW - 5th WWW Workshop on Linked Data on the Web (2012), http://hal.archives-ouvertes.fr/hal-00691256/
5. Cyganiak, R., Harth, A., Hogan, A.: N-Quads: Enxtending N-Triples with Context (2009)
6. Das, S., Sundara, S., Cyganiak, R.: R2RML: RDB to RDF Mapping Language. Candidate Recommendation, W3C (February 2012)
7. Evered, M.: A case study in access control requirements for a health information system. In: Second Workshop on Australasian Information Security (2004)
8. Griffiths, P.P.: An authorization mechanism for a relational database system. ACM Transactions on Database Systems 1(3), 242–255 (1976)

9. Javanmardi, S., Amini, M., Jalili, R., GanjiSaffar, Y.: SBAC: A Semantic-Based Access Control Model. In: 11th Nordic Workshop on Secure IT-systems (NordSec 2006), Linköping, Sweden (2006)
10. Kagal, L., Finin, T.: A policy language for a pervasive computing environment. In: Proceedings of the IEEE 4th International Workshop on Policies for Distributed Systems and Networks, POLICY 2003, pp. 63–74. IEEE Comput. Soc. (2003)
11. Lopes, N., Polleres, A., Straccia, U., Zimmermann, A.: AnQL: SPARQLing Up Annotated RDFS. In: Patel-Schneider, P.F., Pan, Y., Hitzler, P., Mika, P., Zhang, L., Pan, J.Z., Horrocks, I., Glimm, B. (eds.) ISWC 2010, Part I. LNCS, vol. 6496, pp. 518–533. Springer, Heidelberg (2010)
12. Lopes, N., Bischof, S., Decker, S., Polleres, A.: On the Semantics of Heterogeneous Querying of Relational, XML and RDF Data with XSPARQL. In: Moura, P., Nogueira, V.B. (eds.) EPIA 2011 – COLA Track, Lisbon, Portugal (October 2011)
13. Lopes, N., Kirrane, S., Zimmermann, A., Polleres, A., Mileo, A.: A Logic Programming approach for Access Control over RDF. In: Technical Communications of ICLP 2012, vol. 17, pp. 381–392. Schloss Dagstuhl–Leibniz-Zentrum fuer Informatik (2012)
14. McCollum, C.J., Messing, J.R., Notargiacomo, L.: Beyond the pale of MAC and DAC-defining new forms of access control. In: Proceedings of the 1990 IEEE Computer Society Symposium on Research in Security and Privacy, pp. 190–200. IEEE (1990)
15. Qin, L., Atluri, V.: Concept-level access control for the Semantic Web. In: Proceedings of the 2003 ACM Workshop on XML Security - XMLSEC 2003, p. 94. ACM Press (2003)
16. Ryutov, T., Kichkaylo, T., Neches, R.: Access Control Policies for Semantic Networks. In: 2009 IEEE International Symposium on Policies for Distributed Systems and Networks, pp. 150–157. IEEE (July 2009)
17. Sacco, O., Passant, A., Decker, S.: An Access Control Framework for the Web of Data. In: 2011 IEEE 10th International Conference on Trust, Security and Privacy in Computing and Communications, pp. 456–463 (November 2011)
18. Samarati, P., de Capitani di Vimercati, S.: Access control: Policies, models, and mechanisms. In: Focardi, R., Gorrieri, R. (eds.) FOSAD 2000. LNCS, vol. 2171, pp. 137–196. Springer, Heidelberg (2001)
19. Sandhu, R.S.: Role-based access control. In: Advances in Computers, pp. 554–563 (1998)
20. Seaborne, A., Prud'hommeaux, E.: SPARQL Query Language for RDF. W3C Recommendation, W3C (January 2008), http://www.w3.org/TR/rdf-sparql-query/
21. Stephens, S.: The Enterprise Semantic Web. In: Cardoso, J., Hepp, M., Lytras, M.D. (eds.) The Semantic Web: Real-World Applications from Industry. Semantic Web and Beyond Computing for Human Experience, vol. 6, pp. 17–37. Springer (2007)
22. Udrea, O., Recupero, D.R., Subrahmanian, V.S.: Annotated RDF. ACM Trans. Comput. Logic 11(2), 1–41 (2010)
23. Zimmermann, A., Lopes, N., Polleres, A., Straccia, U.: A General Framework for Representing, Reasoning and Querying with Annotated Semantic Web Data. J. Web Sem. 11, 72–95 (2012)

Active Learning of Domain-Specific Distances for Link Discovery

Tommaso Soru and Axel-Cyrille Ngonga Ngomo

Department of Computer Science
University of Leipzig
Augustusplatz 10, 04109 Leipzig
{tsoru,ngonga}@informatik.uni-leipzig.de
http://limes.sf.net

Abstract. Discovering cross-knowledge-base links is of central importance for manifold tasks across the Linked Data Web. So far, learning link specifications has been addressed by approaches that rely on standard similarity and distance measures such as the Levenshtein distance for strings and the Euclidean distance for numeric values. While these approaches have been shown to perform well, the use of standard similarity measure still hampers their accuracy, as several link discovery tasks can only be solved sub-optimally when relying on standard measures. In this paper, we address this drawback by presenting a novel approach to learning string similarity measures concurrently across multiple dimensions directly from labeled data. Our approach is based on learning linear classifiers which rely on learned edit distance within an active learning setting. By using this combination of paradigms, we can ensure that we reduce the labeling burden on the experts at hand while achieving superior results on datasets for which edit distances are useful. We evaluate our approach on three different real datasets and show that our approach can improve the accuracy of classifiers. We also discuss how our approach can be extended to other similarity and distance measures as well as different classifiers.

1 Introduction

Discovering cross-knowledge-base links is of central importance to realize the vision of the Linked Data Web [1]. Over the last years, several frameworks and approaches have been developed to address the two main hurdles of link discovery: the quadratic runtime [16,10] and the discovery of accurate link specifications [18,19]. In both cases, most approaches assume that link discovery can be carried out by devising a similarity function σ such that instances from a source and target knowledge base whose similarity is larger than a certain threshold θ should be linked. In most cases, σ is a combination of several atomic similarity measures, i.e., of measures that compare solely one pair of property values from the source and target instances. The combination (σ, θ) is usually called link specification [16] or linkage rule [10]. So far, the detection of accurate link specification was carried out by using techniques such as linear or Boolean classifier

H. Takeda et al. (Eds.): JIST 2012, LNCS 7774, pp. 97–112, 2013.

Table 1. Example of property values

Source labels	Target labels
Adenomatoid tumor	Adenomatoid tumour
Odontogenic tumor	Odontogenic tumour
Generalised epidermolysis	Generalized epidermolysis
Diabetes I	Diabetes I
Diabetes II	Diabetes II
Anemia Type I	Anemia Type I
Anemia Type II	Anemia Type II
Anemia Type III	Anemia Type III

learning [17] or genetic programming [11,18,19]. However, only generic string similarity measures have been used as atomic measures within these learning approaches. In this paper, we go beyond the state of the art by not only learning classifiers but also learning the similarity metrics used in these classifiers directly from the data. In addition, we go beyond the state-of-the-art in metric learning by applying an active learning approach [24].

The need for our approach is motivated by the exemplary link discovery task in the medical domain showed in Table 1. The source and target contain property values are written in American and British English and thus display minor differences (e.g., "Generali$\underline{z}$ed" vs. "Generali$\underline{s}$ed"). Therewith intertwined are yet also semantic differences (e.g., "Diabetes $\underline{I}$" vs. "Diabetes $\underline{II}$") which correspond to exactly the same edit distance between the property values. Consequently, if a user was to compute links between these two data sets by means of these property values using the edit distance, he/she could either (1) choose to set the distance threshold to 0, which would lead to a precision P = 1 but to a recall of R = 0.625 (F = 0.77) or (2) choose to set the threshold to 1, which would lead to R = 1 but to P = 0.57 (F = 0.73). It is obvious that for this particular example, there is no distance threshold for the standard edit distance that would lead to an F-measure of 1. The same holds for other generic string similarity measures such as Jaccard. Consequently, none of the previous approaches (see e.g. [19,18]) to learning link specifications would be able to detect an optimal link specification for the data at hand. The basic intuition behind this work is the following: If we could model that replacing "s" by"z" yields less semantics that inserting the token "I" for this particular pair of property values, we could assign different weights to these tokens during the computation of the edit distance. It is important to note that this difference would only be valid for this particular pair of properties. For standardized labels such as company names, the substitution of "s" by"z" might yield exactly the same semantics as inserting an "I". Given that most link specifications rely on a complex similarity functions (i.e., on combination of several similarity values) to compute links (e.g., similarity of label, longitude and latitude for cities, etc.), the additional requirement of devising *a domain-specific similarity measure for each of the property pairs used* for linking concurrently is of central importance. Hence, the main goal of this work is to devise a supervised approach for learning link specifications which rely on complex domain-specific similarities learned directly from the input data. As learning such a metric can require a considerable amount of human labeling, we implement our metric learning approach within the pool-based setting of active learning [24] to minimize the burden on human annotators. In the rest of this paper, we focus on learning similarities based on generalized edit distances. In the discussion section, we show how our approach can be extended to learning other measures such as weighted Jaccard and n-gram-based similarities. Our main contributions are thus as follows:

- We present a supervised machine for learning link specifications based on domain-specific similarities.
- To reduce the burden on the user, we learn these similarities by using active learning.
- We show how our framework can be extended to other similarities such as Q-Grams and Jaccard.
- We evaluate the learned similarities against three different data sets and analyze the strengths and weaknesses of our approach [1].

2   Preliminaries and Notation

2.1   Link Discovery as Classification

In this work, we regard link discovery as a classification problem. Let S (source) and T (target) be two sets of resources. Furthermore and without loss of generality, let each resource from S and T be described by a tuple of n property values such that the ith property value of elements of S maps the ith property value of elements of T.[2] The general aim of link discovery is to devise an (mostly with respect to its F-score) accurate classifier $C : S \times T \rightarrow \{-1, +1\}$ that assigns the class $+1$ to a pair (s, t) should be linked and -1 else. Most approaches for link discovery rely on a combination of operators (such as minimum, maximum and linear combinations), similarity measures σ_i (such as Jaccard and Cosine) that compare property values and similarity thresholds θ_i to determine such as classifier. In previous works, only the operators and thresholds were learned [17,11,18,19]. Yet, as shown by our example, such approaches are not always sufficient to achieve a high F-measure. We go beyond the state of the art by learning the similarity measures σ_i. In this work, we focus on learning these measures in combination with weighted linear classifiers, i.e., classifiers such that

$$C(s, t) = +1 \Leftrightarrow \left(\sum_{i=1}^{n} w_i \sigma_i(s, t) - \theta \geq 0 \right), \tag{1}$$

where σ_i is a similarity function that compares the values of the ith property of s and t and aim to learn the values of w_i, θ and σ_i concurrently. Note that other

---

[1] The implementation of our approach is publicly available at
  http://metric-learning.googlecode.com

[2] Note that such pairs can be calculated using schema matching approaches such as those described in [20].

classifiers can be combined with our approach. For a given classifier C, we call (s, t) *positive* if $C(s, t) = +1$ and *negative* else. Also note that while we focus on string similarities in this work, our implementation also supports numerical values by the means of learning Mahalanobis distances.

Learning classifiers within the context of active learning means relying on user feedback after each iteration to improve the classifiers until a termination condition is met. Given the type of classifier we use, we will assume the pool-based sampling setting [24]. We will denote learning iterations with the variable τ. We will denote classifiers, weights, thresholds, matrices and similarities at iteration τ by using superscripts. Consequently, we will label the input generated by the user at iteration τ with Q^τ. The classifier and distance matrix computed at the same iteration will be denoted C^τ resp. M^τ. Note that in the following, we limit ourselves to learning generalized edit similarities σ_i, which we define as $(1 + \delta_i)^{-1}$, where δ_i is a generalized edit distance. We chose to learn similarities instead of distances because they are bound by finite values, making the computation of the classifier more efficient. However, the basic idea behind our approach can be extended to other similarities and distances with ease. We thus present generalization of other types of similarity measures and show how our approach can be extended to them in Section 6.

2.2 Weighted Edit Distance

Let A and B be two strings over an alphabet Σ. Furthermore, let $\Sigma^* = \Sigma \cup \{\epsilon\}$, where ϵ stand for the empty string. The edit distance $\delta(A, B)$ (also called Levenshtein distance [13]) is defined as the minimal number of edit operations (insertion, deletion and substitution) necessary to transform the string A into B. Thus, e.g., $\delta(\text{"Diabetes I"}, \text{"Diabetes II"}) = \delta(\text{"Adenomatoid tumor"}, \text{"Adenomatoid tumour"}) = 1$. Previous work from the machine learning community (see e.g., [3]) has proposed learning the cost of each possible insertion, deletion and substitution. Achieving this goal is equivalent to learning a positive cost matrix M of size $|\Sigma^*| \times |\Sigma^*|$. The rows and columns of M are issued from Σ^*. The entry m_{ij} stands for the cost of replacing the ith character of Σ^* with the jth character of Σ^*. Note that the deletion of a character $c \in \Sigma$ is modeled as replacing c with ϵ, while insertion of c is modeled as replacing ϵ with c. Within this framework, the standard edit distance can be modeled by means of a cost matrix M such that $m_{ij} = 1$ if $i \neq j$ and 0 else. The distance which results from a non-uniform cost matrix[3] is called a weighted (or generalized) edit distance δ_w. The distance $\delta_w(A, B)$ can now be defined as the least expensive sequence of operations that transforms A into B.[4] Note that we also go beyond the state of the art in metric learning by combining metric learning and active learning,

---

[3] For distances, a non-uniform matrix M is one such that $\exists i, j, i', j' : (i \neq j \wedge i' \neq j' \wedge m_{ij} \neq m_{i'j'})$.

[4] Note that while the edit distance has been shown to be a metric, M being asymmetric leads to δ_w not being a metric in the mathematical sense of the term, as $\exists A, B : \delta_w(A, B) \neq \delta_w(B, A)$.

which has been shown to be able to reduce the amount of human labeling needed for learning link specifications significantly [18].

3   The ACIDS Approach

The aim of our approach, ACIDS (Active Learning of Distances and Similarities), is to compute domain-specific similarity measures while minimizing the annotation burden on the user through active learning. Note that while we focus on learning similarity functions derived from the generalized edit distance, we also show how our approach can be generalized to other similarity measures.

3.1   Overview

Our approach consists of the following five steps:

1. $\tau = 0$; get Q^0 and compute $C^0, M^0_{1...n}$;
2. If all elements of $\bigcup\limits_{i=0}^{\tau} Q^\tau$ can be separated by C^τ then goto step 4.
3. Else update the similarity measures σ_i^τ and update the classifier C^τ. If the maximal number of iterations is reached, goto step 4. Else goto step 2.
4. Increment τ. Select the k most informative positive and k most informative negative examples from $S \times T$ and merge these two sets to Q^τ.
5. Require labeling for Q^τ from the oracle. If termination condition reached, then terminate. Else goto step 2.

In the following, we elaborate on each of these steps.

3.2   Initialization

Several parameters need to be set to initialize ACIDS. The algorithm is initialized by requiring k positive and k negative examples $(s,t) \in S \times T$ as input. The positive examples are stored in P^0 while the negative examples constitute the elements of N^0. Q^0 is set to $P^0 \cup N^0$. For all n dimensions of the problem where the property values are strings, $\sigma_i^0(s,t) = (1 + \delta(s,t))^{-1}$ (where δ stands for the edit distance with uniform costs) is used as initial similarity function.

The final initialization step consists of setting the initial classifier C^0. While several methods can be used to define an initial classifier, we focus on linear classifiers. Here, the classifier is a hyperplane in the n-dimensional similarity space S whose elements are pairs $(s,t) \in S \times T$ with the coordinates $(\sigma_1^0(s,t), ..., \sigma_n^0(s,t))$. We set C^0 to

$$\sum_{i=1}^{n} \sigma_i^0(s,t) \geq n\theta_0. \tag{2}$$

This classifier is the hyperplane located at the Euclidean distance $\theta_0\sqrt{n}$ from the origin of S and at $(1 - \theta_0)\sqrt{n}$ from the point of S such that all its coordinates are 1. Note that $\theta_0 \in [0,1]$ is a parameter that can be set by the user.

3.3   Computing Separability

The aim of this step is to check whether there is a classifier that can separate
the examples we know to be positive from those we know to be negative. At
each iteration τ, this step begins by computing the similarities $\sigma_i^\tau(s,t)$ of all
labeled examples $(s,t) \in Q^\tau$ w.r.t. the current distance matrix M^τ. Note that
$\sigma_i^\tau(s,t)$ varies with time as the matrix M^τ is updated. Checking the separability
is carried out by using linear SVMs.[5] The basic idea behind this approach is
as follows: Given a set of n-dimensional similarity vectors $x_1, ..., x_m$ and the
classification $y_i \in \{+1, -1\}$ for each x_i, we define a classifier by the pair (w, b),
where w is a n-dimensional vector and b is a scalar. The vector x is mapped to
the class y such that

$$y(w^T x - \theta) \geq 0. \tag{3}$$

The best classifier for a given dataset is the classifier that minimizes $\frac{1}{2}w^T w$
subject to $\forall i \in \{1, ..., m\}$, $y_i(w^T x_i - \theta) \geq 0$ [6]. In our case, the coordinates x_i
of a point (s,t) at iteration τ are given by $\sigma_i^\tau(s,t)$. The components w_i of the
vector w are the weights of the classifier C^τ, while θ is the threshold. If we are
able to find a classifier with maximal accuracy using the algorithm presented
in [5] applied to linear kernels, then we know the data to be linearly separable.
In many cases (as for our toy data set for example), no such classifier can be
found when using the original edit similarity function. In this case, the basic
intuition behind our approach is that the similarity measures (and therewith
the distribution of positive and negative examples in the similarity space) need
to be updated in such a way that the positive examples wander towards the class
$+1$ of the classifier, while the negative wander towards -1. Altering all M_i^τ in
such a way is the goal of the subsequent step.

3.4   Updating the Similarity Measures

Overview. Formally, the basic idea behind our approach to updating similarity
measures is that when given the two sets $P \subseteq S \times T$ of positive examples and
$N \subseteq S \times T$ of negative examples, a good similarity measure σ_i^τ is one such that

$$\forall (s,t) \in P \; \forall (s',t') \in N \; \sigma_i^\tau(s,t) \geq \sigma_i^\tau(s',t'). \tag{4}$$

Let N^τ be the subset of Q^τ which contains all pairs that were labeled as false
positives by the oracle at iteration τ. Furthermore, let P^τ be set of pairs labeled
as false negatives during the same iteration. In addition, let $\mu_i^\tau : M \times S \times T \to$
$[0,1]$ be a function such that $\mu_i^\tau(m_{ij}, s, t) = 1$ when the substitution operation
modeled by m_{ij} was used during the computation of the similarity $\sigma_i^\tau(s,t)$ (the
approach to computing μ is shown in the subsequent subsection). We update
the weight matrix M_i^τ of σ_i^τ by employing a learning approach derived from
perceptron learning. The corresponding update rule is given by

$$m_{ij}^\tau := m_{ij}^\tau - \sum_{(s,t) \in P^\tau} \eta^+ \mu_i^\tau(m_{ij}^\tau, s, t) + \sum_{(s',t') \in N^\tau} \eta^- \mu_i^\tau(m_{ij}^\tau, s', t'). \tag{5}$$

---

[5] Note that any other type of classifier could be used here.

While it might seem counter-intuitive that we augment the value of m_{ij} for negatives (i.e., points that belong to -1) and decrease it for positives, it is important to remember that the matrix M_i describes the distance between pairs of resources. Thus, by updating it in this way, we ensure that the operation encoded by m_{ij} becomes less costly, therewith making the pairs (s, t) that rely on this operation more similar. Negative examples on the other hand are a hint towards the given operation being more costly (and thus having a higher weight) than assumed so far. The similarity between two strings is then computed as $\sigma_i(s, t) = \frac{1}{1+\delta_i((s,t))}$, where the generic algorithm for computing weighted distances $\delta_i(A, B)$ for two strings A, B is shown in Algorithm 1. The algorithm basically computes the entries of the following matrix:

$$L_{ij}(A, B) = \begin{cases} i & \text{if } j = 0 \wedge 0 \leq i < |A| \\ j & \text{if } i = 0 \wedge 0 < j < |B| \\ \min\{L_{i-1,j}(A, B) + delCost(A[i]), & 0 < i < |A| \wedge 0 < j < |B| \\ L_{i,j-1}(A, B) + insCost(B[j]), \\ L_{i-1,j-1}(A, B) + subCost(A[i], B[j])\} \end{cases}, \quad (6)$$

where the computation of deletion, insertion and substitution costs for the same strings are as follows:

$$delCost(A[i]) := m_{pos(A[i]),pos(\epsilon)}, \quad (7)$$

$$insCost(A[j]) := m_{pos(\epsilon),pos(A[j])} \text{ and} \quad (8)$$

$$subCost(A[i], B[j]) := m_{pos(A[i]),pos(B[j])}. \quad (9)$$

Note that $A[i]$ is the ith character of the string A, $B[j]$ is the jth character of the string B and $pos(char)$ is the index of a character $char$ in the edit distance matrix.

Algorithm 1. Computation of Weighted Levenshtein distance

Require: $A, B \in \Sigma^*$
  $\forall i, j\ L[i, j] \leftarrow 0$
  for $i = 1 \rightarrow |A|$ **do**
    $L[i, 1] \leftarrow i$
  end for
  for $j = 1 \rightarrow |B|$ **do**
    $L[1, j] \leftarrow j$
  end for
  for $i = 1 \rightarrow |A|$ **do**
    for $j = 1 \rightarrow |B|$ **do**
      $L[i, j] \leftarrow min\{L[i - 1, j - 1] + getSubCost(A[i], B[j]),$
        $L[i - 1, j] + getDelCost(A[i]), L[i, j - 1] + getInsCost(B[j])\}$
    end for
  end for
  return $L[|A|, |B|]$

Computation of μ. One of the key steps in our approach is the extension of the edit distance computation algorithm to deliver the sequence of operations that were used for determining the distance between (and therewith the similarity of) pairs (s, t). This computation has often been avoided (see e.g., [3]) as it is very time-consuming. Instead, the generic edit distance algorithm was used and the weights added to the path in a subsequent computation step. Yet, not carrying out this computation can lead to false weights being updated. Consider for example the computation of the edit distance between the two strings A = "BAG" and B = "BAGS" over a weighted matrix where $getInsCost("s") = getSubCost("s", "S") = 0.3$. While the unweighted edit distance approach would carry out the insertion of "S" in A (cost = 1), it would actually be cheaper to carry out two operations: the insertion of "s" and the transformation of "s" to "S" (total costs: 0.6). To achieve the computation of such paths, we used the weighted distance computation algorithm shown in Algorithm 2. This algorithm basically runs the inverse of the dynamic programming approach to computing weighted edit distances and counts when which of the basic operations was used. It returns the sequence of operations that transformed A into B. Another approach to computing the shortest path would have been using Dijkstra's algorithm [7]. This alternative was evaluated by comparing the time complexity of both approaches. The combination of Algorithm 2 and 1 has a complexity of $O(mn) + O(m + n) \simeq O(mn)$ where $m = |A|$ and $n = |B|$. On the other hand, the best implementation of Dijkstra's algorithm, based on a min-priority queue using Fibonacci heap [8], runs in $O(|E| + |V| \log |V|)$. As the number of edges is mn and the number of vertices is $(m + 1)(n + 1)$, the time complexity is $O(mn + (m + 1)(n + 1) \log((m + 1)(n + 1))) \simeq O((m^2 + n^2) \log(m^2 + n^2))$. Without loss of generality, the computational complexities can be reduced to $O(n^2)$ for our approach and $O(n^2 \log n)$ for Dijkstra's algorithm under the assumption that $n \geq m$. Consequently, we chose our less generic yet faster implementation of the computation approach for μ.

Computation of Learning Rates. Given the potential prevalence of negative examples (especially when one-to-one mappings are to be computed), we define η^- as $\max(|S|, |T|)\eta^+$. The intuition behind this definition is that in most cases, valid links only make up a small fraction of $S \times T$ as each s and each t are mostly linked to maximally one other resource. The probability for a random pair (s, t) of being a positive example for a link is then given by

$$P = \frac{\min(|S|, |T|)}{|S| \times |T|} = \frac{1}{\max(|S|, |T|)}. \tag{10}$$

Consequently the ratio of the positive and negative learning rates should be proportional to

$$\frac{\eta^+}{\eta^-} = \frac{P}{1 - P} \tag{11}$$

Given that P is usually very small, the equation reduces to

$$\eta^- \approx \max(|S|, |T|)\eta^+. \tag{12}$$

Algorithm 2. Computation of μ

Require: $L(A, B)$
  $L \leftarrow L(A, B)$
  $[i, j] \leftarrow \|L\| - 1$
  while $i \neq 0 \vee j \neq 0$ **do**
    if $j > 0$ **then**
      $left \leftarrow -L[i][j - 1]$
    else
      $left \leftarrow \infty$
    end if
    if $i > 0$ **then**
      $up \leftarrow -L[i - 1][j]$
    else
      $up \leftarrow \infty$
    end if
    if $i > 0 \wedge j > 0$ **then**
      $upleft \leftarrow -L[i - 1][j - 1]$
    else
      $upleft \leftarrow \infty$
    end if
    if $upleft \leq left \wedge upleft \leq up$ **then**
      $countSub(A[i - 1], B[j - 1])$ // substitution
      $i \leftarrow i - 1$
      $j \leftarrow j - 1$
    else
      if $left < upleft$ **then**
        $countIns(B[j - 1])$ // insertion
        $j \leftarrow j - 1$
      else
        $countDel(A[i - 1])$ // deletion
        $i \leftarrow i - 1$
      end if
    end if
  end while

The main implication of the different learning rates used by our approach is that the false negatives have a higher weight during the learning process as the probability of finding some is considerably smaller than that of finding false positives.

3.5 Determining the Most Informative Examples

Formally, the most informative examples are pairs $(s, t) \notin \bigcup_{i=0}^{\tau} Q^i$ that are such that knowing the right classification for these pairs would lead to the greatest improvement of the current classifier C^τ. As pointed out in previous work [17], these are the pairs whose classification is least certain. For linear classifiers, the

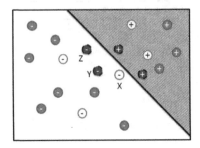

Fig. 1. Most informative examples

pairs (s, t) whose classification is least certain are obviously those elements from $S \times T$ with unknown classification that are closest to the boundary defined by C^τ. Figure 1 depicts this idea. The circles with a dashed border represent the 2 most informative positive and negative examples, the solid disks represent elements from $S \times T$ and the circles are examples that have already been classified by the user. Note that X is closer to the border than Y but is not a most informative example as it has already been classified.

Detecting the most informative examples would thus require computing the distance from all the pairs $(s, t) \in S \times T \backslash \bigcup_{i=0}^{\tau} Q^i$. Yet, this approach is obviously unpractical as it would require a computing almost $|S \times T|$ similarity values, given that $|\bigcup_{i=0}^{\tau} Q^i|$ is only a small fraction of $|S \times T|$. We thus devised the following approximation: Given the classifier defined by $\sum_{i=1}^{n} w_i \sigma_i(s, t) - \theta \geq 0$, we can derive the lower bound α_i for the minimal similarity $\sigma_i(s, t)$ that a pair (s, t) must achieve to belong to $+1$:

$$\sigma_i(s, t) \geq \alpha_i = \theta - \frac{\sum_{j=1, j \neq i}^{n} w_j \sigma_j(s, t)}{w_i}. \tag{13}$$

For each of the n dimensions of the similarity space, the most informative examples are thus likely to be found within the interval $[\alpha_i - \Delta, \alpha_i + \Delta]$, where $\Delta \in [0, 0.5]$. Consequently, we only need to compute the exact similarity scores of pairs (s, t) whose i^{th} coordinates lie within this interval. Several time-efficient approaches such as EdJoin [25] and PassJoin [14] have been developed to compute that set of pairs (s, t) such that $\delta(s, t)$ is smaller than a given threshold. Thus, if we never modified M_i, we could use these approaches to compute the pairs (s, t) with $\sigma_i(s, t) \in [\alpha_i - \Delta, \alpha_i + \Delta]$. Yet, given that we alter M_i and that the aforementioned approaches can only deal with uniform distance matrices M, we cannot only rely on them indirectly. Let $\kappa = \min_{m_{ij} \in M_i} m_{ij}$. Given $m \geq \kappa$ for all entries of m of M_i, the inequality $\delta_i(s, t) \geq \kappa \delta(s, t)$ holds. In addition, we know that by virtue of $\sigma_i = (1 + \delta_i)^{-1}$,

$$\sigma_i(s,t) \geq \alpha_i \Rightarrow \delta_i(s,t) \leq \frac{1 - \alpha_i}{\alpha_i}. \tag{14}$$

Thus, we can approximate the set of pairs (s,t) with $\sigma_i(s,t) \geq \alpha_i - \Delta$ by

$$\sigma_i(s,t) \geq \alpha_i - \Delta \Rightarrow \delta(s,t) \leq \frac{1 - (\alpha_i - \Delta)}{\kappa(\alpha_i - \Delta)}. \tag{15}$$

Consequently, we can run approaches such as EdJoin and PassJoin with the threshold $\frac{1-(\alpha_i-\Delta)}{\kappa(\alpha_i-\Delta)}$ when computing the set of pairs with $\sigma_i(s,t) \geq \alpha_i - \Delta$. Computing the set of pairs with $\sigma_i(s,t) \in [\alpha_i - \Delta, \alpha_i + \Delta]$ can then be achieved by simple filtering. Note that by using this approximation, we can discard a large number of pairs (s,t) for values of κ close to 1. Yet, this approximation is less useful when κ gets very small and is thus only of use in the first iterations of our approach. Devising an approach for the efficient computation of weighted edit distances remains future work.

3.6 Termination

Several termination conditions can be envisaged while running our approaches. Here, we chose to terminate the iteration when the classifier C^τ was able to classify all the entries of $Q^{\tau+1}$ correctly.

4 Experiments and Results

We began our experiments by running our approach on the toy example provided in Table 1. Our approach learned that the costs $getSubCost($"s", "z"$) = getInsCost($"u"$) = 0.3$. Therewith, it was able to achieve an F-measure of 1.0. We then ran our approach on benchmark data as presented in the following.

4.1 Experimental Setup

We evaluated our approach on 3 real data sets (see Table 2)[6]. The first two of these data sets linked the publication datasets ACM, DBLP and Google Scholar. The third dataset linked the products from the product catalogs Abt and Buy [12]. We chose these datasets because they have already been used to evaluate several state-of-the-art machine-learning frameworks for deduplication. In addition, the first two datasets are known to contain data where algorithms which rely on the edit distance can achieve acceptable F-measures. On the other hand, the third dataset, Abt-Buy, is known to be a dataset where the edit distance performs poorly due to the labels of products being written using completely different conventions across vendors. The primary goal of our evaluation was to compare our approach with classifiers based on the generic edit distance. We thus compared our approach with the results achieved by using decision trees and SVMs as implemented in MARLIN [4].

---

[6] The data used for the evaluation is publicly available at http://dbs.uni-leipzig. de/en/research/projects/object_matching/fever/benchmark_datasets_for_ entity_resolution

Table 2. Data sets

Source	Size	Target	Size	Correct Links	# Properties
DBLP	2,616	ACM	2,295	2,224	4
DBLP	2,616	Scholar	64,273	5,349	4
Abt	1,081	Buy	1,092	1,098	3

Table 3. Evaluation of ACIDS on three datasets

	DBLP – ACM		DBLP – Scholar		Abt – Buy	
k	5	10	5	10	5	10
F-score (%)	88.98	**97.92**	70.22	**87.85**	0.40	**0.60**
Precision (%)	96.71	96.87	64.73	91.88	0.20	0.30
Recall (%)	82.40	99.00	76.72	84.16	100.00	100.00
Run time (h)	7.53	7.31	8.45	8.36	12.27	12.01
Iterations	2	2	2	2	5	5

We used the following settings: we used the settings $k = 5, 10$ for the number of examples submitted to the user for labeling. The positive learning rate η^+ was set to 0.1. The boundary for most informative examples Δ was assigned the value 0.1. The maximal number of iterations for the whole algorithm was set to 5. The number of iterations between the perceptron learning and the SVM was set to 50. All experiments were carried out on a 64-bit Linux server with 4 GB of RAM and a 2.5 GHz CPU.

4.2   Results

The results of our evaluation are shown in Table 3 and Figure 2. On all datasets, using $k = 10$ led to better results. The termination condition was met on the first two datasets and the classifiers and measures learned outperformed state-of-the-art tools trained with 500 examples (see Fig. 3 in [12]). Especially, on the first data set, we achieved results superior to the 96.4% resp. 97.4% reported for MAR-LIN(SVM) resp. MARLIN(AD-Tree) when trained with 500 examples. Note that we necessitated solely 40 labeled pairs to achieve these results. In addition, note that we were actually able to outperform all other similarity measures and tools presented in [12] on this particular data set. On the second data set, we outperformed the state of the art (i.e., MARLIN) by more than 7% F-measure while necessitating only 40 labeled pairs. Here both versions of MARLIN achieve less than 80% F-measure when trained with 500 examples. The results on the third data set show the limitations of our approach. Given that the strings used to describe products vary largely across vendors, our approach does not converge within the first five iterations, thus leading to no improvement over the state-of-the art when provided with 100 examples. The main drawback of our approach are the runtimes necessary to compute the most informative examples. A solution to this approach would be to devise an approach that allows for efficiently computing weighted edit distances. Devising such an approach is work in progress.

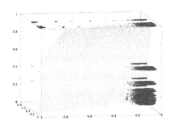

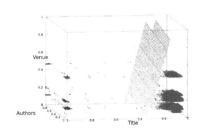

(a) DBLP–ACM dataset with $k = 5$.   (b) DBLP–ACM dataset with $k = 10$.

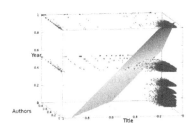

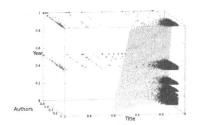

(c) DBLP–Scholar dataset with $k = 5$.   (d) DBLP–Scholar with $k = 10$.

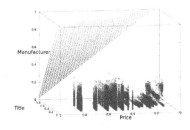

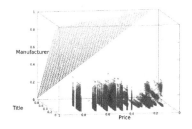

(e) Abt-Buy dataset with $k = 5$.   (f) Abt-Buy dataset with $k = 10$.

Fig. 2. Classifiers learned by ACIDS. Only a subset of all pairs is shown. The red data points are negatives, while the blue are positives.

5   Related Work

Our work is mostly related to Link Discovery and string similarity learning. Several frameworks have been developed with the aim of addressing the quadratic a-priori runtime of Link Discovery. [16,15] presents LIMES, a lossless and hybrid Link Discovery framework, as well as the HYPPO algorithm for the time-efficient computation of links. For example [10] present a lossless approach called Multi-Block that allows to discard a large number of comparisons. Other frameworks include those presented in [21,23]. The second core problem that needs to be addressed while computing links across knowledge bases is the detection of accurate link specifications. The problem has been addressed in manifold ways: The RAVEN approach [17] relies on active learning to learn linear and Boolean classifiers for discovering links. While [11] relies on batch learning and genetic programming to compute link specifications, the approach presented in [18] combines genetic programming and active learning to learn link specifications of high accuracy. In recent work, approaches for the unsupervised computation of link specifications have been devised. For example, [19] shows how link specifications can be computed by optimizing a pseudo-F-measure. Manifold approaches have been developed on string similarity learning (see, e.g., [22,4,3]). [4] for example learns edit distances by employing batch learning and SVMs to record deduplication and points out that domain-specific similarities can improve the quality of classifiers. [3] relies on a theory for good edit distances developed by [2]. An overview of approaches to and applications of learning distances and similarities is given in [9]. To the best of our knowledge, this work is the first to combine active learning and metric learning within the context of link discovery.

6   Conclusion and Future Work

In this paper, we presented an approach for learning string similarities based on edit distances within an active learning setting. We evaluated our approach on three data sets and showed that it yields promising results. Especially, we pointed out that if edit distances are suitable for the properties at hand, then we can actually improve the similarity functions and outperform simple edit distances. Still edit distances are not suitable for use on all datasets. While we focused on how to learn similarity measures based on edit distances, the general idea behind our approach can be easily extended to other string similarity measures. Most of these approaches rely on counting common tokens either in sets or sequences. By replacing the counting of tokens by the addition of weights, we could consequently extend most string similarity measures to weighted measures and learn them within the ACIDS framework. For example, the Jaccard similarity of two strings A and B is defined as $jaccard(A, B) = \frac{|\mathcal{T}(A) \cap \mathcal{T}(B)|}{|\mathcal{T}(A) \cup \mathcal{T}(B)|}$, where $\mathcal{T}(X)$ is the set of tokens (i.e., words) that make up X. Given that all tokens are issued from an alphabet Σ, we can define a weight function $\omega : \Sigma \to [0, 1]$. The weighted Jaccard similarity would then be

$$weighted Jaccard(A, B) = \frac{\sum\limits_{x \in (\mathcal{T}(A) \cap \mathcal{T}(B))} \omega(x)}{\sum\limits_{y \in (\mathcal{T}(A) \cup \mathcal{T}(B))} \omega(y)}. \tag{16}$$

We could then apply the ACIDS learning approach to devise an appropriate weight functions ω. Similarly, we can also assign weights to single n-grams derived from any alphabet and learn weighted n-gram-based similarity functions. Thus instead of computing the trigram similarity as $trigrams(A, B) = \frac{2|\mathcal{T}(A) \cap \mathcal{T}(B)|}{|\mathcal{T}(A)| + |\mathcal{T}(B)|}$, (where $\mathcal{T}(X)$ is the set of trigrams that make up X) we would compute

$$weightedTrigrams(A, B) = \frac{2 \sum\limits_{x \in (\mathcal{T}(A) \cap \mathcal{T}(B))} \omega(x)}{\sum\limits_{y \in \mathcal{T}(A)} \omega(y) + \sum\limits_{z \in \mathcal{T}(B)} \omega(z)}. \tag{17}$$

In addition to carrying out exactly this task, we will devise an approach to improve ACIDS' runtime so as to make it utilizable within interactive settings in future work. Moreover, we will evaluate the effect of the initial classifier and learning rate setting on our approach.

References

1. Auer, S., Lehmann, J., Ngonga Ngomo, A.-C.: Introduction to linked data and its lifecycle on the web. In: Polleres, A., d'Amato, C., Arenas, M., Handschuh, S., Kroner, P., Ossowski, S., Patel-Schneider, P. (eds.) Reasoning Web 2011. LNCS, vol. 6848, pp. 1–75. Springer, Heidelberg (2011)
2. Balcan, M.-F., Blum, A., Srebro, N.: Improved guarantees for learning via similarity functions. In: COLT, pp. 287–298 (2008)
3. Bellet, A., Habrard, A., Sebban, M.: Learning good edit similarities with generalization guarantees. In: Gunopulos, D., Hofmann, T., Malerba, D., Vazirgiannis, M. (eds.) ECML PKDD 2011, Part I. LNCS (LNAI), vol. 6911, pp. 188–203. Springer, Heidelberg (2011)
4. Bilenko, M., Mooney, R.J.: Adaptive duplicate detection using learnable string similarity measures. In: KDD, pp. 39–48 (2003)
5. Chang, C.-C., Lin, C.-J.: LIBSVM: A library for support vector machines. ACM Trans. Intell. Syst. Technol. 2(3), 27:1–27:27 (2011)
6. Cristianini, N., Shawe-Taylor, J.: An introduction to support Vector Machines: and other kernel-based learning methods. Cambridge University Press (2000)
7. Dijkstra, E.W.: A note on two problems in connexion with graphs. Numerische Mathematik 1, 269–271 (1959)
8. Fredman, M.L., Tarjan, R.E.: Fibonacci heaps and their uses in improved network optimization algorithms. J. ACM 34, 596–615 (1987)
9. Hertz, T.: Learning Distance Functions: Algorithms and Applications. PhD thesis, Hebrew University of Jerusalem (2006)
10. Isele, R., Jentzsch, A., Bizer, C.: Efficient Multidimensional Blocking for Link Discovery without losing Recall. In: WebDB (2011)
11. Isele, R., Bizer, C.: Learning linkage rules using genetic programming. In: 6th International Workshop on Ontology Matching, Bonn (2011)

12. Köpcke, H., Thor, A., Rahm, E.: Evaluation of entity resolution approaches on real-world match problems. PVLDB 3(1), 484–493 (2010)
13. Levenshtein, V.I.: Binary codes capable of correcting deletions, insertions, and reversals. Technical Report 8 (1966)
14. Li, G., Deng, D., Wang, J., Feng, J.: Pass-join: A partition-based method for similarity joins. PVLDB 5(3), 253–264 (2011)
15. Ngonga Ngomo, A.-C.: A time-efficient hybrid approach to link discovery. In: Proceedings of OM@ISWC (2011)
16. Ngonga Ngomo, A.-C., Auer, S.: Limes - a time-efficient approach for large-scale link discovery on the web of data. In: Proceedings of IJCAI (2011)
17. Ngonga Ngomo, A.-C., Lehmann, J., Auer, S., Höffner, K.: RAVEN – Active Learning of Link Specifications. In: Sixth International Ontology Matching Workshop (2011)
18. Ngonga Ngomo, A.-C., Lyko, K.: EAGLE: Efficient active learning of link specifications using genetic programming. In: Simperl, E., Cimiano, P., Polleres, A., Corcho, O., Presutti, V. (eds.) ESWC 2012. LNCS, vol. 7295, pp. 149–163. Springer, Heidelberg (2012)
19. Nikolov, A., d'Aquin, M., Motta, E.: Unsupervised learning of link discovery configuration. In: Simperl, E., Cimiano, P., Polleres, A., Corcho, O., Presutti, V. (eds.) ESWC 2012. LNCS, vol. 7295, pp. 119–133. Springer, Heidelberg (2012)
20. Pavel, S., Euzenat, J.: Ontology matching: State of the art and future challenges. IEEE Transactions on Knowledge and Data Engineering 99 (2012)
21. Raimond, Y., Sutton, C., Sandler, M.: Automatic interlinking of music datasets on the semantic web. In: Proceedings of LDoW (2008)
22. Ristad, E.S., Yianilos, P.N.: Learning string-edit distance. IEEE Transactions on Pattern Analysis and Machine Intelligence 20(5), 522–532 (1998)
23. Scharffe, F., Liu, Y., Zhou, C.: Rdf-ai: an architecture for rdf datasets matching, fusion and interlink. In: IK-KR IJCAI Workshop (2009)
24. Settles, B.: Active learning literature survey. Technical Report 1648, University of Wisconsin-Madison (2009)
25. Xiao, C., Wang, W., Lin, X.: Ed-join: an efficient algorithm for similarity joins with edit distance constraints. Proc. VLDB Endow. 1(1), 933–944 (2008)

Interlinking Linked Data Sources
Using a Domain-Independent System

Khai Nguyen[1], Ryutaro Ichise[2], and Bac Le[1]

[1] University of Science, Ho Chi Minh, Vietnam
{nhkhai,lhbac}@fit.hcmus.edu.vn
[2] National Institute of Informatics, Tokyo, Japan
ichise@nii.ac.jp

Abstract. Linked data interlinking is the discovery of every *owl:sameAs* links between given data sources. An *owl:sameAs* link declares the homogeneous relation between two instances that co-refer to the same real-world object. Traditional methods compare two instances by predefined pairs of RDF predicates, and therefore they rely on the domain of the data. Recently, researchers have attempted to achieve the domain-independent goal by automatically building the linkage rules. However they still require the human curation for the labeled data as the input for learning process. In this paper, we present SLINT+, an interlinking system that is training-free and domain-independent. SLINT+ finds the important predicates of each data sources and combines them to form predicate alignments. The most useful alignments are then selected in the consideration of their confidence. Finally, SLINT+ uses selected predicate alignments as the guide for generating candidate and matching instances. Experimental results show that our system is very efficient when interlinking data sources in 119 different domains. The very considerable improvements on both precision and recall against recent systems are also reported.

Keywords: linked data, interlinking, domain-independent, instance matching.

1   Introduction

The linked data is an unsubstitutable component in the generation of semantic web. It provides a mechanism by which the resources are interconnected by structured links. These links not only help the representation of information becomes clearer, but also makes the exploitation of information more efficient. Since then, linked data gets a lot of interest from many of organizations and researchers. Many linked data sources are developed and many support tools are introduced.

Given two linked data sources, data interlinking discovers every *owl:sameAs* links between these sources. An *owl:sameAs* link describes the homogeneity the instances that refer to the same object in the real world. Data interlinking is used in two typical processes: data construction and integration. When publishing linked

H. Takeda et al. (Eds.): JIST 2012, LNCS 7774, pp. 113–128, 2013.

data, an essential task is to declare the links between the instances to ensure the "linked" property of the data. Among these links, finding the *owl:sameAs* is prominently important and challenging [2]. In data integration, because many data sources are independently developed, the consideration of the homogeneous instances is important since it ensures the integrity and consistency of data.

We attempt to develop an interlinking system that is independent with the domain of data. This goal is increasingly interested because linked data is spreading over many areas. Since a linked data instance is represented by a set of RDF triples (subject, predicate, and object), the schema of the data sources is equivalent with the list of used RDF predicates. Frequently, different data sources may use different RDF predicates to describe the same property of the instances. Because the properties are commonly regulated by the domain of the data, difference in domain has the same meaning with difference in schema.

Traditional approaches compare two instances by matching the RDF objects that declared by the corresponding predicates and these alignments are manually mapped by the human [3,8,10]. This approach is inapplicable when the users do not have enough knowledge about the data and manually generating predicate alignments may ignore the hidden useful ones. Recently, researchers have attempted to learn the linkage rules [5,6] for not relying on the domain. However, most approaches require labeled data, which is still involved with human curation.

In this paper, we introduce SLINT+, a training-free system with a domain-independent approach. Our system firstly collects the important predicates of each data sources using the covering and discriminative abilities of predicates. Then, it combines these predicates and selects the most appropriate alignments by considering their confidence. The selected alignments are the guide for comparing instances in the final step, instance matching. While instance matching produces all the *owl:sameAs* links between data sources, the preceding step is the candidate generation, which extracts the potentially homogeneous instances. The basic idea of candidate generation step is quickly comparing the collective information of instances using collected predicates. While the key requirements of an interlinking system are mainly the recall and precision, those of candidate generation are the pair completeness and reduction ratio. These criteria are used to evaluate our system in the experiment. For testing the domain-independent ability, we use 1,6 million *owl:sameAs* links connecting DBpedia[1] and Freebase[2] with 119 different domains. The experimental results are the evidences for the efficiency of SLINT+. Besides, we compare SLINT+ with the previous state-of-the-art systems and report the considerable improvements.

The paper is structured as follows. In the next section, we review some representative linked data interlinking approaches and systems. Section 3 is the detail description about SLINT+. Section 4 reports the experiments as well as the results and analyses. Section 5 closes the paper with the conclusions and future directions.

---

[1] http://dbpedia.org/
[2] http://www.freebase.com/

2   Related Works

One of the first linked data interlinking system is Silk [10]. Silk is a link discovery framework that provides a declarative language for user to define the type of link to discover. The main use of Silk is to find *owl:sameAs* links. Using Silk, users need to declare the pairs of predicates that they want to compare when matching the instances. Besides, the matching threshold is manually configured. AgreementMaker [3] is known as a system that focuses on matching both ontologies and instances. In data interlinking, AgreementMaker use a three steps matching process, including candidate generation, disambiguation, and matching. Candidates are collected by picking the instances that share the same label with others. The disambiguation step divides the candidates into smaller subsets and the matching step verify every pair of instances in each subset to produce the final result. Zhishi.Links [8] is one of the current state-of-the-art systems. This system improves the matching efficiency by using weighting schemes (e.g. TF-IDF, BM25) for RDF objects when generating candidate, as an adoption of pre-matching phase of Silk. However, this system is still depending on the domain of data.

Recently, many domain-independent approaches have been proposed [1,5,6,9]. SERIMI [1] selects the useful predicates and their alignments by considering the entropy and the similarity of RDF objects. SERIMI is the second best system at the OAEI Instance Matching 2011 [4]. Isele and Bizer presented a linkage rule generation algorithm using genetic programming [5]. Nguyen et al. focused on instance matching using learning approach [6]. They build a binary classifier to detect the matched and non-matched pairs of instances. Song and Heflin proposed a domain-independent candidate generation method [9]. They design an unsupervised learning schema to find the most frequent and discriminative predicates. The set of predicates in each data sources are used as the key for candidate generation.

In general, most proposed domain-independent approaches use the labeled data as the replacement for user knowledge about the schema or the domain of the data. Our innovation is developing SLINT+, an interlinking system that does not need any information about both the domain and the matching state of a portion of data sources. SLINT+ is an extension of SLINT [7]. Comparing with SLINT, SLINT+ is similar in the architecture of the system, and different in the use of techniques in some steps. The major improvement of SLINT+ against SLINT is the reduction of many manual thresholds. In the next section we will describe the technical detail of SLINT+.

3   Domain-Independent Linked Data Interlinking System

In this section, we describe the SLINT+ system. The process of interlinking two data sources D_S and D_T is summarized in Fig. 1. There are four ordered steps in this process: predicate selection, predicate alignment, candidate generation, and instance matching. The predicate selection step aims at finding the important

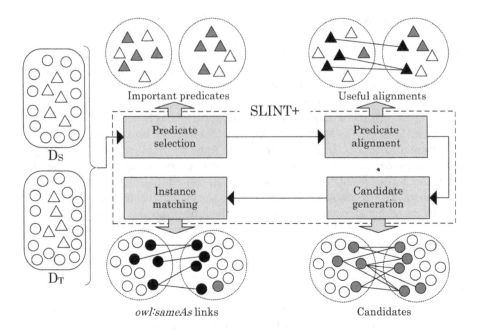

Fig. 1. The interlinking process of SLINT+

predicates, which are illustrated as the dark triangles in the figure. This kind of predicates in both source data D_S and target data D_T are then combined to construct the raw predicate alignments in the next step. We collect the useful alignments by comparing their confidence level with an automatically selected threshold. While the first two steps resolve the domain-independent objective, the last two steps perform the interlinking using the output of the previous ones. The candidate generation step is installed to find the candidates, which are the potentially homogeneous instances and are expected to be very small in quantity compared with all possible pairs of instances. In Fig. 1, the candidates are visualized as the connected pairs of small dark circles in the output of the third step. SLINT+ only conducts the comparison for these candidates to produce the final *owl:sameAs* links. In the following sections, we give the detail of each step.

3.1   Predicate Selection

The predicate selection step is designed to find the important predicates. We assume that important predicates are the ones used to declare the common properties and the distinct information of the object. Therefore, we implement two criteria of an important predicate: the coverage and the discrimination. The coverage of a predicate expresses its frequency while the discrimination represents the variation of the RDF objects described by this predicate. In Eq.1 and Eq.2, we define the coverage $cov(p, D)$ and discrimination $dis(p, D)$ of a predicate p in the data source D.

$$cov(p, D) = \frac{|\{x | \exists < s, p, o > \in x, x \in D\}|}{|D|}.$$ (1)

$$dis(p, D) = HMean(V(p, D), H(p, D))$$

Where

$$V(p,D) = \frac{|\{o | \exists x \in D, < s, p, o > \in x\}|}{|\{< s, p, o > | \exists x \in D, < s, p, o > \in x\}|}$$ (2)

$$H(p,D) = \sum_{w_i \in O} \frac{f(w_i)}{\sum_{w_j \in O} f(w_j)} \log \frac{f(w_i)}{\sum_{w_j \in O} f(w_j)}$$

$$O = \{o | \exists x \in D, < s, p, o > \in x\}.$$

In these equations, x represents an instance, which is a set of RDF triple $<$ $s, p, o >$ (subject, predicate, object). Function f returns the frequency of input RDF object in the interested data source. Using the coverage and discrimination, a predicate is considered to be important if it satisfies the condition of Eq.3.

$$\begin{cases} cov(p, D) & \geq \alpha \\ dis(p, D) & \geq \beta \\ HMean(cov(p, D), dis(p, D)) \geq \gamma. \end{cases}$$ (3)

The coverage of a predicate is the percent of the instances sharing it. This metric is the first requirement of an important predicate because we aim at finding the predicates that are used to describe the common properties of the instances. The number of important predicates is considerably small when being compared with all predicates. For example, in 24,110 instances of *school* domain in DBpedia, there are 2,771 distinct predicates but only 1% of them, 28 predicates, has the coverage that is over 0.5. On another hand, the discrimination is the harmonic mean of the diversity and the entropy of RDF objects. The function H yields the entropy of distinct RDF objects, while V returns the percent of them over total triples. V is a good option to estimate the variation of RDF objects, and H is used to reveal the difference of the predicates that share the same number of distinct RDF objects but differ in the distribution of each value. Clearly, a predicate that describes the ambiguous information of instances should not be important. Therefore, we install the discrimination as the second criterion of an important predicate.

The α and β are automatically configured using the average value of the coverage and discrimination of all predicates, respectively. In another word, α and β are set to α_{mean} and β_{mean}, respectively. The γ is the main requirement when being used to select the predicates having high harmonic mean of coverage and discrimination. Usually, α_{mean} and β_{mean} are quite small because the percent of important predicates is not large. Therefore, γ should have a value larger than α_{mean} and β_{mean} to correctly select the important predicates.

Table 1. Representative extraction function R

Type	Return values of R
string	String tokens
URI	String tokens (separator '/', domain is omitted)
decimal	Rounded values with 2 decimal numbers
integer	Original values
date	Original values

The idea of Eq.3 and function V are inherited from the work of Song and Heffin [9]. We also implemented V as the discrimination function in SLINT [7]. In SLINT+, we extend the discrimination as the combination of V and entropy function H. Besides, we use the α_{mean} and β_{mean} instead of manually configuring them as in [7,9].

For each input data source, SLINT+ collects the important predicates of each source and forwards them into the next step, predicate alignment.

3.2   Predicate Alignment

The aim of this step is to collect the useful alignments of important predicates. A predicate alignment is a pair of two predicates from source data and target data. A useful alignment is expected to be the predicates that describe the same information of the existing instances in each data source. We construct predicate alignments by combining the predicates having the same data type, and after that, we select the ones having high confidence for the result. We categorize RDF predicates into five different types: *string*, *URI*, *decimal*, *integer*, and *date*. The type of a predicate is determined using the major type of RDF objects, which are accompanied by this predicate. For example, predicate p is used to declare the values that can be *string* and *URI* if the frequency of the manner is higher than the latter, then the data type assigned for p will be *string*.

We estimate the confidence of an alignment using the similarity of the representatives of RDF objects. The representatives are the pre-processed values of them. Denoting R as the representative extraction function, the return values of R is given in Table 1. The confidence $conf(p_S, p_T)$ of the alignment between two predicates p_S and p_T is then computed using Eq.4

$$conf(p_S, p_T) = \frac{2 \times |R(O_S) \cap R(O_T)|}{|R(O_S)| + |R(O_T)|}, O_k = \{o | \exists x \in D_k, < s, p_k, o > \in x\}. \quad (4)$$

The idea of Eq.4 starts from the assumption in that the corresponding predicates are used to describe the same properties of the instances. Therefore, the higher value of confidence is, the more useful an alignment is. We use thresholding approach for choosing the useful alignments with the threshold δ is set to the average confidence δ_{mean} of all alignments. Because there are many meaningless alignments when combining the predicates of two data sources, we ignore the low confidences. Therefore, we take the average value of the confidences larger than ϵ, a small value.

Predicate alignment step is the key difference in the interlinking model of SLINT+ against previous systems. Other domain-independent approaches [1,5,6,9] directly generate linkage rules and do not find the corresponding predicates. We attempt a new approach, in which the linkage rules are implicitly defined, and they are the useful alignments.

The two first steps of SLINT+ are the solution for the domain-independent ability. In the next steps, we use their results to perform the candidate generation and instance matching.

3.3   Candidate Generation

When interlinking two data sources D_S and D_T, each instance in D_S should be compared with each instance in D_T. However, conducting all the comparisons is impractical. Candidate generation step aims at collecting all pairs of instances that have a high possibility to be homogeneous. These kind of pair is called *candidate*. Since the basic idea of every set-matching problem is exhausting similitude, candidate generation should be designed not to perform explicit comparisons. There are three consecutive parts in this step: indexing, accumulating, and candidate selection. The first part indexes every instance in each data source with the representative value of RDF objects. The second part builds the weighted co-occurrence matrix to store the co-occurrence value of all pairs of instances. The last part selects the candidate by considering the high elements in the matrix. We put the summary of our candidate generation method in Algorithm 1. In this algorithm, Pr_s and Pr_t represent the predicates that appear in the useful alignments, where Pr_k belongs to D_k. H, M, C, Rp represent the index table, weighted co-occurrence matrix, candidates set, and representative extraction function, respectively. ζ is a factor that will be combined with max to produce the threshold for candidate selection, as written in line 22.

In Algorithm 1, lines 4-11 describe the indexing process. The result of this part is the inverted-index table H. An entry of H is formed by two elements: the key $r.Label$, and the triples $< D, x, r.Value \times sumConf >$. In a triple, the first two elements D and x indicate the identifier of instance $x \in D$ (e.g. index of x in D), which contains $r.Label$ in the representative set of its RDF objects. The last element is the weight of $r.Label$. If an entry of H contains n triples, there will be n instances sharing the same key of this entry. In the design of function Rp, $r.Label$ is extracted as the same manner with function R (Table 1), and $r.Value$ is regulated by the data type of interested predicate. For *string* or *URI*, $r.Value$ is set to the TF-IDF score of the token. For *decimal*, *integer*, or *date*, $r.Value$ is fixed to 1.0.

Lines 12-15 are the building process of co-occurrence matrix M. An element of this matrix is the co-occurrence value of two instances from source data and target data. We use the confidence of useful predicate alignments as the weight for each accumulated value. In addition with the inverted-indexing, the accumulating process improves the speed of candidate generation because each data source and the index table need to be traversed one time.

Algorithm 1. Generating candidates set

Input: D_S, D_T, Pr_S, Pr_T, ζ
Output: Candidate set C
1 $H \leftarrow \emptyset$
2 $M[|D_S|, |D_T|] \leftarrow \{0\}$
3 $C \leftarrow \emptyset$
4 **foreach** $< D, P > \in \{< D_S, Pr_S >, < D_T, Pr_T >\}$ **do**
5   **foreach** $x \in D$ **do**
6   **foreach** $p_i \in P$ **do**
7              $sumConf \leftarrow \sum_{p_j \in \{Pr_S, Pr_T\} \setminus P} conf(p_i, p_j)$
8   **foreach** $r \in Rp(O), O = \{o| < s, p_i, o > \in x\}$ **do**
9   **if** *not H.ContainsKey(r.Label)* **then**
10   H.AddKey$(r.Label)$
11   H.AddValue$(r.Label, < D, x, r.Value \times sumConf >)$

12 **foreach** $key \in H.AllKeys()$ **do**
13   **foreach** $< x_S, v_S > \in H.GetIndices(key, D_S)$ **do**
14   **foreach** $< x_T, v_T > \in H.GetIndices(key, D_T)$ **do**
15             $M[x_S, x_T] \leftarrow M[x_S, x_T] + v_S \times v_T$

16 $\lambda = \text{Mean}(M)$
17 **foreach** $x_S \in D_S$ **do**
18   **foreach** $x_T \in D_T$ **do**
19         $max_S \leftarrow \text{Max}(M[x_S, x_j]), \forall x_j \in D_T$
20         $max_T \leftarrow \text{Max}(M[x_i, x_T]), \forall x_i \in D_S$
21         $max \leftarrow \text{HMean}(max_S, max_T)$
22   **if** $M[x_S, x_T] \geq \lambda$ *and* $M[x_S, x_T] \geq \zeta \times max$ **then**
23             $C \leftarrow C \cup < x_S, x_T >$

24 **return** C

Lines 16-23 are the candidate selection process. We apply an adaptive filtering technique by using the data driven thresholds. λ is installed to warranty there is no assumption about the surjection. This threshold is assigned to the average co-occurrence value of all pairs of instances. This value is usually small because the number of homogeneous pairs is frequently very much lower than that of the non-homogeneous ones. ζ is manually configured and timed with an automatically selected value max to produce the dynamic threshold $\zeta \times max$. This threshold is the main requirement of a homogeneous pair of instances.

Comparing with previous systems, our method is distinct for the weighted co-occurrence matrix and adaptive filtering in candidate selection. While proposed methods compare one or a few pairs of RDF objects, we aggregate multiple similarities from many pairs of RDF objects for improving the quality of "rough" similarity of instances. For candidate selection, SERIMI [1] and Song et al. use traditional *thresholding* [9]. SLINT+ also use this approach as the availability of λ. However, the impact of λ is very small because the value assigned for it is quite low and the key idea of the selection method is adaptive filtering. Silk [10] and Zhishi.Links [8] selects k best correlated candidates for each instance. This approach has advantage in handling the number of candidates, but the fixed value for k seems not reasonable because the ambiguous level are not the same for each instance. While some instances should have small value of k, the others must have larger candidates for still including the homogeneous ones. In the next step, SLINT+ verifies all selected candidates for producing the final *owl:sameAs* links.

3.4   Instance Matching

We estimate the correlation of each candidate and the ones satisfying the process of adaptive filtering will be considered to be homogeneous. The correlation of a candidate, or of two instances, is computed using the similarity of RDF objects of interested instances. These objects are described using useful predicate alignments and the final correlation of two instances is the weighted average of these similarities. The correlation $corr(x_S, x_T)$ of two instances $x_S \in D_S$ and $x_T \in D_T$ is defined in Eq.5.

$$corr(x_S, x_T) = \frac{1}{W} \sum_{<p_S, p_T> \in A} conf(p_S, p_T) \times sim(R(O_S), R(O_T)),$$

$$Where$$

$$O_k = \{o | \exists x \in D_k, < s, p_k, o > \in x\}$$

$$W = \sum_{<p_S, p_T> \in A} conf(p_S, p_T). \tag{5}$$

In this equation, A is the set of useful predicate alignments; R is the representative extraction function, which is the same as that in Table 1; $conf(p_S, p_T)$ is the confidence of interested predicate alignment $< p_S, p_T >$ (Eq.3). The sim function returns the similarity of two set of representatives, and is regulated by the data type of the predicates. For *decimal* and *integer*, we use the variance of representative values. For *date*, we use exact matching and return the value 1 or 0 when the values are totally equal or not, respectively. For *string* and *URI*, we take the cosine similarity with TF-IDF weighting, as given in Eq.6. While cosine is widely used to estimate the similarity of two sets, the TF-IDF weighting very helpful for many disambiguation techniques.

$$sim(Q_s, Q_t) = \frac{\sum_{q \in Q_s \cap Q_t} TFIDF(q, Q_s) TFIDF(q, Q_t)}{\sqrt{\sum_{q \in Q_s} TFIDF^2(q, Q_s) \times \sum_{q \in Q_t} TFIDF^2(q, Q_t)}}. \tag{6}$$

After computing correlation value for every candidate, we apply adaptive filtering to collect homogeneous pairs. An *owl:sameAs* link is created when two instances have the score larger than a threshold, which is dynamically adjusted in accordance with the interested instances. Denoting C as the output candidates of Algorithm 1, we define the set of homogeneous instances I as in Eq.7.

$$I = \{< x_S, x_T > | corr(x_S, x_T) \geq \eta \wedge$$
$$\frac{corr(x_S, x_T)}{max_{\forall < x_m, x_n > \in C, x_m \equiv x_S \vee x_n \equiv x_T} corr(x_m, x_n)} \geq \theta\} \quad (7)$$

We assume that an *owl:sameAs* link connects two instances that have the highest correlation if compared this value with the correlation of other candidate, in which each instance appears. However, θ threshold is necessary to be installed to select the pairs of instance that are co-homogeneous, because we do not assume that there is no duplication in each data sources. In SLINT+, θ is set to a quite large value. We use the average correlation η_{mean} of all candidates for configuring η. Since there are frequently many candidates that are not homogeneous, η_{mean} is usually small. Like λ in Algorithm 1, η ensures there is no assumption about the surjection of given data sources and should be assigned with a low value.

Compared with previous systems, the instance matching step of SLINT+ is distinct in the use of weighted average and adaptive filtering. We use adaptive thresholds for each pair of data sources and the confidence of useful alignments for weighting. Silk [10] also provides a weighted average combination method for similarities, however these weights must be configured manually instead of observing from the data. Zhishi.Links [8] and AgreementMaker[3] select the best correlated candidates, while Silk [10] and SERIMI [1] use traditional threshold-based approach. In the next section, we will report our experiments and the results.

4   Experiments

4.1   Evaluation Metrics

We evaluate the efficiency of the interlinking process of SLINT+ at two main metrics: recall and precision. The recall represents the ability that retains the true *owl:sameAs* links, while the precision indicates the disambiguation ability because it expresses the percent of true elements in all discovered links. In addition to precision and recall, we also report the $F1$ score, the harmonic mean of them. Eq. 8 and 9 are the computation of recall RC and precision PR, respectively.

$$RC = \frac{Number\ of\ correctly\ discovered\ links}{Number\ of\ actual\ links}. \quad (8)$$

$$PR = \frac{Number\ of\ correctly\ discovered\ links}{Number\ of\ all\ discovered\ links}. \quad (9)$$

For evaluating the candidate generation, we also use two main metrics, the pair completeness and reduction ratio. The pair completeness expresses the percent of correct generated candidates and the reduction ratio shows the compactness of candidate set. The aim of candidate generation is to maximize the pair completeness while reserving a very high reduction ratio. Eq. 10 and 11 show the formula of pair completeness PC and reduction ratio RR, respectively. Some studies also use the $F1$ score for PC and RR [9], however we think that it is not equivalent to combine these metrics because RR is frequently very high in compared with PC in most cases.

$$PC = \frac{Number\ of\ correct\ candidates}{Number\ of\ actual\ links}. \tag{10}$$

$$RR = 1 - \frac{Number\ of\ candidates}{Number\ of\ all\ instance\ pairs}. \tag{11}$$

We also report the execution times of SLINT+ in the division of three parts, predication selection and predicate alignment, candidate generation, and instance matching. Every experiment is conducted on a desktop machine with 2.66Ghz of CPU, 8GB of memory, 32-bit Windows 7 operating system. In addition, we use C# language for developing SLINT+.

4.2 Experiment Setup and Datasets

Since SLINT+ contains few manual thresholds, we use the same value for each threshold on every dataset to evaluate the domain-independent ability. We set the value 0.5, 0.1, 0.5, and 0.95 for γ (Eq.3), ϵ (Section 3.2), ζ (Algorithm 1), and θ (Eq. 7), respectively.

The aim of experiment is to evaluate the efficiency of SLINT+, especially the domain-independent goal. Besides, a comparison with existing systems is also necessary. Therefore, we use two datasets DF and OAEI2011. The first dataset, DF is selected from a ground-truth provided by DBpedia[3]. This set contains about 1,6 million *owl:sameAs* links between DBpedia and Freebase, and 141 domains of data. DBpedia and Freebase are the most well-known and are very large data sources in the linked data community. DBpedia and Freebase contains many domains such as people, species, drugs, rivers, songs, films, settlements,... For constructing the DF, we pick up the domains that have the number of *owl:sameAs* links is at most 40,000. After this selection, we have 891,207 links with 119 domains. We divide these links into 119 subsets and each subset is respective to a distinct domain. The smallest subset contains 1,008 links in *architect* domain and the largest subset contains 37,120 links about *plant*. We do not use the domains that contain more than 40,000 instances because the computation of candidate generation requires a large amount of memory, which is the limitation of our testing environment. However, the number 119 is still convinced for verifying the domain-independent ability.

The second dataset, OAEI2011 is the data that was selected for the most recent OAEI 2011 Instance Matching Campaign [4]. In this campaign, participants are asked to build the *owl:sameAs* links between NYTimes[4] to DBpedia, Freebase, and Geonames[5]. NYTimes is a high quality data source while DBpedia,

---

[3] http://downloads.dbpedia.org/3.8/links/freebase_links.nt.bz2
[4] http://data.nytimes.com/
[5] http://www.geonames.org/

Table 2. Domains and numbers of owl:sameAs links in OAEI2011 dataset

ID	Source	Target	Domain	Links
D1	NYTimes	DBpedia	Locations	1920
D2	NYTimes	DBpedia	Organizations	1949
D3	NYTimes	DBpedia	People	4977
D4	NYTimes	Freebase	Locations	1920
D5	NYTimes	Freebase	Organizations	3044
D6	NYTimes	Freebase	People	4979
D7	NYTimes	Geonames	Locations	1789

Table 3. Number of predicates and predicate alignments in DF dataset

	Pr_S	Pr_T	P_S	P_T	A	K
Min	228	9	7	4	12	6
Max	2711	1388	30	21	158	40
Average	468	315	14	8	45	16

Freebase, and Geonames are very large ones. Geonames is the data sources focusing on geography domain and currently contains over 8,0 million geographic names. The OAEI2011 dataset has 7 subsets and 3 domains, which are *location*, *organization*, and *people*. Denoting the 7 subsets as D1 to D7, the overview of this dataset is given in Table 2.

In the next sections, we report the results and the discussions for each step of SLINT+.

4.3   Results and Analyses

4.3.1 Discussion on Predicate Selection and Predicate Alignment

Table 3 gives the number of all predicates Pr in each data source, important predicates P, all predicate alignments A and the useful alignments K in DF dataset. S and T stand for DBpedia and Freebase, respectively. According to this table, the numbers of all predicates are very high when being compared with the numbers of important predicates at the threshold $\gamma = 0.5$, which means that the requirements for covering and discriminative abilities are around 0.5. If we consider the percent of important predicates with this threshold, the maximum values are 25.00% and 17.24% in DBpedia and Freebase, respectively; and the averages are only 4.95% and 3.70% in DBpedia and Freebase, respectively. The low ratio of important predicates indicates the high heterogeneity of the predicates in the schema of data sources. The predicate selection step is therefore necessary to detect the useful predicates. As our observation, the selected predicates are frequently the predicates describing the name or the label of the instances, and some important predicates that rely on the domain of the data. For example, in *city* domain, beside the predicate declaring the name, the useful predicates are also about the longitude and latitude of the city. This example also indicates the *string* data type is not the only important one although this assumption is right in many cases.

Table 4. Results of candidate generation on DF dataset

	NC	PC	RR
Min	1238	0.9652	0.9969
Max	389034	1.0000	0.9999
Weighted average	87089	0.9890	0.9996

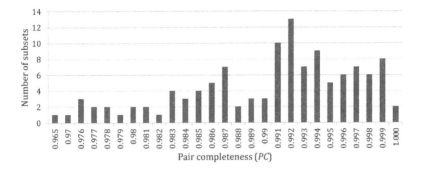

Fig. 2. Histogram of subsets on their pair completeness

For predicate alignment, since the number of important predicates is reduced in the first step, the number of all alignments is not large when the maximum value is only 158. In detail, the selected alignments, which are higher than the average confidence, occupy 42.70% of all alignments at average, 91.30% at the maximum, and 12.50% at the minimum. As our observation, the useful predicate alignments are very appropriate when the predicates describing the similar property are always combined. For example, in *philosopher* domain, the selected alignments contains 3 interesting ones: the combination of 3 different DBpedia predicates describing full name, surname, and given name with 1 Freebase predicate describing the name of the person. It is not easy for a user to point out every useful alignment, especially in the case that useful alignments can reach the number of 16, as the average value in Table 3.

There is currently no dataset for clearly evaluating the efficiency of predicate selection and predicate alignment. However, the high results of the candidate generation and the whole interlinking process are the very clear evidences for the efficiency of these tasks.

4.3.2 Candidate Generation

In this section, we report the results of candidate generation on DF dataset. Table 4 summarizes the candidate generation results. In this table, NC is the number of candidates, and the weight of weighted average values are the numbers of *owl:sameAs* links in each subset. The pair completeness is very high when candidate generation step reserves over 96.52% of correct candidates. Fig. 2 is the histogram of subsets when their PC is rounded into the nearest value in the horizontal axis. According to this figure, most PCs are distributed in the range above 0.991 and reach the highest number at 0.992. The subsets having

Table 5. Results of data interlinking on DF dataset

	RC	PR	$F1$
Min	0.9002	0.8759	0.8985
Max	0.9973	1.0000	0.9986
Weighted average	0.9763	0.9645	0.9702

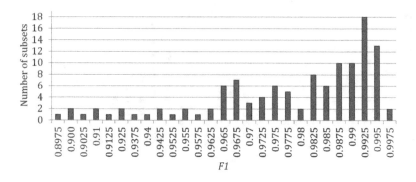

Fig. 3. Histogram of subsets on their F1 score

the PC that is higher than 0.991 are 73 and cover 61.34% of all the subsets in DF. The reduction ratio RR in Table 4 is very impressive when always higher than 0.9969 on every subset. The adaptive filtering technique is also verified through the high result of pair completeness and reduction ratio. In general, the candidate generation step has successfully performed its role on over 119 subsets.

4.3.3 The Whole Interlinking Process

For the whole interlinking process, we report the result of recall RC, precision PR, and $F1$ score. As summarized in Table 5, both recall and precision are very considerable when the weighted average value for $F1$ is very high at 0.9702. Fig. 3 show the histogram of subsets when considering their $F1$ scores. There are 82.35% of subsets having the $F1$ higher than 0.965, the middle value of horizontal axis. These subsets cover 79.40% of all $owl:sameAs$ links in the DF dataset. The high precision and recall not only express the efficiency of interlinking, but also reveal that the predicate selection and predicate alignment produced useful predicates and appropriate alignments.

Fig. 4 shows the runtime of predicate selection and predicate alignment, candidate generation, and instance matching in accordance with the size of the subsets. There are a few points that the runtime does not consistently increase because beside the size of data, the number of predicates also affects the speed of each step. On large subsets, the time-variation of the each step is clearer and we can see that the major time of interlinking process is the candidate generation. If we consider the weighted average runtime, this step occupies 54.45% of total while the first two steps and the instance matching cover only 21.32% and 24.23%, respectively. In general the speed of SLINT+ is very high when it takes

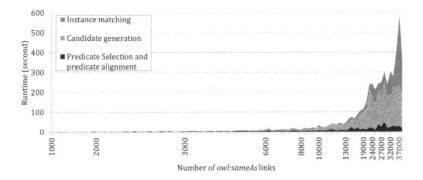

Fig. 4. Execution time of interlinking steps

Table 6. Comparison with previous interlinking systems on OAEI2011 dataset

Dataset	SLINT+			Agree.Maker			SERIMI			Zhishi.Links		
	PR	RC	$F1$	PR	RC	$F1$	PR	RC	$F1$	PR	RC	$F1$
D1	**0.96**	**0.97**	**0.97**	0.79	0.61	0.69	0.69	0.67	0.68	0.92	0.91	0.92
D2	**0.97**	**0.95**	**0.96**	0.84	0.67	0.74	0.89	0.87	0.88	0.90	0.93	0.91
D3	**0.99**	**0.99**	**0.99**	0.98	0.80	0.88	0.94	0.94	0.94	0.97	0.97	0.97
D4	**0.95**	**0.95**	**0.95**	0.88	0.81	0.85	0.92	0.90	0.91	0.90	0.86	0.88
D5	**0.97**	**0.96**	**0.96**	0.87	0.74	0.80	0.92	0.89	0.91	0.89	0.85	0.87
D6	**0.99**	**0.99**	**0.99**	0.97	0.95	0.96	0.93	0.91	0.92	0.93	0.92	0.93
D7	**0.99**	**0.99**	**0.99**	0.90	0.80	0.85	0.79	0.81	0.80	0.94	0.88	0.91
H.Mean	**0.97**	**0.97**	**0.97**	0.92	0.80	0.85	0.89	0.88	0.89	0.93	0.92	0.92

below 10 minutes to interlink the largest subset with 37,120 links and nearly 1
second for the smallest one with 1,008 links.

4.4 Comparison with Previous Systems

We compare SLINT+ with AgreementMaker [3], SERIMI [1], and Zhishi.Links
[8]. These systems recently participated the OAEI 2011 Instance Matching Campaign [4]. Among these systems, SERIMI is the only one that is domain-
independent while AgreementMaker and Zhishi.Links are not. Table 6 shows the
results of SLINT+ and the others. According to this table, SLINT+ is very much
higher if compared with other systems on both precision and recall. Zhishi.Links
is the second best in this comparison but is still 0.05 lower than SLINT+ in
overall. Particularly, SLINT+ performs very well on D4 and D5 datasets, which
are seem to be difficult for Zhishi.Links to interlink.

The improvement of SLINT+ against compared systems is the confirmation
of the robustness of domain-independent approach over domain-dependent ones,
such as AgreementMaker and Zhishi.Links. The limitation of SERIMI is that
this system solely uses the string measurement and only focus on short strings.
Therefore, this system is lower in precision and recall than those of SLINT+ on
every subset.

5   Conclusion

In this paper, we present SLINT+, a domain-independent linked data inter-linking system. Our system interlinks different data sources by comparing the instances using collective information and weighted similarity measure. The information to be compared are extracted by RDF predicates which are automatically selected using their covering and discriminative abilities. The key idea to remove the requirement of labeled matched instances is the concept of the confidence of a predicate alignment. The candidate generation step is also investigated and evaluated. Experimental results show that SLINT+ is outstanding comparing to previous famous systems on OAEI2011 dataset. The domain-independent capability can be said to be achieved when SLINT+ perform very well on 119 different domains of data.

In the future, we will study on improving the scalability of SLINT+ in order to match very large data sources. Besides, a cross-domain interlinking system, which aims at matching the sources containing multiple domains, is a very interesting objective to be explored.

References

1. Araujo, S., Tran, D., de Vries, A., Hidders, J., Schwabe, D.: SERIMI: Class-based disambiguation for effective instance matching over heterogeneous web data. In: SIGMOD 2012 15th Workshop on Web and Database, pp. 19–25 (2012)
2. Bizer, C., Heath, T., Berners-Lee, T.: Linked data - the story so far. Semantic Web and Information Systems 4(2), 1–22 (2009)
3. Cruz, I.F., Antonelli, F.P., Stroe, C.: AgreementMaker: efficient matching for large real-world schemas and ontologies. VLDB Endow. 2, 1586–1589 (2009)
4. Euzenat, J., Ferrara, W., Hage, A., Hollink, L., Meilicke, C., Nikolov, A., Scharffe, F., Shvaiko, P., Stuckenschmidt, H., Zamazal, O., Trojahn, C.: Final results of the ontology alignment evaluation initiative 2011. In: ISWC 2011 6th Workshop on Ontology Matching, pp. 85–113 (2011)
5. Isele, R., Bizer, C.: Learning linkage rules using genetic programming. In: ISWC 2011 6th Workshop on Ontology Matching, pp. 13–24 (2011)
6. Nguyen, K., Ichise, R., Le, B.: Learning approach for domain-independent linked data instance matching. In: KDD 2012 2nd Workshop on Minning Data Semantics, pp. 7:1–7:8 (2012)
7. Nguyen, K., Ichise, R., Le, B.: SLINT: A schema-independent linked data inter-linking system. In: ISWC 2012 7th Workshop on Ontology Matching (2012)
8. Niu, X., Rong, S., Zhang, Y., Wang, H.: Zhishi.links results for OAEI 2011. In: ISWC 2011 6th Workshop on Ontology Matching, pp. 220–227 (2011)
9. Song, D., Heflin, J.: Automatically generating data linkages using a domain-independent candidate selection approach. In: Aroyo, L., Welty, C., Alani, H., Taylor, J., Bernstein, A., Kagal, L., Noy, N., Blomqvist, E. (eds.) ISWC 2011, Part I. LNCS, vol. 7031, pp. 649–664. Springer, Heidelberg (2011)
10. Volz, J., Bizer, C., Gaedke, M., Kobilarov, G.: Discovering and maintaining links on the web of data. In: Bernstein, A., Karger, D.R., Heath, T., Feigenbaum, L., Maynard, D., Motta, E., Thirunarayan, K. (eds.) ISWC 2009. LNCS, vol. 5823, pp. 650–665. Springer, Heidelberg (2009)

Instance Coreference Resolution
in Multi-ontology Linked Data Resources

Aynaz Taheri and Mehrnoush Shamsfard

Computer Engineering Department, Shahid Beheshti University, Tehran, Iran
ay.taheri@mail.sbu.ac.ir, m-shams@sbu.ac.ir

Abstract. Web of linked data is one of the main principles for realization of semantic web ideals. In recent years, different data providers have produced many data sources in the Linking Open Data (LOD) cloud upon different schemas. Isolated published linked data sources are not themselves so beneficial for intelligent applications and agents in the context of semantic web. It is not possible to take advantage of the linked data potential capacity without integrating various data sources. The challenge of integration is not limited to instances; rather, schema heterogeneity affects discovering instances with the same identity. In this paper we propose a novel approach, SBUEI, for instance co-reference resolution between various linked data sources even with heterogeneous schemas. For this purpose, SBUEI considers the entity co-reference resolution problem in both schema and instance levels. The process of matching is applied in both levels consecutively to let the system discover identical instances. SBUEI also applies a new approach for consolidation of linked data in instance level. After finding identical instances, SBUEI searches locally around them in order to find more instances that are equal. Experiments show that SBUEI obtains promising results with high precision and recall.

Keywords: Linked Data, Coreference Resolution, Ontology, Schema, Instance, Matching.

1 Introduction

Linked data is a new trend in the semantic web context. Nowadays increasing the amount of linked data in Linking Open Data project is not the only challenge of publishing linked data; rather, matching and linking the linked data resources are also equally important and can improve the effective consuming of linked data resources. Linked data integration is one of the main challenges that become more important considering development of linked data. Without these links, we confront with isolated islands of datasets. The fourth rule of publishing linked data in [2] explains the necessity of linking URIs to each other. Therefore, extension of datasets without interlinking them is against the Linked Data principles.

In the web of linked data with so large scale, there are obviously many different schemas in the various linked data sources. Considering that there is no compulsion for data providers in utilizing specific schema, we confront with the problem of

H. Takeda et al. (Eds.): JIST 2012, LNCS 7774, pp. 129–145, 2013.
© Springer-Verlag Berlin Heidelberg 2013

schema heterogeneity in data sources. This issue is considerable in instance corefe-rence resolution. Paying attention to schemas in linked data consolidation has many advantages. When we are going to discover instances with unique identity in two data sources, it is a complicated process to compare all the instances of two data sources in order to find equivalents. Processing all of the instances has harmful effects on execu-tion time and needs more computing power. However, if we know about schema matching of two data sources, it is not necessary to look up all the instances. Rather, it is enough to search only instances of two matched concepts of schemas, so perfor-mance would become better. In addition, ignoring the schema may cause precision decrease in instance matching. In many cases, the internal structure and properties of instances do not have enough information to distinguish distinct instances and this increases the possibility of wrong recognition of co-referent instances that are apparently similar in some properties in spite of their different identities.

Although ontology/schema matching can be beneficial for instance matching, it could be detrimental if it is done inefficiently. In [18] effects of ontological mis-matches on data integration (in instance level) are described. They divide all types of mismatches into two groups: conceptual mismatches and explication mismatches. They represent that these kinds of mismatches such as conceptual mismatches could be harmful for instance matching and could decrease the amount of precision by wrong matching of concepts in the schema level. They do ontology matching at the first step and instance matching at the second step. Because of this sequential process, the errors of the first step can propagate into the next step.

In this paper, we propose a solution, SBUEI, to deal with the problem of instance matching and schema matching in linked data consolidation. SBUEI proposes an interleaving of instance and schema matching steps to find coreferences or unique identities in two linked data sources. SBUEI, unlike systems such as [13, 21, 27] - which uses just instance matching- or systems such as [15, 24] -which use just schema matching- exploits both levels of instance and schema matching. The main difference between SBUEI and other systems like [19], which exploit both levels, is that SBUEI exploits an interleaving of them while [19] exploits them sequentially one after the other (starts instance matching after completing schema matching). SBUEI utilizes schema matching results in instance matching and use the instance matching results in order to direct matching in schema level. SBUEI also has a new approach for instance matching.

This paper is structured as follows: section 2 discusses some related work. Sec-tion3 explains the instance coreference resolution algorithm at the first phase, and section 4 describes the schema matching algorithm at the second phase. Section 5 demonstrates the experimental results of evaluating SBUEI. Finally, section 6 con-cludes this paper.

2   Related Work

We divide related works in the area of entity coreference resolution in the context of semantic web into four groups:

Some related works deal with specific domains. Raimond et al. in [25] proposed a method for interlinking two linked data music-related datasets that have similar ontologies and their method was applicable on that specific ontology. In [10], authors described how they interlink a linked data source about movies with other data sources in LOD by applying some exact and approximate string similarity measures. In [32] a method for linking WordNet VUA (WordNet 3.0 in RDF) to DBpedia is proposed. The methodology of this project is customized for only these two datasets. Finding identical instances of foaf:person at social graph is explained in [26] by computing graph similarity.

Some pieces of related works follow the challenge of linked data consolidation so that their methods are not constrained to special domains but their methods are based on the assumption that schemas of different data sets are equal and their approach only concentrate on consolidation of data in instance level. In [11], capturing similarity between instances is based on applying inverse functional properties in OWL language. In [21], authors used a similarity measure for computing similarity of instance matching between two datasets with the same ontology. LN2R [27] is a knowledge based reference reconciliation system and combines a logical and a numerical method. LN2R requires manual alignment at first and then turns to the reconciliation of instances.

Another group of approaches, which focus on consolidating linked data, claims that their approach does not depend on schema information and can identify same instances in heterogeneous data sources without considering their ontologies. Hogan et al. [12] proposed a method for consolidation of instances in RDF data sources that is based on some statistical analysis and suggests some formula in order to find "quasi" properties. ObjectCoref [13] is a self-training system based on a semi supervised learning algorithm and tries to learn discriminative property-value pair based on a statistical measurement. Song et al. [31] proposed an unsupervised learning algorithm in order to find some discriminable properties as candidate selection key. SERIMI [1] is an unsupervised method and has a selection phase in order to find some specific properties and a disambiguating phase. Zhishi.links [20] is a distributed instance matching system. It does not follow any special techniques for schema heterogeneity. It uses an indexing process on the names of instances and uses string similarity to filter match candidates.

Some other approaches of instance matching in linked data sources take advantage of schema in data sets. HMatch(τ) [4] is an instance matcher and use HMatch 2.0 for TBox matching and then tries to capture the power of properties at instance identification [9]. RiMOM [33] , ASMOV [16] and AgreementMaker [5] are three ontology matching systems that recently equipped with instance matchers and participated in instance matching track of OAEI 2010. CODI [22] is also a system for ontology and instance matching and is based on markov logic. Seddiqui et al. [30] proposed an instance matching algorithm and Anchor-Flood algorithm [29] for ontology matching at first and then begin instance matching. Nikolov et al. [18] discussed the effects of ontology mismatches on instance resolution and proposed Knofuss architecture in [19]. Linked Data Integration Framework (LDIF) [28] has two main components: Silk Link Discovery Framework [14] and R2R Framework [3] for identity resolution and vocabulary normalization respectively.

From the four above-mentioned groups, approaches in the last two groups are not dependent on any schemas or domains. It seems that the forth group have more advantages in comparison with third group, because the approaches in the third group have deprived themselves of utilizing useful information in the schema level. As we said in the section 1, paying attention to schema has beneficial effects on linked data consolidation. Our proposed approach, SBUEI, belongs to the forth group. What distinguish SBUEI from the aforementioned approaches in the forth group are its consecutive movements between schema and instance level and the matching algorithm in the instance level. SBUEI exploits the instance level information in order to find accurate proposal about schema matching and then applies schema matching in order to do instance resolution with high precision and recall.

3   Instance Coreference Resolution In Linked Data Sources

The instance coreference resolution algorithm has two phases that are executed iteratively. The first phase needs to receive an anchor as input. Anchor is defined as a pair of similar concepts across ontologies [23, 29]. As the first and second phases are executed in a cycle, for the first round, the user should provide this input, but in the next times the input of the first phase (the anchors) is provided by the output of the second phase.

The first phase starts by getting anchor: $X \equiv \langle C_1, C_2 \rangle$. C_1 and C_2 are two similar concepts from ontologies O_1 and O_2 respectively. It comprises three steps explained in the following in details.

3.1   First Step: Create Linked Instances Cloud

In the first step, we introduce a new construction that is called Linked Instances Cloud (LIC), as the basis of our matching algorithm.

For anchors $\langle C_1, C_2 \rangle$, we must create LICs. For each instance of concepts C_1 and C_2 we make one LIC. C_1 has some instances. For example the URI of one of the C_1 instances is called 'i' .We explain how to create LIC for instance 'i'. For creating this LIC, SBUEI extracts all of the triples whose subjects are instance 'i' and adds them to the LIC. Then, in the triples that belong to LIC, we find neighbors of instance 'i'. If instance 'j' is one of the neighbors of instance 'i', SBUEI repeats the same process for instance 'j'. Triples whose subjects are instance 'j', are added to LIC and the process is repeated for neighbors of the instance 'j'. This process is actually like depth first search among neighbors of instance 'i'. To avoid falling in a loop and to eliminate the size of search space, the maximum depth of search is experimentally set to 5. The LIC that is created for instance 'i' is called LIC_i. Starting point for this LIC is instance 'i'. The process of creating LICs is done for all of the instances of the two concepts C_1 and C_2.

Sometimes the identities of instances are not recognizable without considering the instances that are linked to them, and neighbors often present important information about intended instances. In some cases in our experiments, we observed that even discriminative property-value pairs for an instance may be displayed by its neighbors. Fig. 1 shows an illustration about an instance that its neighbors describe its identity. This example is taken from IIMB dataset in OAEI 2010. Fig. 1 shows $LIC_{item21177}$. '*Item21177*' is the starting point of this *LIC* and is an Actor, Director and a character-creator. Each instance in the neighborhood of '*Item21177*' describes some information about it. For example '*Item74483*' explains the city that '*Item21177*' was born in and '*Item27054*' explains the name of the '*Item21177*'.

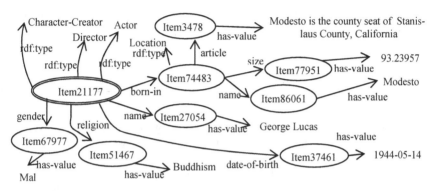

Fig. 1. An Illustration of LIC

Creating *LICs* not only helps us in discovering identities of instances, but also it helps us to find locally more similar instances. This issue is explained in section 3.3.

3.2   Second Step: Compute Similarity between LICs

In the previous step, SBUEI created the LICs of two concepts C_1 and C_2. In this step, we must compare them. Each LIC from concept C_1 should be compared with all LICs of concept C_2 in order to find similar LICs. Starting points of two similar LICs would be equal. Therefore, the triples of two LICs should be compared so that two LICs can be compared. In this process, only triples whose objects are data type values (and not instances) would participate in the comparison. Properties values are very important in comparison. This does not mean that properties (predicates in triples) are effectless in similarity computation. However, finding equal properties in two different ontologies is not usually easy, especially in very heterogeneous ontologies. Therefore, values of properties have more importance for SBUEI. However, if we could find similar properties, the process of similarity computation between properties values would be easier and more effective in increasing of similarity value.

SBUEI computes similarity of LICs in three separate parts as the followings.

Part 1: Normalization of Properties Values
SBUEI applies some normalization techniques on properties values for improving the result of string comparison. These techniques consist of: case normalization, blank normalization, punctuation elimination and removing stop words.

Part 2: Compare Triples in Two LICs

In this part, all of the triples whose objects are values of data type will be compared in two *LICs*. However, before comparing values of properties, properties itself will be compared. Similarity computation for properties are done regarding to three different aspects: range similarity, lexical similarity and properties hierarchies. Equation (1) describes the function that computes properties similarities.

HierSim function computes hierarchal similarity between two properties p_1 and p_2. Properties in the hierarchy are examined by two functions *LexicalSim* and *RanSim*. *RanSim* function compares ranges of two properties p_1 and p_2. *LexicalSim* function computes similarity between the labels of two properties p_1 and p_2. Section 4.1 explains these functions in detail.

$$PropertySim(p_1, p_2)$$
$$= \frac{(RanSim(p_1, p_2) + LexicalSim(p_1, p_2) + HierSim(p_1, p_2, H^P))}{3} \tag{1}$$

If SBUEI finds similar properties in two *LICs*, then objects of triples that have equal properties will be compared. Other triples in two *LICs* that we could not find equal properties for them, will be compared considering only their values of properties. Our experiments show that often in most cases SBUEI does not find similar properties because of lexical, structural and semantical heterogeneity in schemas. Thus, we focus on values of properties when similar properties are not found.

We use Edit Distance method for computing similarity of properties values. Considering v_1 and v_2 as values of two properties p_1 and p_2, SBUEI computes their similarity according to (2).

$$EditDisSim(v_1, v_2) = \begin{cases} EditDistance(v_1, v_2), & if \ EditDistance(v_1, v_2) \geq \delta \\ 0, & if \ EditDistance(v_1, v_2) < \delta \end{cases} \tag{2}$$

Similarity values of triples objects -obtained by *EditDisSim* function- are added together for obtaining similarity value of two *LICs*. In (2) we applied a threshold equal to 0.6 for edit distance method. This threshold was found by making a benchmark and execution of edit distance algorithm based on the benchmark. Equation (2) removes similarity values that are less than threshold. This prevents accumulating small similarity values in the sum of similarity values of triples objects.

Some properties are comments about instances. For similarity computation of these kinds of properties, we do not use (2). SBUEI in (3) applies a specific method for finding comment similarity. It is based on the number of common tokens. Again, we made a benchmark, found a threshold (δ) equal to 0.7.

$$ComSim(x, x') = \begin{cases} CommentSim(x, x'), & if \ CommentSim(x, x') \geq \delta \\ 0, & if \ CommentSim(x, x') < \delta \end{cases}$$
$$CommentSim(x, x') = \frac{2 \times (|x \cap x'|)}{|x| + |x'|} \tag{3}$$

After calculating similarity of properties values, SBUEI computes similarity of two *LICs*. Similarity of two *LICs* is dependent on similarity of their properties values. Triples in two *LICs* have specific importance depend on the depth of their subjects (instances that triples belong to) in the *LICs*. We noted in section 3.1 that depth of instances are estimated towards the starting point of *LIC*. When depth of instances in

LIC increases, their effectiveness in similarity computation of *LICs* decreases. The following triples belong to $LIC_{item21177}$ in Fig. 1.

1 ('Item67977', has-value, Male)
2 ('Item37461', has-value, 1944-05-14)
3 ('Item77951', has-value, 93.23957)
4 ('Item3478', has-value, Modesto is the county seat of Stanislaus County, California)

The above triples describe some information about the starting point of $LIC_{item21177}$. Two first triples explain that '*item21177*' has male gender and date of his birth is 1994-05-14. Instances in the subjects of these two triples have depth equal to two. Two second triples explain that '*item21177*' has born in a city that its size is 93.23957 and also is the county seat of Stanislaus County, California. Instances in the subjects of these two triples have depth equal to three. As you can see, the first two triples have more important role for determining the identity of '*item21177*' than the second two triples. Gender of a person and date of his birth is more important than some comments about the city that he lives in. Nevertheless, this does not mean that existence of instances with greater depth are not beneficial in the *LICs*; rather, they are less important in identity recognition of the starting point of the *LIC* than those with less depth. In addition, we utilize such instances in step three for finding more instances that are similar.

In this regard, similarities of properties values are added with an particular coefficient. SBUEI uses a weighted sum for computing similarity of *LICs*. The coefficients (4) in this sum have inverse relations to the depth of the subject of triples in *LIC* .

$$S_n = \frac{1}{n}$$

$$(4)$$

$$S = \{S_n : n \in N\} \quad n \text{ is the depth of the subject } s$$

SBUEI normalizes the sum of similarities of properties values in two *LICs* into a range of 0 and 1 by dividing the result to the sum of the numbers of triples in two *LICs*.

In (5) LIC_1 and LIC_2 have n and m triples, respectively. We find for each object of LIC_1 the most similar object in LIC_2 via *LexicalPropValSim* function. *LexicalPropValSim* works based on (2) and (3). Each of the two objects with most similarity; have a coefficient equal to inverse of the depth of the triple that belongs to LIC_1. We do the same process for LIC_2 and then add obtained values of similarity LIC_1 and LIC_2. Normalization is done by the sum of the triples numbers in LIC_1 and LIC_2 (In (5) this value is m+n), but each triple has a coefficient according to what mentioned in (4).

$$LIC_1 = \{t_1, t_2, \dots, t_n\}$$
$$LIC_2 = \{t'_1, t'_2, \dots, t'_m\}$$

$$LICSimilarity(LIC_1, LIC_2)$$
$$= \frac{\left(\sum_{i=1}^{n} \frac{1}{Depth(t_i)} max_{1 \leq j \leq m} \left(LexicalPropValSim \left(Obj(t_i), Obj(t'_j) \right) \right) + \atop \sum_{j=1}^{m} \frac{1}{Depth(t'_j)} max_{1 \leq i \leq n} \left(LexicalPropValSim \left(Obj(t'_j), Obj(t_i) \right) \right) \right)}{\left(\sum_{i=1}^{n} \frac{1}{Depth(t_i)} + \sum_{j=1}^{m} \frac{1}{Depth(t'_j)} \right)}$$

$$(5)$$

Part 3: Choose Two Similar LICs

In the previous part, we computed the similarity of two *LICs*. Now we confront with a complex challenge. The main challenge is determining the value of the threshold for deciding whether two *LICs* are equal or not. This threshold is the final approver about equality of two *LICs* based on their similarity value. Our experiments show that the value of threshold is completely variable depending on the data sources and their characteristics. The amounts of threshold have considerable differences in a range between zero and one. For example results of several tests on two data sources, indicated that the best value for similarity threshold of *LICs* are 0.7. While on two other data sources this value was 0.2. Wide range of obtained values for threshold led us to have a dynamic selection for threshold of *LICs* similarity.

Dynamic selection of threshold means that after creating *LICs* and computing their similarities, we choose the value of threshold depending to the concepts that *LICs* belong to. LIC_1 and LIC_2 belong to two concepts C_1 and C_2 respectively. We calculate a specific threshold for two concepts C_1 and C_2. Afterward we apply acquired threshold for comparing *LICs* from these two concepts. For all *LICs* that their starting points belong to two particular concepts from two ontologies, SBUEI use a specific threshold.

The purpose of threshold value variation and the necessity of dynamicity in threshold selection can be justified because of the heterogeneity of ontologies viewpoints regarding to concepts. Two ontologies may have information considering different aspects of concepts and therefore their instances even with the same identities have diverse information with little overlap. In such cases, ontologies are semantically different. For example, one ontology has some properties such as name, director, characters for describing a film and another ontology has name, actors, character-creator properties for describing a film. Hence, equal instances of these two ontologies have low similarities despite the fact that they have the same identities.

SBUEI uses a heuristic for solving this problem. After finding the most similar *LICs* of two concepts C_1 and C_2, SBUEI calculates the similarity average of the most similar *LICs* of two concepts. Instances of concept C_1 have almost the same properties for describing individuals and instances of concept C_2 are also the same way. Hence, the similarity amounts of instances of these two concepts are approximately predictable and are the same.

SimLIC in (6) is the set of most similar *LICs* of two concepts C_1 and C_2. Each member of the set is a triple: the first element is the intended *LIC* from concept C_1, the second element is the most similar *LIC* from concept C_2 and the third element is the value of their similarity. *LICSimThreshold* computes the threshold of two concepts C_1 and C_2.

$$SimLIC = \{(L_1, L_1', S_1), (L_2, L_2', S_2), \ldots, (L_n, L_n', S_n)\}$$

$$LICSimThreshold = \frac{\sum_{i=1}^{n} S_i}{n} \qquad (6)$$

3.3   Third Step: Determine New Equal Instances in Two Similar LICs

In this step, we continue the process of matching on those LICs of the previous step that led to discovering equal instances or in the other words, those LICs that have equal starting points. The strategy in this step is searching locally around the identical instances in order to find new equal instances. In [29] an algorithm for ontology matching is created, and their algorithm is based on the idea that if two concepts of two ontologies are similar, then there is a high possibility that their neighbors are similar too. We use this idea but in instance level. This means that if two instances are identical, then there is possibility that their neighbors are similar too.

Suppose that 'i' and 'j' are two instances of two concepts C_1 and C_2 and they are detected identical in the previous step. Their LICs are called LIC_i and LIC_j. In this step we describe how SBUEI finds more identical instances in LIC_i and LIC_j. This step is composed of three parts as following.

Part 1: Create Sub-Linked Instances Cloud
We define a new construction called 'Sub-LIC'. SBUEI makes one LIC for each instance in LIC_i and LIC_j. Each LIC has some Sub-LICs in itself as many as the number of its instances.

Part 2: Discover Similar Sub-LICs
For discovering similar instances in LIC_i and LIC_j, we compare their *Sub-LICs*. Similarity of *Sub-LICs* is computed same as *LICs* with the same threshold value. *Sub-LICs* that have similarities more than the threshold are equal and their starting points are considered as identical instances.

Finding identical instances of two concepts initially costs a lot because of considering all neighbors of an instance; later we can find locally more identical instances by paying low computational cost.

Part 3: Compute Concept Similarities in Schema Level
After finding identical instances in *Sub-LICs*, now it is time to arrange for moving to schema level. In this part instance matcher gives feedback to schema matcher. Concepts whose instances are the starting points of equal *Sub-LICs* are candidates to be similar. Suppose that $Sub\_LIC_{i_1}$ and $Sub\_LIC_{j_1}$ are two *Sub-LICs* of LIC_i and LIC_j respectively and are detected similar in the previous part. Starting points of $Sub\_LIC_{i_1}$ and $Sub\_LIC_{j_1}$ belong to two concepts C_3 and C_4 from ontologies O_1 and O_2 respectively. We can conclude that C_3 and C_4 are probably similar because they have identical instances.

SBUEI repeats three above parts for all LICs of concepts C_1 and C_2 that are detected similar in step 2. We define the following measure for similarity estimation between two concepts C_3 and C_4: 'ratio of equal Sub-LICs of two concepts C_1 and C_2 that their starting points belong to concepts C_3 and C_4 to the total numbers of LICs of two concepts C_1 and C_2'. Instance matcher utilizes this measure for helping schema matcher in order to find equal concepts.

4   Schema Matching in Linked Data Sources

The second phase is done by a schema matcher. It receives feedback from the first phase, which contains some similarities between concepts from the viewpoint of instance matcher. At this time, schema matcher begins the process of matching in schema level by applying some ontology matching algorithms. SBUEI compares all of these similarity values that are proposed by instance matcher or obtained by schema matcher, and choose a pair of concepts that have the most similarity. SBUEI repeats these two phases consecutively until all of similar concepts that are given feedback to schema level or are detected by schema matcher, to be selected and schema matcher could not find any similar concepts.

4.1   Compute Similarity between Concepts

When SBUEI wants to do ontology matching, it considers to the concepts that are proposed as equal concepts in the previous iterations and the process of ontology matching starts in the neighborhood of these concepts.

We applied the definition of concept neighborhood in [29]. In this definition, neighbors of a concept include its children, grandchildren, parents, siblings, grandparents and uncles. Schema matcher utilizes two similarity measures for ontology matching: label-based techniques and structure-based techniques. Therefore, we have two kinds of matchers: lexical matcher and structural matcher.

Compute Lexical Similarity

Lexical similarity of two concepts or two properties are described by *LexicalSim()* function in (8). We compute the similarity of two concepts or two properties with using a string-based method and utilizing a lexical ontology.

Lexical features of a concept are such as id, label and comments. For computing comments similarities between concepts or properties, we use a token-based method, the same as equation (3). For other features of concepts or properties, SBUEI uses Princeton WordNet [8], a lexical ontology, for finding similarities.

Let l_1 and l_2 be two labels or ids for two concepts C_i and C_j respectively, and let *Synset (l_1)* denotes a synset in WordNet that l_1 belongs to it and *Senses(s)* returns all of senses of a synset s, then

$$WNSenses(l_1) = \{x | x \in Senses(Synset(l_1))\} \tag{7}$$

$$LexicalSim(l_1, l_2) = \begin{cases} 1, & \text{if } l_1 = l_2 \\ 1, & \text{if } l_1 \in WNSenses(l_2) \text{ or } l_2 \in WNSenses(l_1) \\ WNStructuralSim(l_1, l_2), & \text{if } l_1 \notin WNSenses(l_2) \text{ AND} \\ & l_2 \notin WNSenses(l_1) \\ EditDistance(l_1, l_2), & \text{if } l_1 \notin WordNet \text{ or } l_2 \notin WordNet \end{cases} \tag{8}$$

l_1 and l_2 have the most lexical Similarity value, if they are equal or l_1 is one of the senses of the synset that l_2 belongs to or vice versa. *WNStructuralSim(l_1, l_2)* in (8) computes the similarity of two synsets that l_1 and l_2 are one of their senses respectively. The similarity of those synsets is computed regarding to their positions in the hierarchy of WordNet. We used the Wu-Palmer measure presented in [34]. If l_1 or l_2 is not in WordNet we use Edit Distance method for computing their similarity.

Compute Structural Similarity

In addition to the names, ids, comments and all other lexical features of concepts, their structure in ontologies are also important for calculating concepts similarities. In [6] structure based techniques are divided into two groups based on the internal structure and relational structure. SBUEI utilizes internal and relational structures for computing similarities between concepts.

Internal Similarity.

Internal similarity compares properties of two concepts from two different ontologies. Let p_i and p_j be two properties belong to two concepts C_i and C_j respectively, *InternalStructuralSim()* function in (9) is calculated using the following formula:

$$InternalStructuarlSim(C_i, C_j)$$

$$= \frac{\left(2 \times \left(\sum_{p_i \in Property_{(C_i)}} Max_{p_j \in Property(C_j)} \left(\begin{array}{c} RanSim(p_i, p_j) + \\ LexicalSim(p_i, p_j) + \\ CardSim(p_i, p_j) \end{array} \right) \right) \right)}{\left(|Property(C_i)| + |Property(C_j)| \right)} \tag{9}$$

We defined *LexicalSim* previously. *CardSim* compares cardinality of properties. It will be 1.0 if Maximum and Minimum values of properties are equal and 0 otherwise. *RanSim* computes similarity value between data types of two properties range. Utilizing hierarchy of XML schema data types is proposed in [6] for computing data type similarity. SBUEI also uses XML schema data type hierarchy but applies Wu-Palmer method on the hierarchy in order to find the similarity of two data types regarding to their positions in the hierarchy.

Relational Similarity

Relational similarity in SBUEI computes similarity of hierarchies that two concepts from two different ontologies belong to. SBUEI considers taxonomic structure and pays more attention to rdfs:subClassOf for relational similarity computation. Parents and children of a concept are compared with the parents and children of another concept. For comparing parents and children in (10), we use *LexicalSim* and *InternalStructuralSim* functions.

$$RelationalStructuralSim(C_i, C_j)$$

$$= \left(\frac{\sum_{n_i \in Parent(C_i)} Max_{n_j \in Parent(C_j)} \left(\begin{array}{c} LexicalSim(n_i, n_j) + \\ InternalStructuarlSim(n_i, n_j) \end{array} \right)}{\left(|Parent(C_i)| + |Parent(C_j)| \right)} \right)$$

$$+ \left(\frac{\sum_{n_i \in Children(C_i)} Max_{n_j \in Children(C_j)} \left(\begin{array}{c} LexicalSim(n_i, n_j) + \\ InternalStructuarlSim(n_i, n_j) \end{array} \right)}{\left(|Children(C_i)| + |Children(C_j)| \right)} \right) \tag{10}$$

Similarity Aggregation

SBUEI uses a weighted linear aggregation process to compose the results of different similarity values .The weights have been determined experimentally. We obtained values as: $w_{lexical}=0.4$, $w_{internal}=0.3$ and $w_{relational}=0.3$

5   Evaluation

The experimental results are downloadable at our website[1] and some statistics of SBUEI operations in comparison with other systems are prepared[2].

For evaluating SBUEI, we use the datasets in OAEI [7] , a benchmarking initiative in the area of semantic web. We report the experimental results of our proposed approach on four different datasets: PR in OAEI 2010, IIMB in OAEI 2010, IIMB in OAEI 2011 and TAP-SWETO datasets in OAEI 2009. In the last experiment, we found an interesting use case that takes advantage from our approach.

5.1   Person-Restaurants Benchmark

Person-Restaurants benchmark is one of the tasks in instance matching track in OAEI 2010 campaign. This benchmark is composed of three datasets: Person1, Person2 and Restaurants. Each of them has two different owl ontologies and two sets of instances. Person1 and Person2 contain instances about some peoples, and Restaurants has instances about some restaurants. Sizes of these datasets are small. Reference alignments are provided in OAEI 2010 and all of participants in this task must compare their results with the reference alignments. Five systems have participated in this task and we compare our results with the results of other systems.

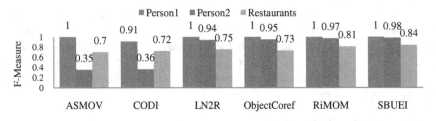

Fig. 2. Result of OAEI'10 Person-Restaurant Benchmark

F-measure values of all systems on the Person-Restaurant benchmark are shown in Fig. 2. Performance of all systems (including SBUEI) are quite good on Person1 dataset. Collected statistics of all the participants show that SBUEI has the best values of F-measure and precision on Person2 and Restaurants datasets. According to obtained values, we conclude that SBUEI performed well in matching at both schema and instance levels of these three datasets.

---

[1] http://nlp.sbu.ac.ir/sbuei/result.rar
[2] http://nlp.sbu.ac.ir/sbuei/statistics.html

5.2   IIMB'10 Track

The second part of our experiments includes IIMB task of OAEI 2010. IIMB composed of 80 test cases. Each test case has OWL ontology and a set of instances. Information of test cases in IIMB track is extracted from Freebase dataset. IIMB divided test cases in four groups. Test cases from 1 to 20 have data value transformations, 21 to 40 have structural transformations, test cases from 40 to 60 have data semantic transformations and 61 to 80 have combination of these three transformations. All of these 80 test cases must be matched against a source test case. We choose IIMB 2010 test cases for the evaluation because this track of OAEI has a good number of participants and its test cases have all kinds of transformations and we could compare all aspects of our system against the other systems.

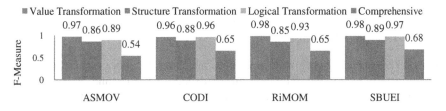

Fig. 3. Results of OAEI'10 IIMB Track

The results of four systems on F-measure are depicted in Fig. 3. We could observe that all four systems have good values of F-measure on datasets with data value transformation. SBUEI and ASMOV have the best values for F-measure. All participating systems have weaker results in test cases with structural value transformation.

SBUEI has better operations than others in these datasets. This means that SBUEI is more stable in modifications such as removing, adding and hierarchal changing of properties. Systems have better results in test cases with semantic value transformation against structure value transformation. SBUEI has the best F-measure in such cases. Four systems do not have desirable results in datasets with combination all kinds of transformations.

5.3   IIMB'11 Track

In this part of our experiments, we are going to show the results of SBUEI on IIMB track of OAEI 2011. This track has also 80 test cases with four kinds of transformations just like the last year IIMB track. The purpose of selecting this track as one of our experiments is that the size of IIMB 2011 has increased greatly compared to last year and is more than 1.5 GB. Increased amount of the dataset size lets us evaluate scalability of our approach. Unfortunately, there has been just one participant in this track, CODI, with which we will compare our results. This shows the scalability difficulties in systems performance at large scale datasets. We observe in Fig. 4 that the recall values of SBUEI in four kinds of transformations are better than CODI but this is not always true for precision value. The operations of SBUEI is clearly better than CODI in datasets with structure transformation considering three aspects of precision, recall and F-measure.

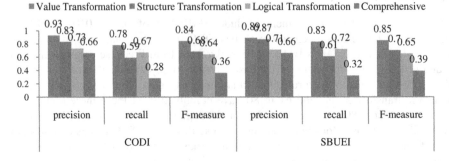

Fig. 4. Results of OAEI'11 IIMB Track

5.4   TAP-SWETO Datasets

OAEI 2009 created a benchmark in instance matching track that was created according to different ontologies. This benchmark includes SWETO and TAP datasets. In [18], authors used these two datasets and explained the effects of ontology mismatches in instance matching on these two datasets. We explain again one of the problems that Nikolov pointed and describe how SBUEI overcame some difficulties in instance matching of these two datasets. Unfortunately, this track was cancelled due to the lack of participants and there are no reference alignments or any other systems to be compared with our results.

Therefore, we mention only how SBUEI performed the process of matching. There are some instances about computer researchers in two datasets. TAP ontology has a concept that is called 'ComputerScientist' and on the other hand SEWTO ontology has a concept that is called 'ComputerScienceResearcher'. These two concepts have also some structural similarities in addition to lexical similarities, because their positions in the hierarchy of ontologies are similar. The first one has Sub-class relation with 'Scientist' concept and the second has Sub-class relation with 'Reasercher' concept. More over both 'Scientist' and 'Reasercher' concepts are children of 'Person' concept in their ontologies. Therefore, 'ComputerScientist' and 'ComputerScience Researcher' are good candidates to be matched by structural and lexical matchers. If we confirm this matching and then start instance resolution process between their instances, we confront with a few numbers of matched instances. This is because of conceptualization mismatches in schema. 'ComputerScientist' in TAP includes only famous computer scientist and 'ComputerScienceResearcher' has more general sense and includes who have a paper in computer science research areas.

SBUEI removed this problem and had a good instance resolution process. SBUEI began the process of matching after receiving a pair of concepts as anchor: ('ResearchPaper' , 'Scientific_Publication'). After acquiring these two concepts, SBUEI is confident about similarity of instances between the concepts. SBUEI created LICs around the instances of two concepts 'ResearchPaper' and 'Scientific_Publication' and computed similarity of LICs and then selected equal LICs with similarities more than threshold. SBUEI found more similar instances in equal LICs of these two concepts. In the neighborhood of LICs starting points, SBUEI found some equal instances that are authors of papers. These newly founded instances belonged to concepts such as: '*CMUPerson*' and '*W3CPerson*' from TAP and '*ComputerScienceRe-searcher*' from SEWTO. Thus, SBUEI discovered similarity between '*CMUPerson*' and '*W3CPerson*' concepts of TAP and '*ComputerScienceResearcher*' concept of SWETO. Then SBUEI started the process of instance resolution and found more matched instances compared to the previous time. In particular, TAP has divided computer researchers considering to their place of their works. These kinds of mismatches in schema level could not be discovered easily by ontology matchers. Ontology matchers and instance matchers can improve their performance cooperatively.

6   Conclusion

In this paper, we proposed a new approach, SBUEI, for linked data consolidation. This approach is applicable in with heterogeneous schemas. SBUEI pays attention to matching in both schema and instance level. Instance resolution process starts in SBUEI after getting two equal concepts as input by instances matcher. Instance matcher creates LICs around the instances of two equal concepts and then compares these clouds. After discovering instances with the same identity in *LICs*, instance matcher utilizes them and sends some similarity feedback to the schema matcher. Schema matcher receives feedback, applies some ontology matching algorithms, determines two equivalent concepts in ontologies and gives them to instance matcher as input. This process continues consecutively. Our experiments showed that our approach achieved high precision and recall and outperforms other systems.

Our future target includes utilizing some methods that are proposed at the third group of section 2 in our approach. This means that we are going to use some learning algorithms to find discriminable properties in the *LICs*. This will help us to find similar *LICs* efficiently. Considering that SBUEI is a recently created approach, does not have appropriate user interface. Therefore, it is important to make a powerful user interface for SBUEI.

References

1. Araujo, S., Hidders, J., Schwabe, D., de Vries, A.P.: SERIMI - Resource Description Similarity, RDF Instance Matching and Interlinking. CoRR, abs/1107.1104 (2011)
2. Bizer, C., Heath, T., Berners-Lee, T.: Linked Data-The Story So Far. Int. J. Semantic Web Inf. Syst. 5(3), 1–22 (2009)
3. Bizer, C., Schultz, A.: The R2R Framework: publishing and discovering mapping on the web. In: 1st International Workshop on Consuming Linked Data, China (2010)
4. Castano, S., Ferrara, A., Montanelli, S., Lorusso, D.: Instance matching for ontology population. In: 16th Symposium on Advanced Database Systems, Italy (2008)
5. Cruz, I.F., Store, C., Caimi, F., Fabiani, A., Pesquita, C., Couto, F.M., Palmonari, M.: Using AgreementMaker to align ontologies for OAEI 2011. In: 6th International Workshop on Ontology Matching, Germany (2011)
6. Euzenat, J., Shvaiko, P.: Ontology Matching, 1st edn. Springer, Heidelberg (2007)

7. Euzenat, J., Meilicke, C., Stuckenschmidt, H., Shvaiko, P., Trojahn, C.: Ontology Alignment Evaluation Initiative: Six Years of Experience. In: Spaccapietra, S. (ed.) Journal on Data Semantics XV. LNCS, vol. 6720, pp. 158–192. Springer, Heidelberg (2011)
8. Fellbaum, C.: WordNet: An Electronic Lexical Database. MIT Press, Cambridge (1998)
9. Ferrara, A., Lorusso, D., Montanelli, S.: Automatic identity recognition in the semantic web. In: 1st ESWC Workshop on Identity and Reference on the Semantic Web, Spain (2008)
10. Hassanzadeh, O., Consense, M.: Linked movie data base. In: 2nd Link Data on the Web, Spain (2009)
11. Hogan, A., Harth, A., Decker, S.: Performing object consolidation on the semantic web data graph. In: 1st Identity, Identifiers, Identification Workshop, Canada ((2007)
12. Hogan, A., Polleres, A., Umbrich, J., Zimmermann, A.: Some entities are more equal than others: statistical methods to consolidate linked data. In: 4th International Workshop on New Forms of Reasoning for the Semantic Web, Greece (2010)
13. Hu, W., Chen, J., Qu, Y.: A Self-training Approach for Resolving Object Coreference Semantic Web. In: 20th International World Wide Web Conference, India (2011)
14. Isele, R., Jentzsch, A., Bizer, C.: Silk server- adding missing links while consuming linked data. In: 1st International Workshop on Consuming Linked Data, China (2010)
15. Jain, P., Hitzler, P., Sheth, A.P., Verma, K., Yeh, P.Z.: Ontology Alignment for Linked Open Data. In: Patel-Schneider, P.F., Pan, Y., Hitzler, P., Mika, P., Zhang, L., Pan, J.Z., Horrocks, I., Glimm, B. (eds.) ISWC 2010, Part I. LNCS, vol. 6496, pp. 402–417. Springer, Heidelberg (2010)
16. Jean-Mary, Y.R., Shironoshita, E.P., Kabuka, M.R.: ASMOV: Results for OAEI 2010. In: 5th International Workshop on Ontology Matching, China (2010)
17. Jean-Mary, Y.R., Shironoshita, E.P., Kabuka, M.R.: Ontology matching with semantic verification. J. Web Sem. 7(3), 235–251 (2009)
18. Nikolov, A., Uren, V., Motta, E.: Toward data fusion in a multi-ontology environment. In: 2nd Linked Data on the Web Workshop, Spain (2009)
19. Nikolov, A., Uren, V., Motta, E., de Roeck, A.: Overcoming schema heterogeneity between linked semantic repositories to improve coreference resolution. In: Gómez-Pérez, A., Yu, Y., Ding, Y. (eds.) ASWC 2009. LNCS, vol. 5926, pp. 332–346. Springer, Heidelberg (2009)
20. Niu, X., Rong, S., Zhang, Y., Wang, H.: Zhishi.links results for OAEI 2011. In: 6th International Workshop on Ontology Matching, Germany (2011)
21. Noessner, J., Niepert, M., Meilicke, C., Stuckenschmidt, H.: Leveraging terminological structure for object reconciliation. In: Aroyo, L., Antoniou, G., Hyvönen, E., ten Teije, A., Stuckenschmidt, H., Cabral, L., Tudorache, T. (eds.) ESWC 2010, Part II. LNCS, vol. 6089, pp. 334–348. Springer, Heidelberg (2010)
22. Noessner, J., Niepert, M.: CODI : Combinatorial Optimization for Data Integration – Results for OAEI 2010. In: 5th International Workshop on Ontology Matching, China (2010)
23. Noy, N., Musen, M.: Anchor-PROMPT: using non-local context for semantic matching. In: Ontologies and Information Sharing Workshop, USA (2001)
24. Parundekar, R., Knoblock, C.A., Ambite, J.L.: Linking and building ontologies of linked data. In: Patel-Schneider, P.F., Pan, Y., Hitzler, P., Mika, P., Zhang, L., Pan, J.Z., Horrocks, I., Glimm, B. (eds.) ISWC 2010, Part I. LNCS, vol. 6496, pp. 598–614. Springer, Heidelberg (2010)
25. Raimond, Y., Sutton, C., Sandler, M.: Automatic Interlinking of music datasets on the semantic web. In: 1st Link Data on the Web, China (2008)

26. Rowe, M.: Interlinking Distributed Social Graphs. In: 2nd Linked Data on the Web Workshop, Spain (2009)
27. Sais, F., Niraula, N., Pernelle, N., Rousset, M.: LN2R a knowledge based reference reconciliation system: OAEI 2010 results. In: 5th International Workshop on Ontology Matching, China (2010)
28. Schultz, A., Matteini, A., Isele, R., Bizer, C., Becker, C.: LDIF-Linked data integration framework. In: 2nd International Workshop on Consuming Linked Data, Germany (2011)
29. Seddiqui, M.H., Aono, M.: An Efficient and Scalable Algorithm for Segmented Alignment of Ontologies of Arbitrary Size. J. Web Sem. 7(4), 344–356
30. Seddiqui, M.H., Aono, M.: Ontology Instance Matching by Considering Semantic Link Cloud. In: 9th WSEAS International Conference on Applications of Computer Engineering, Russia (2009)
31. Song, D., Heflin, J.: Automatically generating data linkages using a domain-independent candidate selection approach. In: Aroyo, L., Welty, C., Alani, H., Taylor, J., Bernstein, A., Kagal, L., Noy, N., Blomqvist, E. (eds.) ISWC 2011, Part I. LNCS, vol. 7031, pp. 649–664. Springer, Heidelberg (2011)
32. Taheri, A., Shamsfard, M.: Linking WordNet to DBpedia. In: Proceedings of the 6th Global WordNet Conference, Japan (2012)
33. Wang, Z., Zhang, X., Hou, L., Zhao, Y., Li, J., Qi, Y., Tang, J.: RiMOM Results for OAEI 2010. In: 5th International Workshop on Ontology Matching, China (2010)
34. Wu, Z., Palmer, M.: Verb Semantics and Lexical Selection. In: 32nd Annual Meeting of the Association for Computational Linguistics, Las Cruces (1994)

The Dynamic Generation of Refining Categories in Ontology-Based Search

Yongjun Zhu[1], Dongkyu Jeon[1], Wooju Kim[1,*], June S. Hong[2], Myungjin Lee[1], Zhuguang Wen[1], and Yanhua Cai[1]

[1] Dept. of Information and Industrial Engineering, Yonsei University, Seoul, Korea
{izhu,jdkclub85,wkim,xml}@yonsei.ac.kr,
{zhuguang1027,helencai0708}@gmail.com
[2] Division of Business Administration, Kyonggi University, Kyonggi, Korea
Junehong@kyonggi.ac.kr

Abstract. In the era of information revolution, the amount of digital contents is growing explosively with the advent of personal smart devices. The consumption of the digital contents makes users depend heavily on search engines to search what they want. Search requires tedious review of search results from users currently, and so alleviates it; predefined and fixed categories are provided to refine results. Since fixed categories never reflect the difference of queries and search results, they often contain insensible information. This paper proposes a method for the dynamic generation of refining categories under the ontology-based semantic search systems. It specifically suggests a measure for dynamic selection of categories and an algorithm to arrange them in an appropriate order. Finally, it proves the validity of the proposed approach by using some evaluative measures.

Keywords: Ontology, Dynamic Classification, Categorization.

1 Introduction

For the last twenty years, the landscape of information technology has drastically changed. According to Hilbert and Lopez [1], the volume of information has grown exponentially: particularly in the period of 1986 to 2007, from six hundred exabytes to three zettabytes for the capacity of the broadcasted information. The portion of the digitally broadcasted information increased from 7% to 24% from 2000 to 2007, which reflects the explosive growth of the production, the storage and the consumption of digital contents. The consumption of digital contents depends heavily on search engines in the Internet for users to find what they want. The current state of search systems, however, requires users to review long list of items from search results laboriously. A traditional solution to alleviate it is the provision of refining categories. Search engines provide categories that classify items in the search result into several groups, and users may refine their searches by selecting a category relevant to

---

* Corresponding author.

H. Takeda et al. (Eds.): JIST 2012, LNCS 7774, pp. 146–158, 2013.

what they want to find. A problem of this solution is that categories are predefined and fixed regardless of the difference in queries and search results. Fixed categories do not reflect the context of searches and the information relevant to search results, and may provide insensible information. To present information relevant to search results better and more sensibly, categories should be generated dynamically according to search results.

The objective of this paper is to propose a method for generating dynamically categories refining search results in ontology-based search systems. It is thereby assumed that items resulted in search are structured data with their associated properties. Another assumption made in this paper is that search results contain only one type of exact item since ontology-based search systems aim at finding items that represent the user's intention and such items belong to an identical type. For example, if a search result is on books and one of its items is "The Old Man and the Sea", it has properties in the book ontology such as "Author", "Publication Date", and "Publisher". The properties for a given object have their own values that describe the characteristics of the object. On the other hand, book stores may group fictions according to their author's names. Importantly, properties may be used for categorizing objects as well as for describing. In this research, categories are properties whose values classify items in search results; items with the same property value go into the same group. Hence, the dynamic generation of categories can be rephrased to selecting properties that classify items depending on search results; different search results induce different categorizations. The category selection requires a criterion that is applied in the selection process. The ordering of categories is another issue. Some orderings of categories may result in a better classification than others. For example, when items from a search consist of books with the same publisher, it is not a good idea to classify them with the property "Publisher" because they all would go into the same group, and this is meaningless in categorization. This research, therefore, focuses on developing a measure for a criterion in the category selection and an algorithm for the category ordering.

The presentation is organized after this selection as follows. Section 2 provides some background knowledge and information with related works. Section 3 presents the detailed method for the dynamic generation of categories: a selection measure and an algorithm for the ordering of selected categories. Section 4 demonstrates an evaluation for some experimentation. The conclusion comes in the last section.

2   Related Works

This research assumes that a search system is based on ontology. Ontology is a specification of a representational vocabulary for a shared domain of discourse (definitions of classes, relations, and functions with other objects) [2].

In generation categories dynamically, a measure needs to be used in selection on property against another. A fundamental concept that is necessary to construct such a measure is Shannon's entropy [3]. To define the measure, another conceptual measure is needed, called information gain [4]. According to [5], it is a non-symmetric measure of the difference between two probability distributions, and also the change in information entropy from a prior state to a state that takes some additional information.

The dynamic category generation has not been studied well, and it is hard to find an article on it in the literature. Park et al. [6], one of the rare studies, proposed an e-mail classification agent using a category generation method. Their agent generates categories dynamically based on the information of titles and contents in e-mail messages. Since titles and contents in e-mail messages contain unstructured information that is hard to classify, categories cannot be prebuilt before search and selected from the prebuilt. Thus, they are generated on the spot, and because of that, users cannot anticipate what categories will be generated. This implies that it is very hard to avoid meeting odd and unintuitive categories that are not easily understandable. Search results considered in this research, however, are structured because search is based on structured ontology databases; ontologies have predefined properties which can be used as categories that classify search results. That structuredness also enables precise analysis in search results and may generate more intuitive and understandable categories. Khoshnevisan [7] proposed a method that identifies a set of search categories based on category preference obtained from search results. Items in the search result are ranked according to the relevancy to query and each has category preference information. The preference of a category is determined by the order of items found from that category. On the other hand, all items of search results in this paper become correct answers to the given query because ontology search is used; items in a search result cannot be ranked because they have no concept of priority and have the same relevancy to query. All items in an ontology search result are evaluated together, rather than dealt separately, in generating categories. Also, the generated categories may provide more appropriate classification than the method proposed by Khoshnevisan because this research uses an ontology-based complete search while the Khoshnevisan's uses the conventional text-based search.

3   The Dynamic Category Generation

3.1   A Measure for the Category Selection

The goal of this research is to propose a method to generate dynamically categories depending on search results. Figure 1 shows the overall process taken in this research. An ontology and resources as a search result under the ontology become inputs to the process. The process analyzes the resources with the knowledge of the structure of the ontology and outputs a classification of categories, also called properties, according to the resources.

The dynamic category generation in the semantic search compels selections of appropriate properties predefined in a given ontology. A natural arising question is then in what criterion to select properties, and a notion is introduced to answer that question. *Property A* explains *Property B* well if the category made by *Property A* contains items homogeneous in the value of *Property B*. Table 1 lists ten movie titles with three relevant properties: "Director", "Genre" and "Country". If the "Genre" of a

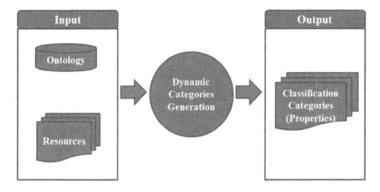

Fig. 1. Overview of the dynamic category generation

movie is known to be "Action", it is easily deduced from Table 1 that the movie was released in "Korea" and directed by "Gyeongtaek Kwak". If a movie is known to be directed by "Changdong Lee", its "Genre" is "Drama" and it is a Korean movie. This deduction is not always available because the knowledge of "Korea" gives no hints on properties of "Genre" and "Director". The main reason for this unavailability is that movies with the value "Korea" of property "Country" have many different values in the other properties.

When movies in Table 1 are classified in the property "Country", eight movies belong to one category of value "Korea", but their similarity within the category is very small because they have different values in properties "Director" and "Genre". This is the same on the items in the category of value "USA". They have almost no similarity except that they are released in the same countries. If the movies are classified in the property "Genre", the extent of similarity increases. Movies in "Action" and "Thriller" categories are all directed by "Gyeongtaek Kwak" and "Junho Bong", respectively. Even though the "Drama" category still has different values in the property "Director", the others have the same value. Classification in "Director" also increases similarity in comparison with the classification in "Country". When several classifiers are available in categorization of objects, a question arises: which classifier is the best for categorization? It is conventionally stated that a good classifier classifies objects similar in values of other properties or features into the same category, and different objects into different categories [8]. This question naturally leads to the requirement of a measure of similarity within each category, made by a given classifier. If one property explains the others better than any other does in such a measurement, the property should become the best classifier. In the example of Table 1, the measures of the properties, "Director", "Genre" and "Country", represent their abilities to explain "Genre" and "Country", "Director" and "Country", and "Director" and "Genre", respectively.

Table 1. Movies with three properties

Movie	Director	Genre	Country
Friend	Kwak Gyeongtaek	Action	Korea
Typhoon	Kwak Gyeongtaek	Action	Korea
The Host	Bong Junho	Thriller	Korea
Mother	Bong Junho	Thriller	Korea
Tokyo!	Bong Junho	Drama	Korea
SILENCED	Hwang Donghyeok	Drama	Korea
Secret Sunshine	Lee Changdong	Drama	Korea
Poetry	Lee Changdong	Drama	Korea
The Beaver	Jodie Foster	Comedy	USA
Battleship	Peter Berg	War	USA

The *TotalGainRatio* is used in this paper for the measurement of the extent that a property explains the others. Every property has the value of *TotalGainRatio*; if its values is bigger, it explains the other properties well. *TotalGainRatio* is calculated as follows:

$$TotalGainRatio(P^*, P_j) = \sum_{P_i \in P^*} GR(P_i, P_j) \tag{1}$$

where P is the set of properties, P_j is the jth property of P, P^* is the set of properties except P_j.

The process of calculating *TotalGainRatio* can be divided into ten steps. Input value of the process is ontology and the set of resources R in search results. The ten steps are as follows:

1. Obtain the type of R and query to ontology to find property set P which have the type of R as a domain. For example, if search results include "The Old Man and the Sea", and so on. The type of R is "Book" and the properties of "Book" such as "title" and "author" are property set P.
2. For all property $P_i(P_i \in P)$, obtain $vals(R, P_i)$ which is value set without duplication of P_i and $values(R, P_i)$ which is all value set of P_i. Some resources may have same values for a property; we count both the values with duplication and without duplication.
3. Calculate Entropy H of all property $P_i(P_i \in P)$.

$$H(P_i) = -\sum_{a \in vals(R, P_i)} \frac{size\ of\ values(R, P_i|P_i^V=a)}{size\ of\ values(R, P_i)} \log_2 \frac{size\ of\ values(R, P_i|P_i^V=a)}{size\ of\ values(R, P_i)} \tag{2}$$

4. Calculate Conditional Entropy of property P_i ($P_i \in P$), and property P_j ($P_j \in P, P_i \neq P_j$) when P_j has the value b.

$$H(P_i|P_j^V = b) =$$
$$-\sum_{a \in vals(R|P_j^V=b, P_i)} \frac{size\ of\ values(R|P_j^V=b, P_i|P_i^V=a)}{size\ of\ values(R|P_j^V=b, P_i)} \log_2 \frac{size\ of\ values(R|P_j^V=b, P_i|P_i^V=a)}{size\ of\ values(R|P_j^V=b, P_i)} \tag{3}$$

This is about calculating Entropy of a property when we know another property has a certain value.

5. Calculate Conditional Entropy of property P_i ($P_i \in P$), and property P_j ($P_j \in P, P_i \neq P_j$).

$$H(P_i|P_j) = \Sigma_{b \in vals(R|P_j)} \frac{size\ of\ values(R|P_j^V = b,\ P_i)}{\Sigma_{c \in vals(R|P_j)}\ size\ of\ values(R|P_j^V = c,\ P_i)} \times H(P_i|P_j^V = b) \quad (4)$$

This is about calculating Entropy of a property when we know all values of another property.

6. Calculate Information Gain of property $P_i(P_i \in P)$, and property $P_j(P_j \in P, P_i \neq P_j)$.

$$IG(P_i, P_j) = H(P_i) - H(P_i|P_j) \quad (5)$$

We calculated the Entropy of P_i and the Conditional Entropy of P_i when we know the values of P_j, the difference of two is the amount of information of P_i we get as the result of knowing the value of P_j. A property may have many values and Information Gain of that property is high, so as a penalty of this we introduce the concept of Classification Information in Step 7.

7. Calculate Classification Information of property $P_i(P_i \in P)$, and property $P_j(P_j \in P, P_i \neq P_j)$.

$$CI(P_i, P_j) = \frac{vals(R|P_j)}{\Sigma_{c \in vals(R|P_j)}\ size\ of\ values(R|P_j^V = c,\ P_i)} \quad (6)$$

Classification Information represents the characteristic of properties. If a property has a lot of values, but they are all different, Classification Information is big, which will make the Information Gin smaller. Though a property has a lot of values and only small amount of them are duplicated, it is not an ideal property.

8. Calculate Gain Ratio of property $P_i(P_i \in P)$, and property $P_j(P_j \in P, P_i \neq P_j)$.

$$GR(P_i, P_j) = \frac{IG(P_i, P_j)}{CI(P_i, P_j)} \quad (7)$$

Take Classification Information into consideration, Gain Ratio is calculated. It is a better way than using Information Gain directly.

9. Calculate $TotalGainRatio$ P_j ($P_j \in P$) and P^*.

After Gain Ratio between a property and all the other properties are calculated, we sum them up and obtain $TotalGainRatio$.

10. Calculate $TotalGainRatio$ of all properties and ordering them.

Properties are ordered by the score of $TotalGainRatio$, property with high score of $TotalGainRatio$ can be a good classifier.

3.2   The Procedure of the Category Selection

In section 3.1, we introduced the methodology for selecting categories. In this section, we will introduce how to apply the methodology to our overall process. In Figure2, class *Film* has properties and *Actor* which is the range of a property of *Film* also has properties. We consider not only the direct properties of *Film* but also all properties that are not directly connected to *Film*, but can connect to *Film* within several steps. If class *Education* has properties, they are also considered.

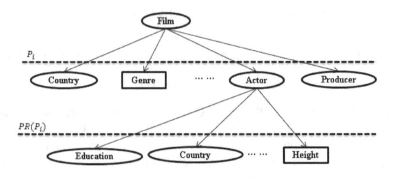

Fig. 2. The structure of a sample ontology

The procedure of selecting categories is as follows:

1. Select and order properties of resources in search results based on score of *TotalGainRatio*. Properties with high score of *TotalGainRatio* are top-ranked.
2. If the range of $P_i(P_i \in P)$ has no properties, property P_i is ranked to its original order.
3. If the range of $P_i(P_i \in P)$ has properties, ($PR(P_i)$ denotes that properties) replace $P_j\left(P_j \in PR(P_i)\right)$ with P_i and calculate *TotalGainRatio* of P_j.
4. Choose $P_j\left(P_j \in PR(P_i)\right)$ which has the highest score of *TotalGainRatio* and compare it with that of P_i.
5. If the *TotalGainRatio* of P_j is smaller than that of P_i, P_i is ranked to its original order, else P_j replaces the order of P_i.

By considering all the properties that can connect to resources in search results, we select and order best properties that can represent the characteristic of search results well.

3.3   Reordering Categories

Through previous process, we selected and ordered properties. We consider this ordering as Alternative 1 and in this section, we will introduce another methodology for ordering selected properties and we call this Alternative 2. Alternative 1 and 2 has same properties but properties have different order. When ordering properties, we considered similarity of classification results of properties. The algorithm is in Figure 3. Where FP is the set of properties that ordered and UFP is the set of properties that do not ordered. They have relationships such as FP ∪ UFP=P and FP∩UFP=∅.

The algorithm fixes first property and calculates similarity between classification results of first property and other properties. Second property is the property which records the smallest similarity with first property. Other properties are also ordered by calculating the sum of similarity between them and every property in FP. Figure 4 is the algorithm calculating similarity of classification results of two properties. Groups classified by property are represented by vector of 0 and 1. Vectors are multiplied and the result is the similarity.

```
1    FP = ∅
2    UFP = P
3    add firstProperty to FP
4    remove firstProperty from UFP
5    while UFP is not empty
6        for all P_j (P_j ∈ UFP)
7            similarity(P_j, FP) = 0
8                for all P_i (P_i ∈ FP)
9                    similarity(P_j, FP) += sim(P_j, P_i)
10           selectedProperty= the property that has Min(similarity(P_j, FP))
11               add selectedProperty to FP
12           remove selectedProperty from UFP
13   end
```

Fig. 3. Property Reordering

```
sim(P_j, P_i)
1    C(p_j) is the groups classified by P_j
2    C_n[ ] is the vector of C_n that represented by 0 and 1
2    totalScore = 0
3    for all C_m (C_m ∈ C(P_j))
4        score = Max Int
5        for all C_n (C_n ∈ C(P_i))
6            tempScore = C_m[ ] × C_n[ ]
7            if tempScore < socre
8                score = tempScore
9                temp = C_n
10           remove temp from C(P_i)
11           totalScore += socre
12       remove C_m from C(P_j)
13   return totalScore
```

Fig. 4. Similarity Calculation

4   Evaluation

In order to evaluate the proposed methods, Film ontology was manually constructed. All data was collected from the Web site of DAUM movies [9]. One thousand movie ID's were randomly generated and searched. Property information, such as directors, actors and so on, were also collected. The ontology thereby contains one thousand movie instances and around nine thousand person instances. Class *Film* has fifteen properties, the classes of type *Person*, such as *Director*, *Actor*, *Producer* and so forth has five properties, and class *Country* has two properties.

Three evaluative measures were used in the experimentations: Recall, Precision, and mean absolute error (MAE). Recall and Precision are widely used in the fields of pattern recognition and information retrieval. Recall is the fraction of relevant instances that are relevant [10]. They are conceptually different in that Recall is counted out of all relevant instances, some of which may not be retrieved, and Precision is the ratio out of all retrieved instances. These two measures were applied, though they come from information retrieval, because of no suggested evaluation measures in literature of category generation. Recall and Precision are defined as follows:

$$recall = \frac{the\ number\ of\ relevant\ catetoreis\ the\ system\ generates}{the\ number\ of\ all\ relevant\ categories} \tag{8}$$

$$precison = \frac{the\ number\ of\ relevant\ catetoreis\ the\ system\ generates}{the\ number\ of\ all\ categories\ the\ system\ generates} \tag{9}$$

Mean absolute error (MAE) is a quantity used to measure how close forecasted or predicted data are to the eventual outcomes [11]. It is defined as follows:

$$MAE = \frac{1}{n}\sum_{i=1}^{n}|SR(P_i) - R(P_i)| \tag{10}$$

where n denotes the number of properties, $SR(P_i)$ is the desired order of the property P_i and $R(P_i)$ the order of the property P_i in the categories generated by the system. Since class *Film* has fifteen properties, the system generates fifteen categories. The values of Recall, Precision, and MAE vary depending on the number of categories to select. As for standard measures, the first one to fifteen properties were selected, the values of Recall, Precision, and MAE were calculated in each of fifteen cases, and the values were finally averaged over the cases. This is a reasonable way to make a standard measure because the suggested measures increase in their values, as the number of properties increases, and are not possible to compare with different numbers of categories are obtained.

Search results were generated by an ontology-based semantic search system. The search result forms structured data consisting of ontology classes with their properties. Table 2 shows two different category lists obtained by two different methods described before under a query "Movie". The list by Method 1 was obtained without considering similarity of properties while that of Method 2 considered it. A survey was conducted to have referential categories according to users' preference. Eleven experts on movie were invited and participated in the survey. The list of fifteen properties, selected by system, with the information of items classified by each property, was given. The orders of properties were hidden to those experts so that they could not recognize the system-generated orderings. They were asked to arrange the properties according to their preference. From the collected data, the average ordering value was calculated for each property in survey and the final ordering of the properties was determined by the comparison of their average ordering values. Table 3 shows the category preference ordering.

The category orderings from the two methods were compared in the reference of the user preference ordering. Values of Recall, Precision, and MAE for Method 1 and Method 2 were calculated for the cases of 1 to 15 properties, as mentioned before. For the case of the calculation of Recall and Precision, a category was counted as a relevant one when in the preference ordering. When k properties were considered in evaluation, the first k properties for the two methods were compared with the first k of the preference ordering. If a property in the list of either method existed in the list of the preference ordering, it was taken as a relevant category. When MAE was considered, the categories in the preference ordering worked as the desired order of the property. The average MAE was obtained by dividing MAE with the number of properties. After the calculation of the measures for all cases, the averages over these 15 cases were also calculated. The values of Recall, Precision, and MAE for Method 1 and 2 are in Table 4. Recall and Precision vary according to the number of properties. Their averages are 0.56 in Method 1. The average of the average MAE is 4.7 in Method 1, which shows that if we choose more than 2 properties, it is hard to anticipate the order of properties. The average of Recall and Precision for Method 2 is about 20% higher than those for Method 1. The values of Recall and Precision for Method 2 are 0.67. This means that more than half of all properties were included in the list of the user preference ordering. The average of MAE for Method 2 is smaller than that of Method 1. That implies that the ordering by Method 2 is the closer to the user preference ordering. From the data, Method 2 is recognized to be better than Method 1. The dynamic generation of categories is thereby more efficient than using fixed categories since fixed categories cannot adapt themselves to represent user preference.

Table 2. Categories of query "Movie" by Method 1 and Method 2

Order	Method 1	Method 2
1	film_genre	film_genre
2	has_continent	film_editor_country
3	film_distributioncompany	film_musicdirector_country
4	film_editor_country	film_writer_country
5	film_musicdirector_country	film_storywriter_country
6	film_cinematographer_country	film_director_country
7	film_artdirector_country	film_artdirector_country
8	film_costumedesigner_country	film_costumedesigner_country
9	film_actor_country	initial_release_date
10	film_director_country	film_producer_country
11	film_productioncompany	film_distributioncompany
12	film_writer_country	film_productioncompany
13	initial_release_date	has_continent
14	film_producer_country	film_actor_country
15	film_storywriter_country	film_cinematographer_country

Table 3. Orders of property of survey

Order	Properties	No.1	No.2	No.3	No.4	No.5	No.6	No.7	No.8	No.9	No.10	No.11	Average
1	film_genre	2	2	2	1	1	1	1	1	1	1	1	1.272727
2	film_director_country	6	6	6	4	2	6	3	7	5	7	5	5.181818
3	film_actor_country	8	10	1	2	5	10	10	3	4	2	9	5.818182
4	film_writer_country	7	7	1	13	3	8	8	2	8	5	3	5.909091
5	has_continent	5	12	14	8	6	2	2	6	15	3	4	7
6	film_storywriter_country	9	9	12	7	4	7	9	10	7	8	2	7.636364
7	film_producer_country	10	4	5	10	7	9	7	5	12	6	11	7.818182
8	film_productioncompany	3	11	7	5	14	3	11	8	3	10	13	8
9	film_editor_country	11	3	9	9	9	9	6	4	13	13	6	8.363636
10	initial_release_date	1	15	3	3	15	5	15	14	2	4	15	8.363636
11	film_distributioncompany	4	13	8	6	11	4	12	9	14	12	14	9.727273
12	film_cinematographer_country	13	8	4	12	12	9	5	15	6	14	12	10
13	film_costumedesigner_country	15	5	11	11	13	9	4	11	11	15	10	10.45455
14	film_artdirector_country	14	1	13	14	10	9	14	13	10	11	8	10.63636
15	film_musicdirector_country	12	14	10	15	8	9	13	12	9	9	7	10.72727

Table 4. Measures for categories by Method 1 and Method 2

No. of Properties	Method 1				Method 2			
	Recall	Precision	MAE	Avg. MAE	Recall	Precision	MAE	Avg. MAE
1	1.00	1.00	0	0.00	1.00	1.00	0	0.00
2	0.50	0.50	3	1.50	0.50	0.50	4	2.00
3	0.33	0.33	11	3.67	0.33	0.33	15	5.00
4	0.25	0.25	16	4.00	0.50	0.50	15	3.75
5	0.40	0.40	26	5.20	0.40	0.40	23	4.60
6	0.33	0.33	32	5.33	0.67	0.67	24	4.00
7	0.29	0.29	39	5.57	0.57	0.57	27	3.86
8	0.25	0.25	44	5.50	0.50	0.50	31	3.88
9	0.44	0.44	50	5.56	0.56	0.56	38	4.22
10	0.50	0.50	58	5.80	0.70	0.70	39	3.90
11	0.64	0.64	61	5.54	0.73	0.73	39	3.55
12	0.75	0.75	69	5.75	0.75	0.75	42	3.50
13	0.85	0.85	72	5.54	0.85	0.85	47	3.62
14	0.93	0.93	79	5.64	0.93	0.93	54	3.86
15	1.00	1.00	88	5.87	1.00	1.00	66	4.40
Average	0.56	0.56	-	4.70	0.67	0.67	-	3.61

5   Conclusion

This paper proposed a method for dynamic generation of categories refining ontology search results. This approach of the dynamic category generation has advantages in that it provides different categories for users according to their search contexts. If users query with different key words, they might assume different contexts and natu-rally categories provided with the results should be adapted to their situations. The proposed approach evaluates search results with a measure and shows different cate-gories according to the results. This is significant because this categorization leads to different paths of search experience that suit for users' search contexts, and users eventually reach the items with ease that they want to search for. It can save much time for users and improve the efficiency in search systems. While this approach is targeting the ontology-based search, it can also be extended to any search systems based on structured data. Together with the measure, which analyses information of search results to generate categories, this paper extended the study to an algorithm to improve the categories obtained from the measure. This algorithm is based on the measure of similarity. The experimentation showed that the outcome from the algorithm is more efficient than that without applying the algorithm.

This research has some limitations; First of all, it is that the quality of ontology affects the efficiency of our method. In this paper, we assume that most instances in ontology are well inputted with proper values. Since every ontology is not well constructed, it is our future work to ameliorate our method to be able to deal with all ontologies. Secondly, it is difficult to find studies in the literature related to the issues we discussed. It is even harder to find evaluative methods for the proposed approach. That is why the methodology in the field of information retrieval was used. It is very desirable to make an appropriate evaluative methodology which may be a future work. Furthermore, many groups classified by categories tend to contain few or even a single instance since instances that share values in properties are sporadic. It caused a larger number of groups that surveyees ought to review in the survey in order to arrange categories. The small number of surveyees with that fact made very close ties in survey scores of some properties. This might be alleviated by considering an ontology with more instances in experimentation. Since this research used one thousand instances, close to the limit in the current PC environment, more instances may require a large scale computing environment.

Acknowledgement. This work (research) is financially supported by the Ministry of Knowledge Economy(MKE) and Korea Institute for Advancement in Technology (KIAT) through the Workforce Development Program in Strategic Technology.

References

1. Hilbert, M., Lopez, P.: The world's technological capacity to store, communicate, and compute information. Science 332(60), 60–65 (2011)
2. Gruber, T.R.: A translation approach to portable ontology specifications. Knowledge Acquisition 5(2), 199–200 (1993)
3. Shannon, C.E.: A mathematical theory of communication. ACM SIGMOBILE Mobile Computing and Communications Review 5(1), 3–55 (2001)
4. Gray, R.M.: Entropy and Information Theory. Springer (2011)
5. Kullback, S., Leibler, R.A.: On information and sufficiency. Annals of Mathematical Statistics 22(1), 76–86 (1951)
6. Park, S., Park, S.-H., Lee, J.-H., Lee, J.-S.: E-mail classification agent using category generation and dynamic category hierarchy. In: Kim, T.G. (ed.) AIS 2004. LNCS (LNAI), vol. 3397, pp. 207–214. Springer, Heidelberg (2005)

 7. Khoshnevisan, C.: Dynamic selection and ordering of search categories based on relevancy information, Technical Report US 7698261B1, U.S. Patent (2010)
 8. Duda, R.O., Hart, P.E., Stork, D.G.: Pattern Classification. John Wiley & Sons (2001)
 9. http://movie.daum.net
10. Baeza-Yates, R., Ribeiro-Neto, B.: Modern Information Retrieval. Addison-Wesley (1999)
11. Hyndman, R.J., Koehler, A.B.: Another look at measures of forecast accuracy. International Journal of Forecasting 22(4), 679–688 (2006)

Keyword-Driven Resource Disambiguation over RDF Knowledge Bases

Saeedeh Shekarpour, Axel-Cyrille Ngonga Ngomo, and Sören Auer

Department of Computer Science, University of Leipzig, Johannisgasse 26,
04103 Leipzig, Germany
{lastname}@informatik.uni-leipzig.de

Abstract. Keyword search is the most popular way to access information. In this paper we introduce a novel approach for determining the
correct resources for user-supplied queries based on a hidden Markov
model. In our approach the user-supplied query is modeled as the observed data and the background knowledge is used for parameter estimation. We leverage the semantic relationships between resources for
computing the parameter estimations. In this approach, query segmentation and resource disambiguation are mutually tightly interwoven. First,
an initial set of potential segments is obtained leveraging the underlying
knowledge base; then, the final correct set of segments is determined after
the most likely resource mapping was computed. While linguistic analysis (e.g. named entity, multi-word unit recognition and POS-tagging)
fail in the case of keyword-based queries, we will show that our statistical approach is robust with regard to query expression variance. Our
experimental results reveal very promising results.

1 Introduction

The Data Web currently amounts to more than 31 billion triples[1] and contains
a wealth of information on a large number of different domains. Yet, accessing
this wealth of structured data *remains* a key challenge for lay users. The same
problem emerged in the last decade when users faced the huge amount of information available of the Web. Keyword search has been employed by popular Web
search engines to provide access to this information in a user-friendly, low-barrier
manner. However, keyword search in structured data raises two main difficulties:
First, the *right segments of data items* that occur in the keyword queries have
to be identified. For example, the query *'Who produced films starring Natalie
Portman'* can be segmented to (*'produce'*, *'film'*, *'star'*, *'Natalie Portman'*) or
(*'produce'*, *'film star'*, *'Natalie'*, *'Portman'*). Note that the first segmentation
is more likely to lead to a query that contain the results intended for by the
user. Second, these segments have to be disambiguated and mapped to the right
resources. Note that the resource ambiguity problem is of increasing importance

---

[1] See http://www4.wiwiss.fu-berlin.de/lodcloud/state/ (May 23th, 2012).

H. Takeda et al. (Eds.): JIST 2012, LNCS 7774, pp. 159–174, 2013.

as the size of knowledge bases on the Linked Data Web grows steadily. Considering the previous example[2], the segment *'film'* is ambiguous because it may refer to the class dbo:Film (the class of all movies in DBpedia) or to the properties dbo:film or dbp:film (which relates festivals and the films shown during these festivals). In this paper, we present an automatic query segmentation and resource disambiguation approach leveraging background knowledge. Note that we do not rely on training data for the parameter estimation. Instead, we leverage the semantic relationships between data items for this purpose. While linguistic methods like named entity, multi-word unit recognition and POS-tagging fail in the case of an incomplete sentences (e.g. for keyword-based queries), we will show that our statistical approach is robust with regard to query expression variance. This article is organized as follows: We review related work in Section 2. In Section 3 we present formal definitions laying the foundation for our work. In the section 4 our approach is discussed in detail. For a comparison with natural language processing (NLP) approaches section 5 introduces an NLP approach for segmenting query. Section 6 presents experimental results. In the last section, we close with a discussion and an outlook on potential future work.

2   Related Work

Our approach is mainly related to two areas of research: text and query segmentation and entity disambiguation. Text segmentation has been studied extensively in the natural language processing (NLP) literature, and includes tasks such as noun phrase chunking where the task is to recognize the chunks that consist of noun phrases (see e.g., [16]). Yet, such approaches cannot be applied to query segmentation since queries are short and composed of keywords. Consequently, NLP techniques for chunking such as part-of-speech tagging [4] or name entity recognition [7,5] cannot achieve high performance when applied to query segmentation. Segmentation methods for document-based Information Retrieval can be categorized into statistical and non-statistical algorithms. As an example of none statistical methods, [15] addresses the segmentation problem as well as spelling correction. Each keyword in a given query is first expanded to a set of similar tokens in the database. Then, a dynamic programming algorithm is used to search for the segmentation based on a scoring function. The statistical methods fall into two groups, namely supervised and unsupervised methods. For example, the work presented in [19] proposes an unsupervised approach to query segmentation in Web search. Yet this technique can not be easily applied to structured data. Supervised statistical approaches are used more commonly. For example, [25] presents a principled statistical model based on Conditional Random Fields (CRF) whose parameters are learned from query logs. For detecting named entities, [9] uses query log data and Latent Dirichlet Allocation. In addition to query logs, various external resources such as Webpages, search

---

[2] The underlying knowledge base and schema used throughout the paper for examples and evaluation is DBpedia 3.7 dataset and ontology.

```
     Data: q: n-tuple of keywords, knowledge base
     Result: SegmentSet: Set of segments
 1   SegmentSet=new list of segments;
 2   start=1;
 3   while start <= n do
 4   │   i = start;
 5   │   while S_(start,i) is valid do
 6   │   │   SegmentSet.add(S_(start,i));
 7   │   │   i++;
 8   │   end
 9   │   start++;
10   end
```

Algorithm 1. Naive algorithm for determining all valid segments taking the order of keywords into account

result snippets and Wikipedia titles have been used [17,20,3]. Current segmentation algorithms are not applicable to our segmentation problem for several reasons. First, because they mostly are not intended for search on structured data, it is not guaranteed that the segments they retrieve are actually part of the underlying knowledge base. Another problem with these segmentation algorithms is that they ignore the semantic relationships between segments of a segmentation. Thus, they are likely to return sub-optimal segmentations.

An important challenge in Web search as well as in Linked Data Search is entity disambiguation. Keyword queries are usually short and inherently lead to significant keyword ambiguity as one query word may represent different information needs for different users [21]. There are different ways for tackling this challenge; firstly, query clustering [6,23,2] applies unsupervised machine learning to cluster similar queries. The basic insight here is that it has been observed that users with similar information needs click on a similar set of pages, even though the queries they pose may vary. Other approaches apply query disambiguation, which tries to find the most appropriate sense of each keyword. To achieve this goal, one way is involving the user in selecting the correct sense [11,10,24]. Another technique for disambiguation is personalized search by using a history of the user activities to tailor the best choice for disambiguation [1,18,26]. Still, the most common approach is using context for disambiguation [14,8,13]. Albeit context has been defined vaguely (with various definitions), herein we define context as information surrounding the given query which can be employed for augmenting search results. In this work, resource disambiguation is based on a different type of context: we employ the structure of the knowledge at hand as well as semantic relations between the candidate resources mapped to the possible segmentations of the input query.

3   Formal Specification

RDF data is modeled as a directed, labeled graph $G = (V, E)$ where V is a set of nodes i.e. the union of entities and property values, and E is a set of directed edges i.e. the union of object properties and data value properties.

The user-supplied query can be either a complete or incomplete sentence. However, after removing the stop words, typically set of keywords remains. The order in which keywords appear in the original query is partially significant. Our approach can map adjacent keywords to a joint resource. However, once a mapping from keywords to resources is established the order of the resources does not affect the SPARQL query construction anymore. This is a reasonable assumption, since users will write strongly related keywords together, while the order of only loosely related keywords or keyword segments may vary. The input query is formally defined as an n-tuple of keyword, i.e. $Q = (k_1, k_2, ..., k_n)$. We aim to transform the input keywords into a suitable set of entity identifiers, i.e. resources $R = \{r_1, r_2...r_m\}$. In order to accomplish this task the input keywords have to be grouped together as segments and for each segment a suitable resource should be determined.

Definition 1 (Segment and Segmentation). *For a given query Q, a segment $S_{(i,j)}$ is a sequence of keywords from start position i to end position j which is denoted as $S_{(i,j)} = (k_i, k_{i+1}, ..., k_j)$. A query segmentation is an m-tuple of segments $SG_q = (S_{(0,i)}, S_{(i+1,j)}, ..., S_{(m,n)})$ where the segments do not overlap with each other and arranged in a continuous order, i.e. for two continuous segments $S_x, S_{x+1} : Start(S_{x+1}) = End(S_x) + 1$. The concatenation of segments belonging to a segmentation forms the corresponding input query Q.*

Definition 2 (Resource Disambiguation). *Let the segmentation $SG' = (S^1_{(0,i)}, S^2_{(i+1,j)}, ..., S^x_{(m,n)})$ be a suitable segmentation for the given query Q. Each segment is mapped to multiple candidate resources from the underlying knowledge base, i.e. $S^i \rightarrow R^i = \{r_1, r_2...r_h\}$. The aim of disambiguation is to choose an x-tuple of resources from the Cartesian product of sets of candidate resources $(r_1, r_2, ..., r_x) \in \{R^1 \times R^2 \times ...R^x\}$ for which each r_i has two important properties. First, it is among the highest ranked candidates for the corresponding segment with respect to the similarity as well as popularity and second it shares a semantic relationship with other resources in the x-tuple.*

When considering the order of keywords, the number of segmentations for a query Q consisting of n keywords is $2^{(n-1)}$. However, not all these segmentations contain valid segments. A valid segment is a segment for which at least one matching resource can be found in the underlying knowledge base. Thus, the number of segmentations is reduced by excluding those containing invalid segments. Algorithm 1 is an extension of the greedy approach presented in [25]. This naive approach finds all valid segments when considering the order of keywords. It starts with the first keyword in the given query as first segment, then it includes the next keyword into the current segment as a new segment and checks whether adding the new keyword would make the new segment no longer valid. We repeat this process until we reach the end of the query. As a running example, lets assume the input query is *'Give me all video games published by Mean Hamster Software'*. Table 1 shows the set of valid segments based on naive algorithm along with some samples of the candidate resources. Note that the suitable segmentation is *('video games', 'published', 'Mean Hamster Software')*.

Table 1. Generated segments and samples of candidate resources for a given query

Segments	Samples of Candidate Resources
video	1. `dbp:video`
video game	1. `dbo:VideoGame`
game	1. `dbo:Game` 2. `dbo:games` 3. `dbp:game` 4. `dbr:Game` 5. `dbr:Game_On`
publish	1. `dbo:publisher` 2. `dbp:publish` 3. `dbr:Publishing`
mean	1. `dbo:meaning` 2. `dbp:meaning` 3. `dbr:Mean` 4. `dbo:dean`
mean hamster	1. `dbr:Mean_Hamster_Software`
mean hamster software	1. `dbr:Mean_Hamster_Software`
hamster	1. `dbr:Hamster`
software	1. `dbo:Software` 2. `dbp:software`

Resource Disambiguation Using a Ranked List of Cartesian Product Tuples: A naive approach for finding the correct $x - tuple$ of resources is using a ranked list of tuples from the Cartesian product of sets of candidate resources $\{R^1 \times R^2 \times ...R^n\}$. The n-tuples from the Cartesian product are simply sorted based on the aggregated relevance score (e.g. similarity and popularity) of all contained resources.

4 Query Segmentation and Resource Disambiguation Using Hidden Markov Models

In this section we describe how hidden Markov models are used for query segmentation and resource disambiguation. First we introduce the concept of hidden Markov models and then we detail how we define the parameters of a hidden Markov model for solving the query segmentation and entity disambiguation problem.

4.1 Hidden Markov Models

The Markov model is a stochastic model containing a set of states. The process of moving from one state to another state generates a sequence of states. The probability of entering each state only depends on the previous state. This memoryless property of the model is called *Markov property*. Many real-world processes can be modeled by Markov models. A hidden Markov model is an extension of the Markov model, which allows the observation symbols to be emitted from each state with a finite probability. The main difference is that by looking at the observation sequence we cannot say exactly what state sequence has produced these observations; thus, the state sequence is *hidden*. However, the probability of producing the sequence by the model can be calculated as well as which state sequence was most likely to have produced the observations.

A hidden Markov model (HMM) is a quintuple $\lambda = (X, Y, A, B, \pi)$ where:

- X is a finite set of states, Y denotes the set of observed symbols;
- $A : X \times X \to \mathbb{R}$ is the transition matrix that each entry $a_{ij} = Pr(S_j|S_i)$ shows the transition probability from state i to state j;

- $B : X \times Y \to \mathbb{R}$ represents the emission matrix, in which each entry $b_{ih} = Pr(h|S_i)$ is associated with the probability of emitting the symbol h from state i;
- π denoting the initial probability of states $\pi_i = Pr(S_i)$.

4.2   State Space and Observation Space

State Space. A state represents a knowledge base entity. Each entity has an associated *rdfs:label* which we use to label the states. The actual number of states X is potentially high because it contains theoretically all RDF resources, i.e. $X = V \cup E$. However, in practice we limit the state space by excluding irrelevant states. A relevant state is defined as a state for which a valid segment can be observed. In other words, a valid segment is observed in an state if the probability of emitting that segment is higher than a certain threshold θ. The probability of emitting a segment from a state is computed based on a similarity scoring which we describe in the section 4.3. Therefore, the state space of the model is pruned and contains just a subset of resources of the knowledge base, i.e. $X \subset V \cup E$. In addition to these candidate states, we add an **unknown entity state** to the set of states. The *unknown entity* (UE) state comprises all entities, which are not available (anymore) in the pruned state space. The *observation space* is the set of all valid segments found in the input user query (using e.g. the Algorithm 1). It is formally is defined as $O = \{o | o \text{ is a valid segment}\}$.

4.3   Emission Probability

Both the labels of states and the segments contain sets of words. For computing the emission probability of the state i and the emitted segment h, we compare the similarity of the label of state i with the segment h in two levels, namely string-similarity level and set-similarity level: (1) The *set-similarity level* measures the difference between the label and the segment in terms of the number of words using the *Jaccard similarity*. (2) The *string-similarity level* measures the string similarity of each word in the segment with the most similar word in the label using the *Levenshtein distance*. Our similarity scoring method is now a combination of these two metrics. Consider the segment $h = (k_i, k_{i+1}, ..., k_j)$ and the words from the label l divided into a set of keywords M and stopwords N, i.e. $l = M \cup N$. The total similarity score between keywords of a segment and a label is then computed as follows:

$$b_{ih} = Pr(h|S_i) = \frac{\sum\limits_{t=i}^{j} argmax_{\forall m_i \in M}(\sigma(m_i, k_t))}{|M \cup h| + 0.1 * |N|}$$

This formula is essentially an extension of the *Jaccard similarity coefficient*. The difference is that in the numerator, instead of using the cardinality of intersections the sum of the string-similarity score of the intersections is computed. As in the Jaccard similarity, the denominator comprises the cardinality of the union of two sets (keywords and stopwords). The difference is that the number of stopwords have been down-weighted by the factor 0.1 to reduce their influence (since they do not convey much meaningful information).

4.4   Hub and Authority of States

Hyperlink-Induced Topic Search (HITS) is a link analysis algorithm for ranking Web pages [12]. Authority and hub values are defined in terms of one another and computed in a series of iterations. In each iteration, hub and authority values are normalized. This normalization process causes these values to converge eventually. Since RDF data is graph-structured data and entities are linked together, we employed a weighted version of the HITS algorithm in order to assign different popularity values to the states in the state space. For each state we assign a hub value and an authority value. A good hub state is one that points to many good authority states and a good authority state is one that is pointed to from many good hub states. Before discussing the HITS computations, we define the edges between the states in the HMM. For each two states i and j in the state space, we add an edge if there is a path in the knowledge base between the two corresponding resources of maximum length k. Note, that we also take property resources into account when computing the path length. The path length between resources in the knowledge base is assigned as weight to the edge between corresponding states. We use a weighted version of the HITS algorithm to take the distance between states into account. The authority of a state is computed as: For all $S_i \in S$ which point to S_j : $auth_{S_j} = \sum_{\forall i} w_{i,j} * hub_{S_i}$ And the hub value of a state is computed as: For all $S_i \in S$ which are pointed to by S_j : $hub_{S_j} = \sum_{\forall i} w_{i,j} * auth_{S_i}$ The weight $w_{i,j}$ is defined as $w_{i,j} = k - pathLength(i,j)$, where $pathLength(i,j)$ is the length of the path between i and j. These definitions of hub and authority for states are the foundation for computing the transition probability in the underlying hidden Markov model.

4.5   Transition Probability

As mentioned in the previous section, each edge between two states shows the shortest path between them with the length less or equal to k-hop. The edges are weighted by the length of the path. Transition probability shows the probability of going from state i to state j. For computing the transition probability, we take into account the connectivity of the whole of space state as well as the weight of the edge between two states. The transition probability values decrease with the distance of the states, e.g. transitions between entities in the same triple have higher probability than transitions between entities in triples connected through extra intermediate entities. In addition to the edges recognized as the shortest path between entities, there is an edge between each state and the *Unknown Entities* state. The transition probability of state j following state i denoted as $a_{ij} = Pr(S_j|S_i)$. For each state i the condition $\sum_{\forall S_j} Pr(S_j|S_i) = 1$ should be held. The transition probability from the state i to *Unknown Entity* (UE)

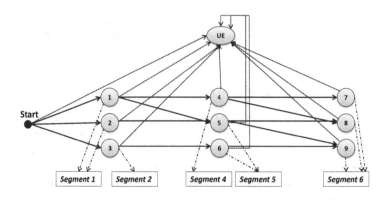

Fig. 1. Trellis representation of Hidden Markov Model

state is defined as: $a_{iUE} = Pr(UE|S_i) = 1 - hub_{S_i}$ And means a good hub has less probability to go to UE state. Thereafter, the transition probability from the state i to state j is computed as: $a_{ij} = Pr(S_j|S_i) = \dfrac{auth_{S_j}}{\sum\limits_{\forall a_{ik} > 0} auth_{S_k}} * hub_{S_i}$.

Here, the edges with the low distance value and higher authority values are more probable to be met.

4.6 Initial Probability

The initial probability π_{S_i} is the probability that the model assigns to the initial state i in the beginning. The initial probabilities fulfill the condition $\sum\limits_{\forall S_i} \pi_{S_i} = 1$. We denote states for which the first keyword is observable by $InitialStates$. The initial states are defined as follows:

$$\pi_{S_i} = \frac{auth_{S_i} + hub_{S_i}}{\sum\limits_{\forall S_j \in InitialStates} (auth_{S_j} + hub_{S_j})}$$

In fact, π_{S_i} of an initial state depends on both hub and authority values. Figure 1 illustrates an instantiated hidden markov model. The set of hidden states are represented by circles. The state UE refers to the absent resources in the model and other hidden states are relevant resources. Each segment box represents a possible observation. The arrows show a transition from one state to another state and the dashed arrows shows an emitted observation associated with a specific state.

4.7 Viterbi Algorithm for the K-Best Set of Hidden States

The optimal path through the Markov model for a given sequence (i.e. input query keywords) reveals disambiguated resources forming a correct segmentation. The *Viterbi algorithm* or *Viterbi path* is a dynamic programming approach for finding the optimal path through the markov model for a given sequence. It discovers the most likely sequence of underlying hidden states that might have generated a given sequence of observations. This discovered path has the maximum joint emission and transition probability of involved states. The sub paths of this most likely path also have the maximum probability for the respective sub sequence of observations. The naive version of this algorithm just keeps track of the most likely path. We extended this algorithm using a tree data structure to store all possible paths generating the observed query keywords. Therefore, in our implementation we provide a ranked list of all paths generating the observation sequence with the corresponding probability. After running the Viterbi algorithm for our running example, the disambiguated resources are: { *dbo:VideoGame, dbo:publisher, dbr:Mean-Hamster-Software* } and consequently the reduced set of valid segments is: { *VideoGam, publisher, Mean-Hamster-Software* } .

5 Query Segmentation Using Natural Language Processing

Natural language processing (NLP) techniques are commonly used for text segmentation. Here, we use a combination of named entity and multi-word unit recognition services as well as POS-tagging for segmenting the input-query. In the following, we discuss this approach in more detail.

Detection of Segments: Formally, the detection of segments aims to transform the set of keywords $K = \{k_1, .., k_n\}$ into a set of segments $\mathcal{T} = \{t_1, ..., t_m\}$ where each k_i is a substring of exactly one $t_j \in \mathcal{T}$. Several approaches have already been developed for this purpose, each with its own drawbacks: Semantic lookup services (e.g., *OpenCalais*[3] and *Yahoo! SeoBook*[4] as used in the current implementation) allow to extract *named entities* (NEs) and *multi-word units* (MWUs) from query strings. While these approaches work well for long queries such as *"Films directed by Garry Marshall starring Julia Roberts"*, they fail to discover noun phrases such as "highest place" in the query *"Highest place of Karakoram"*. We remedy this drawback by combining lookup services and a simple noun phrase detector based on POS tags. This detector first applies a POS tagger to the query. Then, it returns all sequences of keywords whose POS tags abide by the following right-linear grammar:

1. $S \rightarrow adj\ A$ 2. $S \rightarrow nn\ B$ 3. $A \rightarrow B$
4. $B \rightarrow nn$ 5. $B \rightarrow nn\ B$

where S is the start symbol, A and B are non-terminal symbols and nn (noun) as well as adj (adj) are terminal symbols. The compilation of segments is carried as follows: We send the input K to the NE and MWU detection services as well as to the noun phrase detector. Let $\mathcal{N}$ be the set of NEs, $\mathcal{M}$ the set of MWUs

---

[3] http://viewer.opencalais.com/
[4] http://tools.seobook.com/yahoo-keywords/

and $\mathcal{P}$ the set of noun phrases returned by the system. These three sets are merged to a set of labels $\mathcal{L} = (\mathcal{N} \oplus \mathcal{M}) \oplus \mathcal{P}$, where $\oplus$ is defined as follows:

$$A \oplus B = A \cup B \backslash \{b \in B | \exists a \in A \; overlap(a, b)\} \tag{1}$$

where $overlap(a, b)$ is true if the strings a and b overlap. The operation $\oplus$ adds the longest elements of B to A that do not overlap with A. Note that this operation is not symmetrical and prefers elements of the set A over those of the set B. For example, "river which Brooklyn Bridge crosses" leads to $\mathcal{N}$ = {"Brooklyn Bridge"}, M = {"Brooklyn" , "Brooklyn Bridge"} and $\mathcal{P}$ = {"Brooklyn Bridge"}. Thus, $\mathcal{L} = (\mathcal{N} \oplus \mathcal{M}) \oplus \mathcal{P}$ = {"Brooklyn Bridge"}. The final set of segments $\mathcal{T}$ is computed by retrieving the set of all single keywords that were not covered by the approaches above and that do not occur in a list of stopwords. Thus, for the query above, $\mathcal{T}$ = {"Brooklyn Bridge", "river", "cross"}.

6   Evaluation

The goal of our experiments was to measure the accuracy of resource disambiguation approaches for generating adequate SPARQL queries. Thus, the main question behind our evaluation was as follows: Given a keyword-based query(KQ) or a natural-language query (NL) and the equivalent SPARQL query, how well do the resources computed by our approaches resemble the gold standard. It is important to point out that a single erroneous segment or resource can lead to the generation of a wrong SPARQL query. Thus, our criterion for measuring the correctness of segmentations and disambiguations was that *all of the recognized segments* as well as *all of the detected resources* had to match the gold standard.

Experimental Setup. So far, no benchmark for query segmentation and resource disambiguation has been proposed in literature. Thus, we created such a benchmark from the DBpedia fragment of the question answering benchmark *QALD-2*[5]. The QALD-2 benchmark data consists of 100 training and 100 test questions in natural-language that are transformed into SPARQL queries. In addition, it contains a manually created keyword-based representation of each of the natural-language questions. The benchmark assumed the generic query generation steps for question answering: First, the correct segments have to be computed and mapped to the correct resources. Then a correct SPARQL query has to be inferred by joining the different resources with supplementary resources or literals. As we are solely concerned with the first step in this paper, we selected 50 queries from the QALD-2 benchmark (25 from the test and 25 from the training data sets) that were such that each of the known segments in the benchmark could be mapped to exactly one resource in the SPARQL query and vice-versa. Therewith, we could derive the correct segment to resource mapping directly from the benchmark[6]. Queries that we discarded include *"Give me all soccer*

---

[5] http://www.sc.cit-ec.uni-bielefeld.de/qald-2

[6] The queries and result of the evaluation and source code is available for download at http://aksw.org/Projects/lodquery

clubs in Spain", which corresponds to a SPARQL query containing the resources
{dbo:ground, dbo:SoccerClub, dbr:Spain }. The reason for discarding this
particular query was that the resource dbo:ground did not have any match in
the list of keywords. Note that we also discarded queries requiring schema information beyond DBpedia schema. Furthermore, 6 queries out of the 25 queries
from the training data set [7] and 10 queries out of 25 queries from the test data
set [8] required a query expansion to map the keywords to resources. For instance, the keyword *"wife"* should be matched with *"spouse"* or *"daughter"* to
"child". Given that the approaches at hand generate and score several possible
segmentations (resp. resource disambiguation), we opted for measuring the *mean
reciprocal rank MRR* [22] for both the query segmentation and the resource disambiguation tasks. For each query $q_i \in Q$ in the benchmark, we compare the
rank r_i assigned by different algorithms to the correct segmentation and to the
resource disambiguation: $MRR(\mathcal{A}) = \frac{1}{|Q|} \sum_{q_i} \frac{1}{r_i}$. Note that if the correct seg-
mentation (resp. resource disambiguation) was not found, the reciprocal rank is
assigned the value 0.

Results. We evaluated our hidden Markov model for resource disambiguation
by combining it with the naive (Naive & HMM) and the greedy segmentation
(Greedy & HMM) approaches for segmentation. We use the natural language
processing (NLP) approach as a baseline in the segmentation stage. For the resource disambiguation stage, we combine ranked Cartesian product (RCP) with
the natural language processing (NLP & RCP) and manually injected the correct
segmentation (RCP) as the baseline. Note that we refrained from using any query
expansion method. The segmentation results are shown in Figure 2. The *MRR*
are computed once with the queries that required expansion and once without.
Figure 2(a), including queries requiring expansion, are slightly in favor of NLP,
which achieves on overage a 4.25% higher MRR than Naive+HMM and a 24.25%
higher MRR than Greedy+HMM. In particular, NLP achieves optimal scores
when presented with the natural-language representation of the queries from the
"train" data set. Naive+HMM clearly outperforms Greedy+HMM in all settings.
The main reason for NLP outperforming Naive+HMM with respect to the segmentation lies in the fact that Naive+HMM and Greedy+HMM are dependent
on matching segments from the query to resources in the knowledge base (i.e. segmentation and resource disambiguation are interwoven). Thus, when no resource
is found for a segment (esp. for queries requiring expansion) the HMM prefers an
erroneous segmentation, while NLP works independent from the disambiguation
phase. However, as it can be observed NLP depends on the query expression.
Figure 2(b) more clearly highlights the accuracy of different approaches. Here,
the *MRR* without queries requiring expansion is shown. Naive+HMM perfectly
segments both natural language and keyword-based queries. The superiority of
intertwining segmentation and disambiguation in Naive+HMM is clearly shown
by our disambiguation results in the second stage in Figure 3. In this stage,

---

[7] Query IDs: 3, 6, 14, 43, 50, 93.
[8] Query IDs: 3, 20, 28, 32, 38, 42, 46, 53, 57, 67.

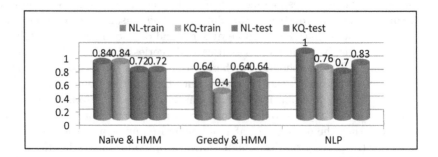

(a) Queries that require query expansion are included.

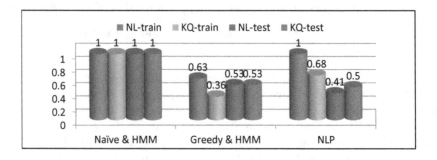

(b) Queries that require query expansion are not included.

Fig. 2. Mean reciprocal rank of query segmentation (first stage)

Naive+HMM outperforms Greedy+HMM, NLP+RCP and RCP in all four experimental settings. Figure 3(a) shows on average 24% higher MRR, although queries requiring expansion are included. In the absence of the queries that required an expansion (Figure 3(b)), Naive+HMM on average by 38% superior to all other approaches and 25% superior to RCP. Note that RCP relies on correct segmentation which in reality is not always a valid assumption. Generally, Naive+HMM being superior to Greedy+HMM can be expected, since the naive approach for segmentation generates more segments from which the HMM can choose. Naive+HMM outperforming RCP (resp. NLP+RCP) is mostly related to RCP (resp. NLP+RCP) often failing to assign the highest rank to the correct disambiguation. One important feature of our approach is, as the evaluation confirms, the robustness with regard to the query expression variance. As shown in Figure 3, Naive+HMM achieves the same MRR on natural-language and the keyword-based representation of queries on both – the train and the test – datasets. Overall, Naive+HMM significantly outperforms our baseline Greedy+HNM as well as state-of-the-art techniques based on NLP. Figure 4 shows the mean of MRR for different values of the threshold θ applied for

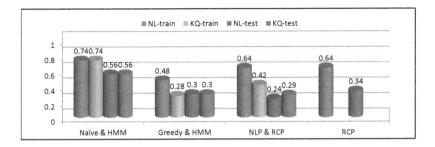

(a) Queries that require query expansion are included.

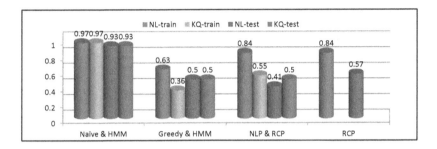

(b) Queries that require query expansion are not included.

Fig. 3. Mean reciprocal rank of resource disambiguation (second stage)

punning the state space. As it can be observed the optimal value of θ is in the range $[0.6, 0.7]$. A high value of θ prevents including some relevant resources and a low value causes a load of irrelevant resources. We set θ to 0.7.

The success of our model relies on transition probabilities which are based on the connectivity of both the source and target node (hub score of source and sink authority) as well as taking into account the connectivity (authority) of all sink states. Especially, employing the HITS algorithm leads to distributing a normalized connectivity degree across the state space. To compare the proposed method for bootstrapping transition probability, we tested two other methods (i.e., normal and Zipfian distribution). Assume the random variable X is defined as the weighted sum of the normalized length of the path (distance) between two states and normalized connectivity degree: $X = \alpha * distance + (1 - \alpha) * (1 - connectivityDegree)$. We bootstrapped the transition probability based on the normal and Zipfian distribution for the variable X. Table 2 shows the MRR of the HMM based on different methods (i.e., normal distribution, Zipfian and the proposed method) employed for bootstrapping the transition probability. The results achieved with the first two methods only led to a low accuracy. The proposed method is superior to the other two methods in all settings.

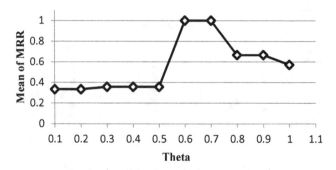

Fig. 4. Mean MRR for different values of θ

Table 2. MRR based on different methods employed in transition probability for 10 queries from train dataset

Query ID	12	15	19	22	23	25	31	33	34	35
Zipf $\alpha = 1$	0.3	0.5	0.25	0.2	0.05	0.05	0.2	0.2	0.2	0.5
Zipf $\alpha = 0.75$	0.3	0.5	0.25	0.2	0.05	0.05	0.16	0.2	0.2	0.5
Zipf $\alpha = 0.5$	0.3	0.5	0.25	0.2	0.05	0.05	0.16	0.2	0.2	0.5
Zipf $\alpha = 0.25$	0.3	0.5	0.25	0.2	0.045	0.05	0.16	0.2	0.2	0.5
Zipf $\alpha = 0$	0.3	0.5	0.25	0.2	0.045	0.05	0.16	0.2	0.2	0.5
The proposed method	1	1	1	1	1	1	1	1	1	1
Normal $\alpha = 1$	1	0.5	0.07	0.25	0.034	0.041	0.3	0.1	1	0.16
Normal $\alpha = 0.75$	1	0.5	0.07	0.3	0.034	0.041	0.125	0.1	1	0.25
Normal $\alpha = 0.5$	1	0.5	0.07	0.3	0.041	0.052	0.14	0.1	1	0.25
Normal $\alpha = 0.25$	1	0.5	0.07	0.3	0.058	0.58	0.2	0.125	1	0.2
Normal $\alpha = 0$	1	0.5	0.1	0.5	0.083	0.045	0.5	0.5	0.5	0.5

7   Discussion and Future Work

We explored different methods for bootstrapping the parameters (i.e. different distributions tested e.g., normal, Zipf) of the HMM. The results achieved with these methods only led to a very low accuracy. The success of our model relies on transition probabilities which are based on the connectivity of both the source and target node (hub score of source and sink authority) as well as taking into account the connectivity (authority) of all sink states. Employing the HITS algorithm leads to distributing a normalized connectivity degree across the state space. More importantly, note that considering a transition probability to the unknown entity state is crucial, since it arranges states with the same emitted segments in a descending order based on their hub scores. Most previous work has been based on finding a path between two candidate entities. For future, we aim to realize a search engine for the Data Web, which is as easy to use as search engines for the Document Web, but allows to create complex queries and returns comprehensive structured query results[9]. A first area of improvements is related to using dictionary knowledge such as hypernyms, hyponyms or co-hyponyms. Query expansion might also, however, result in a more noisy input for

---

[9] A prototype of our progress in this regard is available at http://sina.aksw.org

our model. Thus, a careful extension of our approach and analysis of the results will be required. In addition, we will extend our approach with a query cleaning algorithm. The input query might contain some keywords which semantically are not related to the rest of keywords. Since user usually is looking for information semantically closely related to each other, these unrelated keywords (i.e. noise) should be cleaned.

References

1. Topic-sensitive pagerank: A context-sensitive ranking algorithm for web search. Technical Report 2003-29 (2003)
2. Beeferman, D., Berger, A.: Agglomerative clustering of a search engine query log. ACM Press (2000)
3. Brenes, D.J., Gayo-Avello, D., Garcia, R.: On the fly query entity decomposition using snippets. CoRR, abs/1005.5516 (2010)
4. Brill, E., Ngai, G.: Man* vs. machine: A case study in base noun phrase learning. ACL (1999)
5. Chieu, H.L., Ng, H.T.: Named entity recognition: A maximum entropy approach using global information. In: Proceedings COLING 2002 (2002)
6. Chuang, S.-L., Chien, L.-F.: Towards automatic generation of query taxonomy: A hierarchical query clustering approach. IEEE Computer Society (2002)
7. Collins, M., Singer, Y.: Unsupervised models for named entity classification. In: SIGDAT Empirical Methods in NLP and Very Large Corpora (1999)
8. Finkelstein, L., Gabrilovich, E., Matias, Y., Rivlin, E., Solan, Z., Wolfman, G., Ruppin, E.: Placing Search in Context: the Concept Revisited. In: WWW (2001)
9. Guo, J., Xu, G., Cheng, X., Li, H.: Named entity recognition in query. ACM (2009)
10. Joachims, T., Granka, L.A., Pan, B., Hembrooke, H., Gay, G.: Accurately interpreting clickthrough data as implicit feedback. In: SIGIR. ACM (2005)
11. Kelly, D., Teevan, J.: Implicit feedback for inferring user preference: a bibliography. SIGIR Forum 37(2), 18–28 (2003)
12. Kleinberg, J.M.: Authoritative sources in a hyperlinked environment. J. ACM 46(5) (1999)
13. Kraft, R., Chang, C.C., Maghoul, F., Kumar, R.: Searching with context. In: WWW 2006: 15th Int. Conf. on World Wide Web. ACM (2006)
14. Lawrence, S.: Context in web search. IEEE Data Eng. Bull. 23(3), 25–32 (2000)
15. Pu, K.Q., Yu, X.: Keyword query cleaning. PVLDB 1(1), 909–920 (2008)
16. Ramshaw, L.A., Marcus, M.P.: Text chunking using transformation-based learning. CoRR (1995)
17. Risvik, K.M., Mikolajewski, T., Boros, P.: Query segmentation for web search (2003)
18. Shepitsen, A., Gemmell, J., Mobasher, B., Burke, R.: Personalized recommendation in social tagging systems using hierarchical clustering. ACM (2008)
19. Tan, B., Peng, F.: Unsupervised query segmentation using generative language models and wikipedia. In: WWW. ACM (2008)
20. Tan, B., Peng, F.: Unsupervised query segmentation using generative language models and wikipedia. ACM (2008)
21. Uzuner, A., Katz, B., Yuret, D.: Word sense disambiguation for information retrieval. AAAI Press/The MIT Press (1999)

22. Vorhees, E.: The trec-8 question answering track report. In: Proceedings of TREC-8 (1999)
23. Wen, J.-R., Nie, J.-Y., Zhang, H.-J.: Query Clustering Using User Logs. ACM Transactions on Information Systems 20(1) (2002)
24. White, R.W., Jose, J.M., van Rijsbergen, C.J., Ruthven, I.: A simulated study of implicit feedback models. In: McDonald, S., Tait, J.I. (eds.) ECIR 2004. LNCS, vol. 2997, pp. 311–326. Springer, Heidelberg (2004)
25. Yu, X., Shi, H.: Query segmentation using conditional random fields. ACM (2009)
26. Zhu, Y., Callan, J., Carbonell, J.G.: The impact of history length on personalized search. ACM (2008)

An Automated Template Selection Framework for Keyword Query over Linked Data

Md-Mizanur Rahoman[1] and Ryutaro Ichise[1,2]

[1] Department of Informatics,
The Graduate University for Advanced Studies, Tokyo, Japan
`mizan@nii.ac.jp`
[2] Principles of Informatics Research Division
National Institute of Informatics, Tokyo, Japan
`ichise@nii.ac.jp`

Abstract. Template-based information access, in which templates are constructed for keywords, is a recent development of linked data information retrieval. However, most such approaches suffer from ineffective template management. Because linked data has a structured data representation, we assume the data's inside statistics can effectively influence template management. In this work, we use this influence for template creation, template ranking, and scaling. Our proposal can effectively be used for automatic linked data information retrieval and can be incorporated with other techniques such as ontology inclusion and sophisticated matching to further improve performance.

Keywords: linked data, keyword, information access, data statistics.

1 Introduction

The Linked Open Data [3] initiative opened a new horizon in data usage, where data are connected like a network. The idea is based on link construction and link finding among the data. In linked data, data storage paradigm deviates from the traditional repository centric to open publishing, so that other applications can access and interpret the data. As of September 2011, 295 datasets consisting of over 31 billion RDF triples on various domains were interlinked by approximately 504 million RDF links[1]. It is well understood that linked data presently contain a vast amount of knowledge that underscores the need for good and efficient data access options. Usually information access over linked data requires finding links [9,2,1]. But this access introduces a very basic problem for networked data: it is difficult to find the endpoint of data with a network presentation within a reasonable cost. So finding link's end point on linked data is not easy, especially by general purpose users who have very little knowledge of the linked data's internal structure. Several studies propose using keywords in link data access [5,19,14,10,4], since keywords are considered as easy and familiar

---

[1] `http://www4.wiwiss.fu-berlin.de/lodcloud/state/`

H. Takeda et al. (Eds.): JIST 2012, LNCS 7774, pp. 175–190, 2013.
© Springer-Verlag Berlin Heidelberg 2013

means. Keyword-based linked data or semantic data retrieval is different from other traditional keyword-based data retrieval since it requires adapting semantics with keywords. Since keywords do not hold semantics specifically ontology information, few researchers have proposed automatic ontology inclusion [14]. But, automatic ontology inclusion is a challenge because the machine itself needs to incorporate ontology that is not yet solved. Recently, a few researchers have incorporated templates for keyword searches over linked data [15,13] to solve the above-mentioned problems: link's end point finding, semantics inclusion to keywords. Usually a template is predefined structure which holds some position holders and accomplishs tasks by setting those holders with task specific parameters. From the linked data perspective, these position holders provide some kind of ontology or semantics and the predefined structure assures the finding of link end points. But, current template-based linked data retrieval research lack concrete guidelines for template construction and template ranking, which are highly required for effective adaptation of templates in linked data retrieval. In this work, we address this shortcoming and concentrate on defining guidelines for templates from the perspective of keyword-based linked data retrieval.

In our work, we propose a guided framework for template construction and template ranking, and their adaptation to keyword-based linked data access. We utilize data's inside statistics for template management. We assume that linked data holds a structured representation of data, so the data's inside statistics, such as the positional frequency of a resource, can influence template construction and template ranking. In this assumption, we are motivated by query likelihood model [12]: a traditional information retrieval (IR) model, which argues more frequently required information appears in underlying data, more potential in data retrieval. Our work uses keywords for data retrieval and produces a template that we adapt to the equivalent SPARQL query for accessing real data. For each query question, our intuitive assumption is that keywords are given in order and that order indicates a higher possibility of a relation existing between two adjacent keywords. We deploy a binary progressive approach to obtain the final template, which means we create a template from a pair of keywords, then we use a pair of templates to find the final template. As we follow this approach, the order of keywords in the query question is important from a very practical retrieval aspect. Our proposed framework automatically creates templates from keywords without knowing schema or ontology information , which is an automatic linked data retrieval approach. We assume that future inclusion of that information will improve the system's performance effectively.

The remainder of this paper is outlined as follows. Section 2 describes related work. Section 3 introduces template construction and template management for query questions with two keywords. Section 4 extends our work for query questions with more than two keywords. Section 5 implements our proposal and presents the experimental results. Section 6 concludes our work.

2   Related Work

Information access is an active research field in the linked data research community. In the past few years, researchers have introduced a range of techniques of linked data information retrieval, such as retrieval of an RDF document with sophisticated matching and ranking [6], retrieval with automatic ontology inclusion [17,14], and retrieval by incorporating user feedback [11]. In some approaches, researchers consolidate the benefits of two or more techniques, as in faceted search with explicit query [7,18], or semi-supervised machine learning with user feedback [11]. Since keywords are considered as easy and familiar means for data access, some studies propose keyword based linked data retrieval [5,19] or taking advantage of keywords [14,10,4] in the linked data access paradigm.

In our work, we are also motivated toward keyword-based linked data access. In related approaches, Gheng et al. created virtual documents that provide query-relevant structures on linked object finding [5]. Zhou et al. showed keyword based resource mapping and a graph construction technique [19], whereas automatic ontology inclusion to keywords is another technique proposed in [14,10]. Bicer et al. showed keyword based index creation and query graph construction [4]. Han et al. advocated an easy query learning framework, where keywords are fitted automatically to the query construction using sophisticated matching [8]. So, the linked data research community has been working toward an easy and effective searching technique over linked data, and many of their proposals incorporate keywords as a search option. But keyword based linked data or semantic data retrieval is different from other traditional keyword based data retrieval, since it requires adapting semantics with the keywords. Some researchers propose inclusion of templates to provide a structured framework, which could fill the gap between keywords and linked data semantics [15,13]. Those studies are still not sufficient for devising concrete guidelines for template creation and template ranking, which are highly required for effective adaptation of templates overlying linked data retrieval. For example, Unger et al. proposed a natural language (NL) based QA system for linked data with *Pythia* [16], but a problem is template creation because it employs NL tools that sometimes lead to incorrect template construction [15]. Moreover *Pythia* itself has a scalability issue that is very specific to particular data. Shekarpour et al. proposed an approach similar to our template creation, but their template construction needs to know some schema information such as instance or class type information, which they provide manually [13]. Moreover, their proposal is only able to handle a query with at most two keywords.

3   Construction of Templates

This section describes template construction and SPARQL query generation. We take two adjacent keywords and produce a possible SPARQL query for those keywords. As mentioned, we follow a binary progressive approach for template construction. That is, we use two resources from two different keywords to construct templates. We use this binary approach because we assume this template

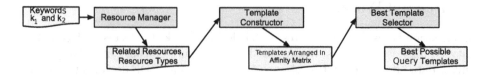

Fig. 1. Template selection process flow for query question with two keywords

construction technique is scalable and can be extended for a query with more than two keywords. Template management for more than two keywords is described in Section 4.

Keyword-based linked data information retrieval is different than other data retrieval. Linked data information retrieval requires a schema or ontology information that is absent in keywords. So, a template is introduced to fill this gap. Usually a template provides a framework that can be fitted to a specific task. More specially, a template provides a predefined structure with a few position holders, and those positions are filled by task-specific parameters. From the linked data retrieval perspective, filling those position holders by resources (i.e., resources from keywords) can be considered as an automatic semantics inclusion technique in the given keywords. This gap-filling process is suggested for semantics inclusion mainly because it can adapt to the linked data's data representation, i.e., $< subject, object, predicate >$ issue. Moreover, linked data is a kind of network data that introduces the option of finding new links but, at the same time, it introduces the very basic problem of network data: link's end point finding. Link's end point finding in network data makes data access task difficult because it can not assure when end point will be found. Template handles this problem by its predefined structure and assures that link's end point finding will not get hanged. So, in link data access, templates solve the problem of link's end point finding and keyword's absent semantics including which are very much required for automatic linked data access.

In our work, we devise a two-keyword based template selection process flow that clearly guides us in template construction and template management. Figure 1 shows this process flow. It starts with two keywords k_1 and k_2, then the process *Resource Manager* manages the resources for those keywords and collects related resources and resource types. Next, the process *Template Constructor* constructs templates arranged in the affinity matrix. Finally, the process *Best Template Selector* selects the best possible query template. We use this best possible query template for constructing the required SPARQL query. In this paper, we describe each process in detail.

3.1   Resource Manager

This process manages resources for given keywords. It takes two adjacent keywords and produces keyword related resources with their classifications. We divide this *Resource Manager* process into three sub-processes. Figure 2 shows the process flow. Sub-process *Resource Extractor* takes keywords and extracts

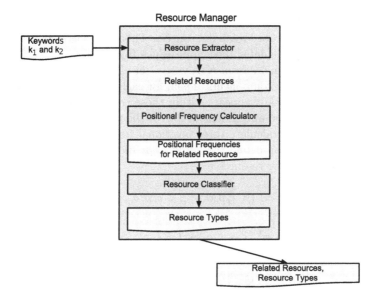

Fig. 2. Details of Resource Manager

related resources, then feeds those related resources to sub-process *Positional Frequency Calculator* to calculate the positional frequencies for related resources, then passes those outputs to sub-process *Resource Classifier* to classify resources into resource types. In the end, process *Resource Manager* provides related resources and resource types. Below we describe each sub-process in detail.

Resource Extractor. We use this sub-process for extracting keyword related resources from a given knowledgebase, where the knowledgebase is defined as KB and is considered as a linked data source with set of RDF triple t. Here, t is represented as $< s, p, o >$, where s, p, and o are considered as the subject, predicate, and object, respectively. We take a keyword as input and find keyword related resources as output. Our keyword related resource for keyword k is defined as follows:

$$RR(k) = \{s \mid \exists < s, p, o > \in KB \wedge (p \in rtag) \wedge (m(o, k) = true)\}$$

where

- $rtag$ is a set of resources representing tags such as label, name, prefLabel, etc.
- $m(o, k)$ is a boolean function between the triple object o and the keyword k, whether they match exactly or not.

We use exact similarity matching to make our process simple. We presume that sophisticated matching will increase system performance in future work.

In summary, the *Resource Extractor* sub-process extracts keyword related resources from knowledgebase KB and forwards them to the next sub-process.

Positional Frequency Calculator. We use this sub-process for counting positional frequencies for keyword related resources. This sub-process takes keyword related resources and calculates their appearance frequencies in KB. To calculate the positional frequencies from KB, we collect all triples that hold keyword related resources as their components. Then, for each keyword related resource, we count how many times it appears within those collected triples. In a triple, because a resource can be an element of any of the three components, *subject, predicate and object*, we calculate three positional frequencies for each keyword related resource. Here, we define the subject, predicate, and object positional frequency, respectively, for resource r:

$$PF_s(r) =| \{< r,p,o >| \exists < r,p,o >\in KB\} |$$

$$PF_p(r) =| \{< s,r,o >| \exists < s,r,o >\in KB\} |$$

$$PF_o(r) =| \{< s,p,r >| \exists < s,p,r >\in KB\} |$$

According to the above definitions, for every keyword related resource r, we collect $PF_s(r)$, $PF_p(r)$, and $PF_o(r)$, and forward those frequencies to the next sub-process for their classification.

Resource Classifier. We use this sub-process for classifying each keyword related resource into two categories: a predicate (PR) or a non-predicate (NP) type resource. We take the three positional frequencies for every keyword related resource and classify it according to its type. To classify each keyword related resource r, we use a type deciding function $uType$, which takes parameter resource r and compares the value of $PF_p(r)$ with the values of $PF_s(r)$ and $PF_o(r)$. Below we define this function:

$$uType(r) = \begin{cases} PR & \text{iff } (PF_p(r) > PF_s(r)) \wedge (PF_p(r) > PF_o(r)) \\ NP & \text{otherwise} \end{cases}$$

To summarize, the *Resource Manager* process collects all keyword related resources for both keywords k_1 and k_2, and performs the classification for those resources. We use these keyword related resources and their types in the next *Template Constructor* process.

3.2   Template Constructor

This process constructs possible templates. We take all related resources along with their types, then construct templates and find a structured arrangement for the templates. We refer to this arrangement as an affinity matrix. As mentioned, the template is a structured pattern with some position holders where template accomplishs a task by setting those holders with task-specific parameters. Considering the above and following our binary progressive approach, we

Table 1. Query templates for resources r_1 and r_2 with their corresponding SPARQL queries (TG: template group, QT: query template)

TG	QT	SPARQL Query for QT
TG_1	$<?uri, r_1, r_2>$	SELECT ?uri WHERE {?uri r_1 r_2.}
	$<r_2, r_1, ?uri>$	SELECT ?uri WHERE {r_2 r_1 ?uri.}
TG_2	$<r_1, ?uri, r_2>$	SELECT ?uri WHERE {r_1 ?uri r_2.}
	$<r_2, ?uri, r_1>$	SELECT ?uri WHERE {r_2 ?uri r_1.}
	$<?uri, ?p_1, r_1> <?uri, ?p_2, r_2>$	SELECT ?uri WHERE {?uri $?p_1$ r_1. ?uri $?p_2$ r_2.}
	$<r_1, ?p_1, ?uri> <r_2, ?p_2, ?uri>$	SELECT ?uri WHERE {r_1 $?p_1$?uri. r_2 $?p_2$?uri.}
	$<r_1, ?p_1, ?uri> <?uri, ?p_2, r_2>$	SELECT ?uri WHERE {r_1 $?p_1$?uri. ?uri $?p_2$ r_2.}
	$<?uri, ?p_1, r_1> <r_2, ?p_2, ?uri>$	SELECT ?uri WHERE {?uri $?p_1$ r_1. r_2 $?p_2$?uri.}

construct a query template with two position holders, where we set up those holders with two resources from two different keywords. Table 1 shows all of our templates for resources r_1 and r_2 with their corresponding SPARQL queries. The first column indicates the template group, the second column shows the query template and the third column shows the query template corresponding SPARQL query. In template construction, if one resource comes as a predicate (PR) and another one comes as a non-predicate (NP), we construct a query template with triples, where both of the resources are in a single triple, the PR type resource is a predicate component, and the NP type resource is either a subject component or a predicate component. We create two query templates for this resource combination and mark them as query template group 1 (TG_1). If both resources come as a non-predicate (NP), we either construct query templates with triples, where both resources sit in a single triple, one resource is a subject component, and the other one is an object component. Or we construct query templates with triples, where each resource belongs to an individual triple and is either a subject component or an object component. We create six query templates for this resource combination and mark them as query template group 2 (TG_2). Another possibility is coming both resources as a predicate (PR), but from the keyword based data access perspective, our intuitive assumption is that there is a small chance of coming two adjacent resources as a predicate, so we do not create any group for this combination (which is also justified in [13]). This resource attachment to the query template is considered as filling position holders for query templates. In query template, other than resource r_1 and r_2, remaining triple components are considered as variable type resources. We mark these variable type resources with the symbol "?".

As is understood, we get several query templates for each two keywords: either the query template construction picks all query templates for a template group, or the query template construction picks query templates for keywords having several resources. To manage these query templates, we arranges them in the affinity matrix (AM). Below we describe the details.

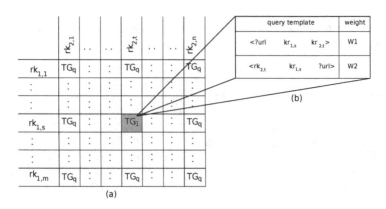

Fig. 3. (a) Affinity matrix filled with template groups ($TG_q \in \{ TG_1 \cup TG_2 \}$). Row elements represent resources from the first keyword and column elements represent resources from the second keyword. (b) Details of an individual cell where the template group is TG_1.

Construction of Affinity Matrix. We construct the AM by putting all keyword related resources for one keyword into columns and second keyword resources into rows, then consider their resource types and decide a template group. Then, according to the template group, we fill each cell with all query templates for that group and calculate the weight for each query template.

Figure 3 (a) shows the structure of an affinity matrix, where it is assumed that k_1 has n number of resources $RR(k_1) = \{rk_{1,1}, .., rk_{1,s}, .., rk_{1,n}\}$ and is stored as a column vector while k_2 has m number of resources $RR(k_2) = \{rk_{2,1}, .., rk_{2,t}, .., rk_{2,m}\}$ and is stored as a row vector and Figure 3 (b) shows an individual cell for that matrix with query templates and their weights.

Calculation of Weight. In weight calculations we consider a query template's frequency and its resources' frequency in a given knowledgebase KB. We base this frequency calculation on the traditional query likelihood IR model [12], which indicates more frequently required information appears in underlying data, and more potentially in data retrieval. In AM construction, as each cell holds all query templates for a template group (either TG_1 or TG_2), we calculate the weights for all of these query templates.

To calculate the weight for query template QT, we use below weight factors:

1. **frequency of query template** $fq(QT)$: number of answers returned by a query template corresponding SPARQL query.
2. **frequency of query template for resource** $fq(r)$: positional frequency of resource r in KB where the position (subject, predicate and object) of r is guided by query template QT.

$$fq(r, qt(r_1, r_2)) = \begin{cases} PF_s(r) & \text{if } r \text{ is on } subject \text{ in } QT \\ PF_p(r) & \text{if } r \text{ is on } predicate \text{ in } QT \\ PF_o(r) & \text{if } r \text{ is on } object \text{ in } QT \end{cases}$$

The final weight $FW(QT)$ of query template QT is as follows:

$$FW(QT) = fq(QT) * fq(r_1, QT) * fq(r_2, QT)$$

For each cell, we calculate all corresponding weights of the query template and save them for later use. In summary, the *Template Constructor* process constructs query templates, store them into the AM and forwards the AM to the next process for selecting the best possible query template.

3.3   Best Template Selector

This process selects best possible query template from all query templates in affinity matrix AM. We select best possible query template by considering its two parameters: depth level and weight. We have already discussed about query template's weight, here we will introduce query template's depth level. Depth level considers how closely resources r_1 and r_2 are attached in triples for a query template. If query template belongs triples where both r_1 and r_2 sit in single triple, the depth level of that query template is one, otherwise depth level is two. In query template table (see Table 1), first four query templates hold depth level value 1, and the remaining query templates hold depth level value 2. In our work, we assume query template with depth level one is more potential than query template with depth level two. So in best possible query template selection, at first, we pick depth level one type query templates with weight greater than zero, sort them with their weight and choose the highest weighted query template as best possible query template. If we do not find any depth level one type query template with weight more than zero, we consider the highest weighted depth level two type query template as best possible query template. In any case, if we get more than one best possible query templates (because of same weight), we pick any of them. Considering the above, process *Best Template Selector* selects best possible query template for keywords k_1 and k_2. And for two keywords query question, this best possible query template related SPARQL query is used for finding our intended result.

4   More Than Two Keywords

In the previous section, we discussed how to create templates for two keywords. In this section, we describe how we make the final query template for a query question with more than two keywords. Again, we follow the binary progressive approach to get our final query template. We take a pair of keywords and a pair of query templates for finding the final query template. We are motivated to this binary approach because we assume keywords are given in order. For a query question with more than two keywords, we follow a query template selection procedure based on four processes. Figure 4 shows this template selection procedure, where it starts with all keywords. Then, process *Best Template Constructor* generates the best possible query templates, the related keywords and weights for each two adjacent keywords. Next, process *Comparator* compares whether it

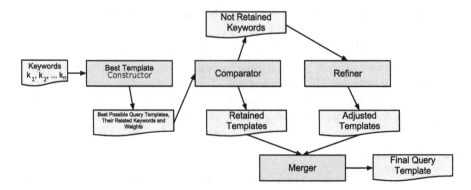

Fig. 4. Template selection process flow for query question with more than two keywords

should retain or adjust these templates, and stores one template as a retained template and finds not retained keyword for adjustment. Process *Refiner* constructs adjusted templates for not retained keywords, and finally process *Merger* merge templates to get the final query template. Below we describe each process in detail.

4.1   Best Template Constructor

This process selects all of the best possible query templates for every two adjacent keywords and stores them along with their keywords and weights. For example, if there are n number of keywords $\{k_1, k_2, k_3, .., k_n\}$, we will get $n-1$ number of best possible query templates, each for two adjacent keywords (i.e., 1st best possible template for keywords k_1 and k_2, 2nd best possible query template for k_2 and k_3, and so on). We keep all of the best possible query templates for two adjacent keywords, their corresponding keywords and weights for the next process. In each two keywords' best possible query template selection, if we do not get any query template with weight more than zero, we consider any of query templates that can be constructed for those two keywords as best possible query template.

4.2   Comparator

This process compares each two adjacent best possible query templates by their depth levels and weights. By this comparision, we retain one query template and forward remaining one to *Refiner* process for further adjustment. Like best possible query template selection in section 3, we select lower depth level and higher weighted query template as *retained template* and find *not retained keyword* (keyword that is not related to *retained template* for each adjacent query template pair) from other query template. In summary, for each two adjacent query templates, process *Comparator* forwards both *retained template* and *not retained keyword* to next processes for their further usage.

Table 2. Modified Query Templates for resources r_1 with their corresponding SPARQL Queries (MTG: modified template group, MQT: modified query template)

MTG	MQT	SPARQL Query for MQT
MTG_1	$<?uri, r_1, ?o_1>$	SELECT ?uri WHERE {?uri r_1 ?o_1.}
MTG_2	$< r_1, ?p_1, ?uri>$	SELECT ?uri WHERE {r_1 ?p_1 ?uri.}
	$<?uri, ?p_1, r_1>$	SELECT ?uri WHERE {?uri ?p_1 r_1.}

4.3   Refiner

This process takes each individual *not retained keyword* and adjusts the query template for those keywords. As earlier query template management was intended for two keywords, it is understood that this time we need to modify query templates to serve a single keyword. To do so, we find keyword related resources for *not retained keyword k* with their type classifications and create modified query templates. This modified query template creation process is similar to the earlier query template creation in Table 1, but we next modify those to some extent to consider only one position holder. Table 2 shows our modified query templates for resource r_1 ($MQT(r_1)$) along with their corresponding SPARQL queries. The first column indicates the modified template group, the second shows the modified query template, and the third column shows the corresponding SPARQL query. We divide the modified query templates into two groups: the first group (MTG_1) for a predicate type resource and the second group (MTG_2) for a non-predicate type resource. For each *not retained keyword* we obtain several modified templates. So, we need a ranking scheme to select the best. We measure the best possible modified query template by using a weight factor called *frequency of modified query template*. This weight factor is similar to the previous *frequency of query template* and is calculated by using the count number of answers returned by the modified query templates with the corresponding SPARQL query.

By checking the above weight factor for all resources of a *not retained keyword*, we select the best possible modified query template, which we call the *adjusted template* for that *not retained keyword*. The *Refiner* process calculates adjusted templates for all *not retained keyword*s and keeps them for later use by the *Merger* process.

4.4   Merger

After the generation of all *retained templates* and *adjusted templates*, this process gives a final query template for all query keywords. We merge all *retained templates* and *adjusted templates* and build a final query template. As shown in the SPARQL query, we collect the query template's results only for ?*uri*. so, in the query template merger we keep the variable resource ?*uri* for all query templates and use a different variable resource identifier for the other cases. For example, if variable resource identifier ?p1 and ?p2 are already used by a previous query template(s), we use a new variable resource identifier (e.g., ?p3) when merging templates. So, this *Merger* process gives us the final query template that we use in the linked data retrieval.

For the final query template, we use same SPARQL query generation technique shown in Table 1 and Table 2. We extend SPARQL query according to the final query template, and this SPARQL query produces our intended result for a query question with more than two keywords.

5   Experiment

We use the QALD-1[2] DBpedia test set in our experiment. The reason for choosing this question set that it uses DBpedia as the underlying dataset. Because DBpedia holds a huge number of triples from multiple domains, we can measure our system in a real linked data implementation. Moreover, Question Answering over Linked Data One (QALD-1) is recognized by the semantic web community as a data retrieval challenge for linked data.

The test set has 50 natural language questions. Out of those 50, we chose 42 questions. The remaining eight questions require a special kind of reasoning between keywords and resources, and so we purposefully discarded them. For example, Q# 9, 17, 18, 23, 24, 44, and 45 require keyword reasoning for quantitative measurement such as "more than 10000", "highest", "more than", etc. and Q# 38 requires keyword reasoning for the word "border", where the resources are not synonymous with the word "border" and so this keyword requires a special kind of reasoning. Out of our 42 questions, we change 3 questions (Q# 10, 30, 35) to their descriptive mode which actually were boolean type questions. We intuitively retrieved keywords for those 42 questions, which we consider were simultaneously balanced in the KB and the query questions. For example, Q# 29 *In which films directed by Garry Marshall was Julia Roberts starring?* We constructed keywords *Film, starring, Julia Roberts, Director, Garry Marshall.* Our KB holds triples from DBPeida, Geoname, and NYtimes. Our proposed framework is not intended for particular datasets and can be implemented with any dataset, so we kept these three datasets together to test our system.

5.1   Result

We categorize our 42 questions into four groups ("2/3/4/5 keyword question") by considering the number of keywords in a question and evaluate our result. We also compare our proposal with GoRelations [8], which also uses the QALD-1 test set for their evaluation. Table 3 and Table 4 show our results.

Table 3 shows the performance of our system. The first column groups query questions into number of keywords, the second column counts the number of questions in each group, and the third, fourth, and fifth columns measure average recall, average precision, and average F1 measure for each group, respectively. In addition, the bottom shows average recall, average precision, and average F1

---

[2] http://www.sc.cit-ec.uni-bielefeld.de/qald-1

Table 3. Proposed system's recall, precision, and F1 measure grouped by number of keywords

No. of Keywords	No. of Questions	Recall (avg)	Precision (avg)	F1 Measure (avg)
2	30	0.933	0.884	0.899*
3	8	0.472	0.442	0.454*
4	3	0.000	0.000	0.000*
5	1	1.000	1.000	1.000
	Average	**0.780**	**0.740**	**0.753***

Table 4. Performance comparison between GoRelations and our System

	Recall	Precision	F1 Measure
GoRelations [8]	0.722	0.687	0.704
Proposed System	**0.780**	**0.740**	**0.753***

measure for all 42 questions. For each group, we calculate the average precision and the average recall by summing up the individual recall and the individual precision and dividing them by the number of questions for each group. But calculation of the average F1 measure in the same way is not possible, because the individual F1 measure cannot be calculated if the recall of that individual question is 0. In that case, we put an individual question's F1 measure as 0, if the recall is 0, and then calculate the average F1 measure for each group of questions. For all 42 questions, we found the overall performance with the average F1 measure to be 0.753*[3] and 28 of them achieved the gold standard.

In the "2 keyword question" (Q# 4, 5, 6, 7, 8, 10, 11, 12, 13, 15, 16, 19, 20, 21, 25, 26, 27, 30, 31, 33, 34, 35, 40, 41, 43, 46, 47, 48, 49, 50) we get an average F1 measure of 0.899*, where most of them (24 out of 30) achieved the gold standard and only two of them (Q# 4 and 16) produced an empty result, which indicates most of our templates correctly captured the possible queries. We only select query templates with the highest weight, which can be considered a very strict choice in data retrieval, but still we get the recall value 0.933 (for two-keyword questions), which indicates system performance can be enhanced by incorporating a flexible choice. For the "2 keyword question", our detailed observation finds that the gold standard had 23 questions (out of 30) that should belong to template group TG_1; our system was successfully able to capture 22 of these to TG_1 query templates, and so we can validate our heuristic on resource type classification.

For the "3 keyword question" (Q# 2, 14, 22, 28, 32, 37, 39, 42) we get relatively low performance (average F1 measure 0.454), in which only three achieved the gold standard. This low performance for primarily keyword based data access as the keyword sometimes lead to a misleading actual meaning. For example, Q# 37 *Who is the daughter of Bill Clinton married to?*, where we retrieve keywords as *Bill Clinton, child, spouse*, the system could not justify whether we asked for

---

[3] As mentioned, we put individual questions as F1 =0, if recall = 0.

Bill Clinton's "Parent" or his daughter's husband? Moreover, by only picking the highest weight template, our system overlooked other possibilities.

In the "4 keyword question" (Q# 1, 3, 36) we encountered template consolidation. In our *Merger* process, except for the variable resource identifier *?uri*, we use different variable resource identifiers, which sometimes leads us to an empty result. To solve this problem, we need to find an attachment strategy for variable identifiers so that we can use the same identifier if they represent the same resource.

We achieved the gold standard for the "5 keyword question" (Q# 29), which justifies our binary progressive approach in finding the final query template. In our proposal, we give higher priority for templates with depth level one than depth level two, which supports us in correct comparision between two adjacent templates and adjust one of them. Getting gold standard result for "5 keyword question" validate our priority giving approach. This approach is validated in the "3 keyword question" as well.

We compared our proposal with GoRelations [8], which also uses the QALD-1 test set in their evaluation. Table 4 shows these results for the recall, precision, and F1 measure. Our system achieved the overall F1 measure 0.753* while GoRelations received the value 0.704 for the same measurement, which indicates our system performs better for current research that also attributes linked data retrieval. In our system, end users are not required to build their query. Instead, they can use a traditional keyword based query, which we believe is a much easier option.

6   Conclusion

Since a keyword is comfortable choice in data retrieval, many researchers try to adapt keywords for linked data access. But keywords do not hold required schema information or semantics, which are very much required in linked data retrieval. A few researchers have proposed a template based retrieval, in which the template fills the gap between keywords and semantics. But those initiatives lack robust guidelines on template establishment. In our work, for automatic keyword based linked data retrieval, we propose a guided outline for template creation and template ranking. We use data's inside statistics for template creation and template ranking. As we rely on the data's inside statistics that are calculated automatically, we consider our system as quite promising in fully automatic information access over linked data. We introduce a binary progressive processing paradigm that is scalable for any number of query keywords. Moreover, our proposal includes matrix-based template management, which is very much required for parallel processing and web-scale implementation. Our system can be fitted for any data set and is suitable for scalable implementation.

We assume our template creation can benefit from other linked data access approaches, such as automatic ontology inclusion, feedback incorporation, and sophisticated matching with keywords. We want to explore this in our future work. Currently, we create a query template (QT) by classifying resources into two types, i.e., a predicate (PR) and a non-predicate (NP), but we assume that *class* information could reduce number of templates effectively, and so we want to investigate this in the future. Because we depend on various statistical parameters, we assume incorporation of off-line processing can increase system performance, which we will also include in our future investigations.

References

1. Akar, Z., Halaç, T.G., Ekinci, E.E., Dikenelli, O.: Querying the Web of Interlinked Datasets using VOID Descriptions. In: Proceedings of WWW 2012 Workshop on Linked Data on the Web (2012)
2. Alexander, K., Cyganiak, R., Hausenblas, M., Zhao, J.: Describing Linked Datasets. In: Proceedings of WWW 2012 Workshop on Linked Data on the Web (2009)
3. Berners-Lee, T.: Linked Data - Design Issues (2006), http://www.w3.org/DesignIssues/LinkedData.html
4. Bicer, V., Tran, T., Abecker, A., Nedkov, R.: KOIOS: Utilizing Semantic Search for Easy-Access and Visualization of Structured Environmental Data. In: Aroyo, L., Welty, C., Alani, H., Taylor, J., Bernstein, A., Kagal, L., Noy, N., Blomqvist, E. (eds.) ISWC 2011, Part II. LNCS, vol. 7032, pp. 1–16. Springer, Heidelberg (2011)
5. Cheng, G., Qu, Y.: Searching Linked Objects with Falcons: Approach, Implementation and Evaluation. International Journal on Semantic Web and Information Systems 5(3), 49–70 (2009)
6. Ding, L., Finin, T.W., Joshi, A., Pan, R., Cost, R.S., Peng, Y., Reddivari, P., Doshi, V., Sachs, J.: Swoogle: A Search and Metadata Engine for the Semantic Web. In: Proceedings of the 13th ACM Conference on Information and Knowledge Management, pp. 652–659 (2004)
7. Ferré, S., Hermann, A.: Semantic Search: Reconciling Expressive Querying and Exploratory Search. In: Aroyo, L., Welty, C., Alani, H., Taylor, J., Bernstein, A., Kagal, L., Noy, N., Blomqvist, E. (eds.) ISWC 2011, Part I. LNCS, vol. 7031, pp. 177–192. Springer, Heidelberg (2011)
8. Han, L., Finin, T., Joshi, A.: GoRelations: An Intuitive Query System for DBpedia. In: Pan, J.Z., Chen, H., Kim, H.-G., Li, J., Wu, Z., Horrocks, I., Mizoguchi, R., Wu, Z. (eds.) JIST 2011. LNCS, vol. 7185, pp. 334–341. Springer, Heidelberg (2012)
9. Hartig, O., Bizer, C., Freytag, J.-C.: Executing SPARQL Queries over the Web of Linked Data. In: Bernstein, A., Karger, D.R., Heath, T., Feigenbaum, L., Maynard, D., Motta, E., Thirunarayan, K. (eds.) ISWC 2009. LNCS, vol. 5823, pp. 293–309. Springer, Heidelberg (2009)
10. Herzig, D.M., Tran, T.: Heterogeneous Web Data Search using Relevance-Based on the Fly Data Integration. In: Proceedings of the 21st World Wide Web Conference, pp. 141–150 (2012)
11. Lehmann, J., Bühmann, L.: AutoSPARQL: Let Users Query Your Knowledge Base. In: Antoniou, G., Grobelnik, M., Simperl, E., Parsia, B., Plexousakis, D., De Leenheer, P., Pan, J. (eds.) ESWC 2011, Part I. LNCS, vol. 6643, pp. 63–79. Springer, Heidelberg (2011)
12. Manning, C.D., Raghavan, P., Schütze, H.: An Introduction to Information Retrieval. Cambridge University Press (2009)
13. Shekarpour, S., Auer, S., Ngomo, A.C.N., Gerber, D., Hellmann, S., Stadler, C.: Keyword-driven SPARQL Query Generation Leveraging Background Knowledge. In: Proceedings of the 10th International Conference on Web Intelligence, pp. 203–210 (2011)

14. Tran, T., Wang, H., Haase, P.: Hermes: Data Web Search on a Pay-As-You-Go Integration Infrastructure. Journal of Web Semantics 7(3), 189–203 (2009)
15. Unger, C., Bühmann, L., Lehmann, J., Ngomo, A.C.N., Gerber, D., Cimiano, P.: Template-Based Question Answering over RDF Data. In: Proceedings of the 21st World Wide Web Conference, pp. 639–648 (2012)
16. Unger, C., Cimiano, P.: Pythia: Compositional Meaning Construction for Ontology-Based Question Answering on the Semantic Web. In: Muñoz, R., Montoyo, A., Métais, E. (eds.) NLDB 2011. LNCS, vol. 6716, pp. 153–160. Springer, Heidelberg (2011)
17. Vallet, D., Fernández, M., Castells, P.: An Ontology-Based Information Retrieval Model. In: Gómez-Pérez, A., Euzenat, J. (eds.) ESWC 2005. LNCS, vol. 3532, pp. 455–470. Springer, Heidelberg (2005)
18. Wang, H., Liu, Q., Penin, T., Fu, L., Zhang, L., Tran, T., Yu, Y., Pan, Y.: Semplore: A Scalable IR Approach to Search the Web of Data. Journal of Web Semantics 7(3), 177–188 (2009)
19. Zhou, Q., Wang, C., Xiong, M., Wang, H., Yu, Y.: SPARK: Adapting Keyword Query to Semantic Search. In: Aberer, K., Choi, K.-S., Noy, N., Allemang, D., Lee, K.-I., Nixon, L.J.B., Golbeck, J., Mika, P., Maynard, D., Mizoguchi, R., Schreiber, G., Cudré-Mauroux, P. (eds.) ISWC/ASWC 2007. LNCS, vol. 4825, pp. 694–707. Springer, Heidelberg (2007)

Leveraging the Crowdsourcing
of Lexical Resources for Bootstrapping
a Linguistic Data Cloud

Sebastian Hellmann⋆, Jonas Brekle⋆, and Sören Auer

Universität Leipzig, Institut für Informatik, AKSW,
Postfach 100920, D-04009 Leipzig, Germany
{hellmann,brekle,auer}@informatik.uni-leipzig.de
http://aksw.org

Abstract. We present a declarative approach implemented in a comprehensive open-source framework based on *DBpedia* to extract lexical-semantic resources – an ontology about language use – from *Wiktionary*. The data currently includes language, part of speech, senses, definitions, synonyms, translations and taxonomies (hyponyms, hyperonyms, synonyms, antonyms) for each lexical word. Main focus is on flexibility to the loose schema and configurability towards differing language-editions of *Wiktionary*. This is achieved by a declarative mediator/wrapper approach. The goal is to allow the addition of languages just by configuration without the need of programming, thus enabling the swift and resource-conserving adaption of wrappers by domain experts. The extracted data is as fine granular as the source data in *Wiktionary* and additionally follows the *lemon* model. It enables use cases like disambiguation or machine translation. By offering a linked data service, we hope to extend DBpedia's central role in the LOD infrastructure to the world of Open Linguistics.

1   Introduction

The exploitation of community-built lexical resources has been discussed repeatedly. *Wiktionary* is one of the biggest collaboratively created lexical-semantic and linguistic resources available, written in 171 languages of which approximately 147 can be considered active[1], containing information about hundreds of spoken and even ancient languages. For example, the English *Wiktionary* contains nearly 3 million words[2]. A *Wiktionary* page provides for a lexical word a hierarchical disambiguation to its language, part of speech, sometimes etymologies and most prominently senses. Within this tree numerous kinds of linguistic properties are given, including synonyms, hyponyms, hyperonyms, example sentences, links to Wikipedia and many more. [13] gave a comprehensive overview on why

---

⋆ Contributed equally to this work.
[1] http://s23.org/wikistats/wiktionaries_html.php
[2] See http://en.wiktionary.org/wiki/semantic for a simple example page.

H. Takeda et al. (Eds.): JIST 2012, LNCS 7774, pp. 191–206, 2013.

this dataset is so promising and how the extracted data can be automatically enriched and consolidated. Aside from building an upper-level ontology, one can use the data to improve NLP solutions, using it as comprehensive background knowledge. The noise should be lower when compared to other automatic generated text copora (e.g. by web crawling) as all information in *Wiktionary* is entered and curated by humans. Opposed to expert-built resources, the openness attracts a huge number of editors and thus enables a faster adaption to changes within the language.

The fast changing nature together with the fragmentation of the project into *Wiktionary language editions* (*WLE*) with independent layout rules, called *ELE guidelines* (Entry Layout Explained, see Section 3.2) poses the biggest problem to the automated transformation into a structured knowledge base. We identified this as a serious problem: Although the value of *Wiktionary* is known and usage scenarios are obvious, only some rudimentary tools exist to extract data from it. Either they focus on a specific subset of the data or they only cover one or two WLE. The development of a flexible and powerful tool is challenging to be accommodated in a mature software architecture and has been neglected in the past. Existing tools can be seen as adapters to single WLE — they are hard to maintain and there are too many languages, that constantly change. Each change in the *Wiktionary* layout requires a programmer to refactor complex code. The last years showed, that only a fraction of the available data is extracted and there is no comprehensive RDF dataset available yet. The key question is: Can the lessons learned by the successful DBpedia project be applied to *Wiktionary*, although it is fundamentally different from Wikipedia? The critical difference is that only word forms are formatted in infobox-like structures (e.g. tables). Most information is formatted covering the complete page with custom headings and often lists. Even the infoboxes itself are not easily extractable by default DBpedia mechanisms, because in contrast to DBpedias *one entity per page* paradigm, *Wiktionary* pages contain information about *several* entities forming a complex graph, i.e. the pages describe the lexical word, which occurs in several languages with different senses per part of speech and most properties are defined *in context* of such child entities. Opposed to the currently employed classic and straight-forward approach (implementing software adapters for scraping), we propose a declarative mediator/wrapper pattern. The aim is to enable non-programmers (the community of adopters and domain experts) to tailor and maintain the WLE wrappers themselves. We created a simple XML dialect to encode the ELE guidelines and declare triple patterns, that define how the resulting RDF should be built. This configuration is interpreted and run against *Wiktionary* dumps. The resulting dataset is open in every aspect and hosted as linked data[3]. Furthermore the presented approach can be extended easily to interpret or *triplify* other MediaWiki installations or even general document collections, if they follow a global layout.

---

[3] http://wiktionary.dbpedia.org/

Table 1. Comparison of existing Wiktionary approaches (ld = linked data hosting). None of the above include any crowd-sourcing approaches for data extraction. The wikokit dump is not in RDF.

name	active	available	RDF	#triples	ld	languages
JWKTL	✓	dumps	✗	-	✗	en, de
wikokit	✓	source + dumps	✓	n/a	✗	en, ru
texai	✗	dumps	✓	~ 2.7 million	✗	en
lemon scraper	✓	dumps	✓	~16k per lang	✗	6
blexisma	✗	source	✗	-	✗	en
WISIGOTH	✗	dumps	✗	-	✗	en, fr
lexvo.org	✓	dumps	✓	~353k	✓	en

2   Related Work

In the last five years, the importance of *Wiktionary* as a lexical-semantic resource has been examined by multiple studies. Meyer et al. ([12,11]) presented an impressive overview on the importance and richness of *Wiktionary*. In [21] the authors presented the *JWKTL* framework to access *Wiktionary* dumps via a Java API. In [13] this JWKTL framework was used to construct an upper ontology called *OntoWiktionary*. The framework is reused within the *UBY project* [4], an effort to integrate multiple lexical resources (besides *Wiktionary* also *Word-Net, GermaNet, OmegaWiki, FrameNet, VerbNet* and *Wikipedia*). The resulting dataset is modelled according to the *LMF ISO standard*[6]. [14] and [18] discussed the use of *Wiktionary* to canonicalize annotations on cultural heritage texts, namely the Thompson Motif-index. Zesch et. al. also showed, that *Wiktionary* is suitable for calculating semantic relatedness and synonym detection; and it outperforms classic approaches [22,20]. Furthermore, other NLP tasks such as sentiment analysis have been conducted with the help of *Wiktionary* [2]. Several questions arise, when evaluating the above approaches: Why are there not more NLP tools reusing the free *Wiktionary* data? Why are there no web mashups of the data[4]? Why has *Wiktionary* not become the central linking hub of lexical-semantic resources, yet?

From our point of view, the answer lies in the fact, that although the above papers presented various desirable properties and many use cases, they did not solve the underlying knowledge extraction and data integration task sufficiently in terms of coverage, precision and flexibility. Each of the approaches presented in Table 1 relies on tools to extract machine-readable data in the first place. In our opinion these tools should be seen independent from their respective usage and it is not our intention to comment on the scientific projects built upon them in any way here. We will show the state of the art and which open questions they raise.

---

[4] For example in an online dictionary from `http://en.wikipedia.org/wiki/List_of_online_dictionaries`

JWKTL is used as data backend of *OntoWiktionary* as well as UBY[5] and features a modular architecture, which allows the easy addition of new extractors (for example *wikokit* [8] is incorporated). The Java binaries and the data dumps in LMF are publicly available. Among other things, the dump also contains a mapping from concepts to lexicalizations as well as properties for part of speech, definitions, synonyms and subsumption relations. The available languages are English, German (both natively) and Russian through *wikokit*. According to our judgement, *JWKTL* can be considered the most mature approach regarding software architecture and coverage and is the current state of the art. *Texai*[6] and *Blexisma*[7] are also Java based APIs, but are not maintained anymore and were most probably made obsolete by changes to the *Wiktionary* layout since 2009. There is no documentation available regarding scope or intended granularity. A very fine grained extraction was conducted using WISIGOTH [17], but unfortunately there are no sources available and the project is unmaintained since 2010. Two newer approaches are the *lexvo.org* service and the algorithm presented in [9]. The *lexvo.org* service offers a linked data representation of *Wiktionary* with a limited granularity, namely it does not disambiguate on sense level. The source code is not available and only the English *Wiktionary* is parsed. As part of the Monnet project[8], McCrae et al. [9] presented a simple scraper to transform *Wiktionary* to the *lemon* RDF model [10]. The algorithm (like many others) makes assumptions about the used page schema and omits details about solving common difficulties as shown in the next section. At the point of writing, the sources are not available, but they are expected to be published in the future. Although this approach appears to be the state of the art regarding RDF modelling and linking, the described algorithm will *not scale to the community-driven heterogeneity* as to be defined in Section 3. All in all, there exist various tools that implement extraction approaches at various levels of granularity or output format. In the next section, we will show several challenges that in our opinion are insufficiently tackled by the presented approaches. Note that this claim is not meant to diminish the contribution of the other approaches as they were mostly created for solving a single research challenge instead of aiming to establish *Wiktionary* as a stable point of reference in computational linguistics using linked data.

3   Problem Description

In order to conceive a flexible, effective and efficient solution, we survey in this section the challenges associated with Wiki syntax, *Wiktionary* and large-scale extraction.

---

[5] http://www.ukp.tu-darmstadt.de/data/lexical-resources/uby/,
http://www.ukp.tu-darmstadt.de/data/lexical-resources/uby/

[6] http://sourceforge.net/projects/texai/

[7] http://blexisma.ligforge.imag.fr/index.html

[8] See http://www.monnet-project.eu/. A list of the adopted languages and dump files can be found at http://monnetproject.deri.ie/lemonsource/ Special:PublicLexica

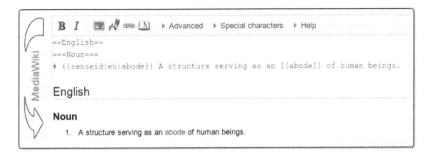

Fig. 1. An excerpt of the *Wiktionary* page *house* with the rendered HTML

3.1 Processing Wiki Syntax

Pages in *Wiktionary* are formatted using the *wikitext* markup language[9]. Operating on the parsed HTML pages, rendered by the *MediaWiki engine*, does not provide any significant benefit, because the rendered HTML does not add any valuable information for extraction. Processing the database backup XML dumps[10] instead, is convenient as we could reuse the DBpedia extraction framework[11] in our implementation. The framework mainly provides input and output handling and also has built-in multi-threading by design. Actual features of the wikitext syntax are not notably relevant for the extraction approach, but we will give a brief introduction to the reader, to get familiar with the topic. A wiki page is formatted using the lightweight (easy to learn, quick to write) markup language *wikitext*. Upon request of a page, the MediaWiki engine renders this to an HTML page and sends it to the user's browser. An excerpt of the *Wiktionary* page *house* and the resulting rendered page are shown in Figure 1.

The markup == is used to denote headings, # denotes a numbered list with * for bullets, [[link label]] denotes links and {{}} calls a template. Templates are user-defined rendering functions that provide shortcuts aiming to simplify manual editing and ensuring consistency among similarly structured content elements. In MediaWiki, they are defined on special pages in the `Template:` namespace. Templates can contain any wikitext expansion, HTML rendering instructions and placeholders for arguments. In the example page in Figure 1, the `senseid` template[12] is used, which does nothing being visible on the rendered page, but adds an id attribute to the HTML `li`-tag, which is created by using #. If the English *Wiktionary* community decides to change the layout of senseid definitions at some point in the future , only a single change to the template definition is required. Templates are used heavily throughout *Wiktionary*, because

---

[9] http://www.mediawiki.org/wiki/Markup_spec
[10] http://dumps.wikimedia.org/backup-index.html
[11] http://wiki.dbpedia.org/Documentation
[12] http://en.wiktionary.org/wiki/Template:senseid

they substantially increase maintainability and consistency. But they also pose a problem to extraction: on the unparsed page only the template name and its arguments are available. Mostly this is sufficient, but if the template adds static information or conducts complex operations on the arguments, which is fortunately rare, the template result can only be obtained by a running MediaWiki installation hosting the pages. The resolution of template calls at extraction time slows the process down notably and adds additional uncertainty.

3.2   Wiktionary

Wiktionary has some unique and valuable properties:

- **Crowd-sourced**
  Wiktionary is community edited, instead of expert-built or automatically generated from text corpora. Depending on the activeness of its community, it is up-to-date to recent changes in the language, changing perspectives or new research. The editors are mostly semi-professionals (or guided by one) and enforce a strict editing policy. Vandalism is reverted quickly and bots support editors by fixing simple mistakes and adding automatically generated content. The community is smaller than Wikipedia's but still quite vital (between 50 and 80 very active editors with more than 100 edits per month for the English *Wiktionary* in 2012[13]).
- **Multilingual**
  The data is split into different Wiktionary Language Editions (WLE, one for each language). This enables the independent administration by communities and leaves the possibility to have different perspectives, focus and localization. Simultaneously one WLE describes multiple languages; only the representation language is restricted. For example, the German *Wiktionary* contains German description of German words **as well as** German descriptions for English, Spanish or Chinese words. Particularly the linking across languages shapes the unique value of *Wiktionary* as a rich multi-lingual linguistic resource. Especially the WLE for not widely spread languages are valuable, as corpora might be rare and experts are hard to find.
- **Feature rich**
  As stated before, *Wiktionary* contains for each lexical word –A lexical word is just a string of characters and has no disambiguated meaning yet– a disambiguation regarding language, part of speech, etymology and senses. Numerous additional linguistic properties exist normally for each part of speech. Such properties include word forms, taxonomies (hyponyms, hyperonyms, synonyms, antonyms) and translations. Well maintained pages (e.g. frequent words) often have more sophisticated properties such as derived terms, related terms and anagrams.

---

[13] http://stats.wikimedia.org/wiktionary/EN/TablesWikipediaEN.htm

semantic

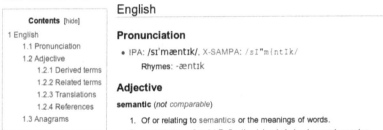

Fig. 2. Example page `http://en.wiktionary.org/wiki/semantic` and underlying schema, only valid for the English *Wiktionary*, as other WLE might look very different

- **Open license**
  All the content is dual-licensed under both the *Creative Commons CC-BY-SA 3.0 Unported License*[14] as well as the *GNU Free Documentation License (GFDL)*.[15] All the data extracted by our approach falls under the same licences.
- **Big and growing**
  English contains 2,9M pages, French 2,1M, Chinese 1,2M, German 0,2 M. The overall size (12M pages) of *Wiktionary* is in the same order of magnitude as Wikipedia's size (20M pages)[16]. The number of edits per month in the English *Wiktionary* varies between 100k and 1M — with an average of 200k for 2012 so far. The number of pages grows — in the English *Wiktionary* with approx. 1k per day in 2012.[17]

The most important resource to understand how *Wiktionary* is organized are the *Entry Layout Explained* (ELE) help pages. As described above, a page is divided into sections that separate languages, part of speech etc. The table of content on the top of each page also gives an overview of the hierarchical structure. This hierarchy is already very valuable as it can be used to disambiguate a lexical word. The schema for this tree is restricted by the ELE guidelines[18]. The entities illustrated in Figure 3.2 of the ER diagram will be called *block* from now on. The schema can differ between WLEs and normally evolves over time.

---

[14] `http://en.wiktionary.org/wiki/Wiktionary:Text_of_Creative_Commons_Attribution-ShareAlike_3.0_Unported_License`

[15] `http://en.wiktionary.org/wiki/Wiktionary:GNU_Free_Documentation_License`

[16] `http://meta.wikimedia.org/wiki/Template:Wikimedia_Growth`

[17] `http://stats.wikimedia.org/wiktionary/EN/TablesWikipediaEN.htm`

[18] For English see `http://en.wiktionary.org/wiki/Wiktionary:ELE`

3.3 Wiki-Scale Data Extraction

The above listed properties that make *Wiktionary* so valuable, unfortunately pose a serious challenge to extraction and data integration efforts. Conducting an extraction for specific languages at a fixed point in time is indeed easy, but it eliminates some of the main features of the source. To fully synchronize a knowledge base with a community-driven source, one needs to make distinct design choices to fully capture all desired benefits. MediaWiki was designed to appeal to non-technical editors and abstains from intensive error checking as well as formally following a grammar — the community gives itself just layout guidelines. One will encounter fuzzy modelling and unexpected information. Editors often see no problem with such "noise" as long as the page's visual rendering is acceptable. Overall, the main challenges can be summed up as (1) the constant and frequent changes to data *and schema*, (2) the heterogeneity in WLE schemas and (3) the human-centric nature of a wiki.

4   Design and Implementation

Existing extractors as presented in Section 2 mostly suffer from their *inflexible* nature resulting from their narrow use cases at development time. Very often approaches were only implemented to accomplish a short term goal (e.g. prove a scientific claim) and only the needed data was extracted in an *ad-hoc* manner. Such evolutionary development generally makes it difficult to generalize the implementation to heterogeneous schemas of different WLE. Most importantly, however, they ignore the community nature of a *Wiktionary*. Fast changes of the data require ongoing maintenance, ideally by the wiki editors from the community itself or at least in tight collaboration with them. These circumstances pose serious requirements to software design choices and should not be neglected. All existing tools are rather monolithic, hard-coded black boxes. Implementing a new WLE or making a major change in the WLE's ELE guidelines will require a programmer to refactor most of its application logic. Even small changes like new properties or naming conventions will require software engineers to align settings. The amount of maintenance work necessary for the extraction correlates with change frequency in the source. Following this argumentation, a community-built resource can only be efficiently extracted by a community-configured extractor. This argument is supported by the successful crowd-sourcing of DBpedia's internationalization [7] and the non-existence of *open* alternatives with equal extensiveness.

Given these findings, we can now conclude four high-level requirements:

- declarative description of the page schema;
- declarative information/token extraction, using a terse syntax, maintainable by non-programmers;
- configurable mapping from language-specific tokens to a global vocabulary;
- fault tolerance (uninterpretable data is skipped).

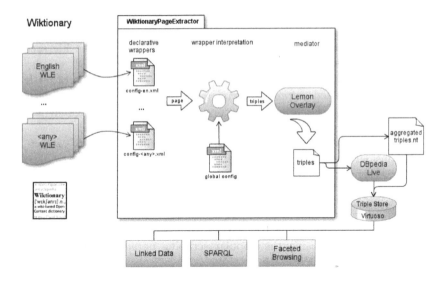

Fig. 3. Architecture for extracting semantics from Wiktionary leveraging the DBpedia framework

We solve the above requirements by proposing an extension to the DBpedia framework (in fact an additional extractor), which follows a rather sophisticated workflow, shown in Figure 3.

The *Wiktionary* extractor is invoked by the DBpedia framework to handle a page. It therefore uses a language-specific configuration file, that has to be tailored to match the WLE's ELE guidelines to interpret the page. At first, the resulting triples still adhere to a language-specific schema, that directly reflects the assumed layout of the WLE. A generic lossless transformation and annotation using the *lemon* vocabulary is then applied to enforce a global schema and reduce semantic heterogeneity. Afterwards the triples are returned to the DBpedia frameworks, which takes care of the serialization and (optionally) the synchronization with a triple store via DBpedia Live[19]. The process of interpreting the declarative wrapper is explained in more detailed in Figure 4.

Fig. 4. Overview of the extractor workflow

---

[19] http://live.dbpedia.org/live

4.1   Extraction Templates

As mentioned in Section 3.2, we define *block* as the part of the hierarchical page that is responsible for a certain entity in the extracted RDF graph. For each *block*, there can be declarations on how to process the page on that level. This is done by so called *extraction templates*(ET) (not to be confused with the templates of *wikitext*). Each possible section in the *Wiktionary* page layout (i.e. each linguistic property) has an ET configured (explained in detail below). The idea is to provide a declarative and intuitive way to encode *what to extract*. For example consider the following page snippet:

```
1   ===Synonyms===
2   * [[building]]
3   * [[company]]
```

Since the goal is to emit a link to each resource per line, we can write the ET in the following style, using the popular scraping paradigms such as regular expressions:

```
1   ===Synonyms===
2   (* [[\$target]]
3   )+
```

Some simple constructs for variables "$target" and loops "(*", ")+" are defined for the ET syntax. If they are *matched against* an actual wiki page, *bindings* are extracted by a matching algorithm. We omit a low-level, technical description of the algorithm — one can think of it like a Regular Expression *Named Capturing Group*. The found *variable bindings* for the above example are {(target->building), (target->company)}. The triple generation rule encoded in XML looks like:

```
1   <triple s="http://some.ns/$entityId" p="http://some.ns/hasSynonym" o="http://some.ns/$target" />
```

Notice the reuse of the $target variable: The data extracted from the page is inserted into a triple. The variable $entityId is a reserved global variable, that holds the page name e.g. the word. The created triples in N-Triples syntax are:

```
1   <http://some.ns/house>    <http://some.ns/hasSynonym>    <http://some.ns/building> .
2   <http://some.ns/house>    <http://some.ns/hasSynonym>    <http://some.ns/company> .
```

The actual patterns are more complex, but the mechanism is consistently used throughout the system.

4.2   Algorithm

The algorithm of processing a page works as follows:

Input: Parsed page obtained from the DBpedia Framework (essentially a lexer is used to split the Wiki Syntax into tokens)

1. Filter irrelevant pages (user/admin pages, statistics, list of things, files, templates, etc.) by applying string comparisons on the page title. Return an empty result on that condition.

2. Build a finite state automaton[20] from the page layout encoded in the WLE specific XML configuration. This schema also contains so called *indicator templates* for each *block*, that — if they match at the current page token — indicate that their respective block starts. So they trigger state transitions. In this respect the mechanism is similar to [9], but in contrast our approach is declarative — the automaton is constructed *on-the-fly* and not hard-coded. The current state represents the current position in the disambiguation tree.

3. The page is processed token by token:

   (a) Check if *indicator templates* match. If yes, the corresponding block is entered. The *indicator templates* also emit triples like in the *extraction template* step below. These triples represent the block in RDF – for example the resource `http://wiktionary.dbpedia.org/resource/semantic-English` represents the English block of the page "semantic".

   (b) Check if any *extraction template* of the current block match.
      – If yes, transform the variable bindings to triples.[21] Localization specific tokens are replaced as configured in the so called *language mapping* (explained in detail in section 4.3).

4. The triples are then *transformed*. In our implementation *transformation* means, that all triples are handed to a static function, which return a set of triples again. One could easily load the triples into a triple store like JENA and apply arbitrary SPARQL Construct and Update transformations. This step basically allows post-processing, e.g. consolidation, enrichment or annotation. In our case, we apply the schema transformation (by the mediator) explained in detail in Section 4.4).

5. The triples are sorted and de-duplicated to remove redundancy in the RDF dumps.

Output: Set of triples (handed back to the DBpedia Framework).

4.3   Language Mapping

The language mappings are a very simple way to translate and normalize tokens, that appear in a WLE. In the German WLE, for example, a noun is described with the German word "*Substantiv*". Those tokens are translated to a shared vocabulary, before emitting them (as URIs for example). The configuration is also done within the language specific XML configuration:

```
1  <mapping from="Substantiv" to="Noun">
2  <mapping from="Deutsch" to="German">
3  ...
```

---

[20] Actually a finite state transducer, most similar to the Mealy-Model.

[21] In our implementation: Either declarative rules are given in the XML config or alternatively static methods are invoked on user-defined classes (implementing a special interface) for an imperative transformation. This can greatly simplify the writing of complex transformation.

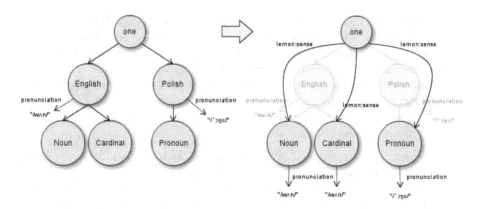

Fig. 5. Schema normalization

4.4   Schema Mediation by Annotation with *lemon*

The last step of the data integration process is the schema normalization. The global schema of all WLE is not constructed in a centralized fashion — instead we found a way to both making the data globally navigable and keeping the heterogeneous schema without loosing information. *lemon* [10] is an RDF model for representing lexical information (with links to ontologies — possibly DBpedia). We use part of that model to encode the relation between *lexical entries* and *lexical senses*. *lemon* has great potential of becoming the *de facto* standard for representing dictionaries and lexica in RDF and is currently the topic of the OntoLex W3C Community group[22]. The rationale is to add *shortcuts* from *lexical entities* to *senses* and propagate properties that are along the intermediate nodes down to the senses. This can be accomplished with a generic algorithm (a generic tree transformation, regardless of the depth of the tree and used links). Applications assuming only a *lemon* model, can operate on the shortcuts and if applied as an overlay — leaving the original tree intact — this still allows applications, to also operate on the actual tree layout. The (simplified) procedure is presented in Figure 5[23]. The use of the *lemon* vocabulary and model as an additional schema layer can be seen as our mediator. This approach is both lightweight and effective as it takes advantage of *multi-schema modelling*.

5   Resulting Data

The extraction has been conducted as a proof-of-concept on four major WLE: The English, French, German and Russian *Wiktionary*. The datasets combined

---

[22] http://www.w3.org/community/ontolex/

[23] Note, that in the illustration it could seem like the information about part-of-speech would be missing in the *lemon* model. This in not the case. Actually from the part-of-speech nodes, there is a link to corresponding language nodes. These links are also propagated down the tree.

Table 2. Statistical comparison of extractions for different languages. XML lines measures the number of lines of the XML configuration files.

language	#words	#triples	#resources	#predicates	#senses	XML lines
en	2,142,237	28,593,364	11,804,039	28	424,386	930
fr	4,657,817	35,032,121	20,462,349	22	592,351	490
ru	1,080,156	12,813,437	5,994,560	17	149,859	1449
de	701,739	5,618,508	2,966,867	16	122,362	671

Table 3. Statistical quality comparison

language	t/w	#wws	s/wws	t/l
en	13.35	591,073	1.39	2.70
fr	7.52	750,206	1.26	1.73
ru	11.86	211,195	1.40	2.25
de	8.01	176,122	1.43	1.06

contain more than 80 million facts. The data is available as N-Triples dumps[24], Linked Data[25], via the *Virtuoso Faceted Browser*[26] or a SPARQL endpoint[27]. Table 2 compares the size of the datasets from a quantitative perspective.

The statistics show, that the extraction produces a vast amount of data with broad coverage, thus resulting in the largest lexical linked data resource. There might be partially data quality issues with regard to missing information (for example the number of *words with senses* seems to be relatively low intuitively), but detailed quality analysis has yet to be done. Instead we defined some simple quality measures that can be automatically computed.

Table 3 gives an assessment of the quality of the language configuration independent from the quality of the underlying source data:
t/w: *Triples per word.* The simplest measure of information density. $\#wws$: *Words with senses.* The number of words, that have at least one sense extracted. An indicator for the ratio of pages for which valuable information could be extracted, but consider stub pages, that are actually empty. s/wws: *Senses per word with sense.* Gives an idea of the average senses per word while ignoring unmaintained pages. t/l: *Triples per line.* The number of triples divided by the number of line breaks in the page source (plus one). Averaged across all pages.

6   Lessons Learned

Making unstructured sources machine-readable creates feedback loops. Although this is not yet proven by empirical data, the argument that extracting structured data from an open data source and making it freely available in turn encourages users of the extracted data to contribute to the source, seems reasonable. The clear

---

[24] http://downloads.dbpedia.org/wiktionary
[25] For example http://wiktionary.dbpedia.org/resource/dog
[26] http://wiktionary.dbpedia.org/fct
[27] http://wiktionary.dbpedia.org/sparql

incentive is to *get the data out again*. This increase in participation besides improving the source, also illustrates the advantages of machine readable data to common Wiktionarians. Such a positive effect from DBpedia supported the current *Wikidata*[28] project.

Suggested changes to Wiktionary. Although it's hard to persuade the community of far-reaching changes, we want to conclude how *Wiktionary* can increase its data quality and enable better extraction.

- **Homogenize Entry Layout across all WLE's.**
- **Use anchors to markup senses:** This implies creating URIs for senses. These can then be used to be more specific when referencing a *word* from another article. This would greatly benefit the evaluation of automatic anchoring approaches like in [13].
- **Word forms:** The notion of word forms (e.g. declensions or conjugations) is not consistent across articles. They are hard to extract and often not given.

7   Discussion and Future Work

Our main contributions are an extremely flexible extraction from *Wiktionary*, with simple adaption to new Wiktionaries and changes via a declarative configuration. By doing so, we are provisioning a linguistic knowledge base with unprecedented detail and coverage. The DBpedia project provides a mature, reusable infrastructure including a public Linked Data service and SPARQL endpoint. All resources related to our *Wiktionary* extraction, such as source-code, extraction results, pointers to applications etc. are available from our project page.[29] As a result, we hope it will evolve into a central resource and interlinking hub on the currently emerging Web of Linguistic Data.

7.1   Next Steps

Wiktionary Live: Users constantly revise articles. Hence, data can quickly become outdated, and articles need to be re-extracted. DBpedia-Live enables such a continuous synchronization between DBpedia and Wikipedia. The WikiMedia foundation kindly provided us access to their update stream, the Wikipedia OAI-PMH[30] live feed. The approach is equally applicable to *Wiktionary*. The *Wiktionary* Live extraction will enable users for the first time ever to query *Wiktionary* like a database in real-time and receive up-to-date data in a machine-readable format. This will strengthen *Wiktionary* as a central resource and allow it to extend its coverage and quality even more.

Wiki Based UI for the WLE Configurations: To enable the crowd-sourcing of the extractor configuration, an intuitive web interface is desirable. Analogue to the mappings wiki[31] of DBpedia, a wiki could help to hide the technical details of the

---

[28] http://meta.wikimedia.org/wiki/Wikidata
[29] http://wiktionary.dbpedia.org
[30] Open Archives Initiative Protocol for Metadata Harvesting,
     cf. http://www.mediawiki.org/wiki/Extension:OAIRepository
[31] http://mappings.dbpedia.org/

configuration even more. Therefore a JavaScript based WYSIWYG XML editor seems useful. There are various implementations, which can be easily adapted.

Linking: Finally, an alignment with existing linguistic resources like WordNet and general ontologies like YAGO or DBpedia is essential. That way *Wiktionary* will allow for the interoperability across a multilingual semantic web.

7.2   Open Research Questions

Publishing Lexica as Linked Data. The need to publish lexical resources as linked data has been recognized recently [16]. Although principles for publishing RDF as Linked Data are already well established [1], the choice of identifiers and first-class objects is crucial for any linking approach. A number of questions need to be clarified, such as which entities in the lexicon can be linked to others. Obvious candidates are entries, senses, synsets, lexical forms, languages, ontology instances and classes, but different levels of granularity have to be considered and a standard linking relation such as `owl:sameAs` will not be sufficient. Linking across data sources is at the heart of linked data. An open question is how lexical resources with differing schemata can be linked and how are linguistic entities to be linked with ontological ones. There is most certainly an impedance mismatch to bridge.

The success of DBpedia as a "crystallization point for the Web of Data" is predicated on the stable identifiers provided by Wikipedia and are an obvious prerequisite for any data authority. Our approach has the potential to drive this process by providing best practices and live showcases and data in the same way DBpedia has provided it for the LOD cloud. Especially, our work has to be seen in the context of the recently published Linguistic Linked Data Cloud[3] and the community effort around the Open Linguistics Working Group (OWLG)[32] and NIF [5]. Our Wiktionary conversion project provides valuable data dumps and linked data services to further fuel development in this area.

Algorithms and Methods to Bootstrap and Maintain a Lexical Linked Data Web. State-of-the-art approaches for interlinking instances in RDF knowledge bases are mainly build upon similarity metrics [15,19] to find duplicates in the data, linkable via `owl:sameAs`. Such approaches are not directly applicable to lexical data. Existing linking properties either carry strong formal implications (e.g. `owl:sameAs`) or do not carry sufficient domain-specific information for modelling semantic relations between lexical knowledge bases.

References

1. Auer, S., Lehmann, J.: Making the web a data washing machine - creating knowledge out of interlinked data. Semantic Web Journal (2010)
2. Chesley, P., Vincent, B., Xu, L., Srihari, R.K.: Using verbs and adjectives to automatically classify blog sentiment. In: AAAI Spring Symposium (2006)
3. Chiarcos, C., Hellmann, S., Nordhoff, S., Moran, S., Littauer, R., Eckle-Kohler, J., Gurevych, I., Hartmann, S., Matuschek, M., Meyer, C.M.: The open linguistics working group. In: LREC (2012)

---

[32] `http://linguistics.okfn.org`

4. Gurevych, I., Eckle-Kohler, J., Hartmann, S., Matuschek, M., Meyer, C.M., Wirth, C.: Uby - a large-scale unified lexical-semantic resource based on lmf. In: EACL 2012 (2012)
5. Hellmann, S., Lehmann, J., Auer, S.: Linked-data aware URI schemes for referencing text fragments. In: ten Teije, A., Völker, J., Handschuh, S., Stuckenschmidt, H., d'Acquin, M., Nikolov, A., Aussenac-Gilles, N., Hernandez, N. (eds.) EKAW 2012. LNCS, vol. 7603, pp. 175–184. Springer, Heidelberg (2012)
6. ISO 24613:2008. Language resource management – Lexical markup framework. ISO, Geneva, Switzerland
7. Kontokostas, D., Bratsas, C., Auer, S., Hellmann, S., Antoniou, I., Metakides, G.: Internationalization of Linked Data: The case of the Greek DBpedia edition. Journal of Web Semantics (2012)
8. Krizhanovsky, A.A.: Transformation of wiktionary entry structure into tables and relations in a relational database schema. CoRR (2010), http://arxiv.org/abs/1011.1368
9. McCrae, J., Cimiano, P., Montiel-Ponsoda, E.: Integrating WordNet and Wiktionary with lemon. In: Chiarcos, C., Nordhoff, S., Hellmann, S. (eds.) Linked Data in Linguistics. Springer (2012)
10. McCrae, J., Spohr, D., Cimiano, P.: Linking Lexical Resources and Ontologies on the Semantic Web with Lemon. In: Antoniou, G., Grobelnik, M., Simperl, E., Parsia, B., Plexousakis, D., De Leenheer, P., Pan, J. (eds.) ESWC 2011, Part I. LNCS, vol. 6643, pp. 245–259. Springer, Heidelberg (2011)
11. Meyer, C.M., Gurevych, I.: How web communities analyze human language: Word senses in wiktionary. In: Second Web Science Conference (2010)
12. Meyer, C.M., Gurevych, I.: Worth its weight in gold or yet another resource — A comparative study of wiktionary, openThesaurus and germaNet. In: Gelbukh, A. (ed.) CICLing 2010. LNCS, vol. 6008, pp. 38–49. Springer, Heidelberg (2010)
13. Meyer, C.M., Gurevych, I.: OntoWiktionary – Constructing an Ontology from the Collaborative Online Dictionary Wiktionary. In: Semi-Automatic Ontology Development: Processes and Resources. IGI Global (2011)
14. Moerth, K., Declerck, T., Lendvai, P., Váradi, T.: Accessing multilingual data on the web for the semantic annotation of cultural heritage texts. In: 2nd Workshop on the MSW, ISWC (2011)
15. Ngonga Ngomo, A.-C., Auer, S.: Limes - a time-efficient approach for large-scale link discovery on the web of data. In: Proceedings of IJCAI (2011)
16. Nuzzolese, A.G., Gangemi, A., Presutti, V.: Gathering lexical linked data and knowledge patterns from framenet. In: K-CAP (2011)
17. Sajous, F., Navarro, E., Gaume, B., Prévot, L., Chudy, Y.: Semi-automatic Endogenous Enrichment of Collaboratively Constructed Lexical Resources: Piggybacking onto Wiktionary. In: Loftsson, H., Rögnvaldsson, E., Helgadóttir, S. (eds.) IceTAL 2010. LNCS (LNAI), vol. 6233, pp. 332–344. Springer, Heidelberg (2010)
18. Mörth, K., Budin, G., Declerck, T., Lendvai, P., Váradi, T.: Towards linked language data for digital humanities
19. Volz, J., Bizer, C., Gaedke, M., Kobilarov, G.: Discovering and maintaining links on the web of data. In: Bernstein, A., Karger, D.R., Heath, T., Feigenbaum, L., Maynard, D., Motta, E., Thirunarayan, K. (eds.) ISWC 2009. LNCS, vol. 5823, pp. 650–665. Springer, Heidelberg (2009)
20. Weale, T., Brew, C., Fosler-Lussier, E.: Using the wiktionary graph structure for synonym detection. In: The People's Web Meets NLP, ACL-IJCNLP (2009)
21. Zesch, T., Müller, C., Gurevych, I.: Extracting Lexical Semantic Knowledge from Wikipedia and Wiktionary. In: LREC (2008)
22. Zesch, T., Müller, C., Gurevych, I.: Using wiktionary for computing semantic relatedness. In: AAAI (2008)

Navigation-Induced Knowledge Engineering by Example

A New Paradigm
for Knowledge Engineering by the Masses

Sebastian Hellmann[1], Jens Lehmann[1], Jörg Unbehauen[1], Claus Stadler[1],
Thanh Nghia Lam[1], and Markus Strohmaier[2]

[1] Universität Leipzig, Institut für Informatik,
AKSW, Postfach 100920, D-04009 Leipzig, Germany
{hellmann,lehmann,unbehauen,cstadler}@informatik.uni-leipzig.de
[2] Graz University of Technology and Know-Center
Inffeldgasse 21a, 8010 Graz, Austria
markus.strohmaier@tugraz.at

Abstract. Knowledge Engineering is a costly, tedious and often time-consuming task, for which light-weight processes are desperately needed. In this paper, we present a new paradigm - Navigation-induced Knowledge Engineering by Example (NKE) - to address this problem by producing structured knowledge as a result of users navigating through an information system. Thereby, NKE aims to reduce the costs associated with knowledge engineering by framing it as navigation. We introduce and define the NKE paradigm and demonstrate it with a proof-of-concept prototype which creates OWL class expressions based on users navigating in a collection of resources. The overall contribution of this paper is twofold: (i) it introduces a novel paradigm for knowledge engineering and (ii) it provides evidence for its technical feasibility.

Keywords: Navigation, Knowledge Engineering, Paradigm, Methodology, Ontology Learning, Search, OWL.

1 Introduction

Over the past years, structured data has increasingly become available on the World Wide Web (WWW). Yet, the actual usage of this data still poses significant barriers for lay users. One of the main drawbacks to the utilization of the structured data on the WWW lies in the blatant cognitive gap between the informational needs of users and the structure of existing knowledge bases. In this paper, we propose a novel paradigm - Navigation-induced Knowledge Engineering by Example (NKE) - which aims to bridge this gap.

Due to the sheer size of large knowledge bases, users can hardly know which identifiers are available or useful for the construction of axioms or queries. As a consequence, users might not be able to express their informational need in a structured form. Yet, users often have a very precise idea of what kind of results

H. Takeda et al. (Eds.): JIST 2012, LNCS 7774, pp. 207–222, 2013.

they would like to retrieve. A historian, for example, searching DBpedia [16] for `ancient Greek law philosophers influenced by Plato` can easily name some examples and - when presented with a selection of prospective results - she will be able to quickly identify correct and incorrect results. However, she might not be able to efficiently construct a formal query adhering to the large DBpedia knowledge base.

In this paper, we will argue and show that Navigation-induced Knowledge Engineering by Example can tackle important parts of the above described information gap problem.

In NKE, the informational need of a user is approximated, e.g. by letting the user formulate preferences or simply by browsing an application. From this interaction, so called *positive and negative examples* can be inferred that are then used as an input to a supervised machine learning algorithm. In a final step, generated knowledge from several user interactions is combined into a taxonomy, which forms the basis for the knowledge engineering process. Following that process, NKE produces structured knowledge as a by-product of users navigating through an information system. Navigation-induced Knowledge Engineering by Example thereby serves several purposes at the same time; it (i) aids users in expressing their informational needs in a structured way (ii) helps them in navigating to resources in a given system and (iii) produces structured knowledge as a result of this process.

Most traditional Knowledge Engineering methodologies heavily rely on a phase-oriented model built on collaboration of a centralized team of domain experts and ontology engineers[21,22,26]. In NKE, web users take the role of domain experts and elicitation is done *en passant* during the navigation process.

The vision of NKE is to enable low-cost knowledge engineering on the largest possible scale - the World Wide Web. The most fundamental consequence of the paradigm is that value is added to data by having a large number of users navigating, using and interacting with it. A reciprocal relation is formed between the informational need of users and the information gained through the created taxonomy. Although structured data is becoming widely available, no other methodology or paradigm – to the best of our knowledge – is currently able to scale up and provide light-weight knowledge engineering for a massive user base. Using NKE, data providers can publish flat data on the World Wide Web without creating a detailed structure *upfront*, but rather observe how structure is created *on the fly* by interested users who navigate the knowledge base.

In summary, this paper makes the following contributions. It

- introduces Navigation-induced Knowledge Engineering by Example (NKE) as a new paradigm for knowledge engineering
- presents a proof-of-concept (HANNE) to demonstrate technical feasibility
- illustrates the new paradigm in an e-commerce context and a query-answering system

The paper is structured as follows: We define Navigation-induced Knowledge Engineering by Example in Section 2 and explain its concepts in detail. To demonstrate the technical feasibility of NKE, we present HANNE – a Holistic

Application for Navigation-induced Knowledge Engineering by Example – in Section 3. HANNE is an Active Machine Learning tool based on Inductive Logic Programming that allows for the extraction of formal definitions (OWL Class Expression) of user-defined concepts based on corresponding examples from arbitrary and possibly large RDF data sets. After we have presented HANNE as a proof-of-concept, we evaluate it in Section4. In Section 5, we review related work on NKE, two fields that we will connect in our work. Finally, we conclude and describe future work.

2   Navigation-Induced Knowledge Engineering by Example - A New Paradigm

In this section, we define NKE and give an explanation of the key concepts and requirements related to this paradigm.

Definition: *Navigation-induced Knowledge Engineering by Example is the manifestation of labeled examples by interpreting user navigation combined with the active correction and refinement of these examples by the user to create an ontology of user interests through supervised active machine learning.*

When a web site is displayed in a browser, links are presented to the user for selection. Users typically select a subset of these links to navigate to a particular resource or set of resources. However, as web sites are heterogeneous and thus present a multitude of heterogeneous links, it is difficult - if not impossible - to make proper assumptions about the users' informational needs that are driving their underlying navigation behavior. If we, however, constrain our focus to web sites serving homogeneous content, such as a list of products, people or bookmarks, it becomes easier to analyze the goal of a user more clearly.

The NKE paradigm focuses on those websites, where objects with some form of defined semantics are available, such as Amazon products or Wikipedia articles. As the user is presented with a list of links to such objects, selecting and clicking on a link can be interpreted as positive feedback. All other links are neglected and can be interpreted as negative feedback. This interpretation is, of course, an oversimplification and often wrong: A user might accidentally click on a link or follow a link and then realize, that the target is not what she was looking for. Furthermore not only the selected item of a list might be of interest, but others as well. In addition, it normally remains hidden to a web system, whether the informational need of a user changes during the course of a visit. As soon as e.g. a product is found, the next user action might be triggered by a different need[1]. In many cases however, especially in more interactive systems, it is feasible to approximate the informational needs of a user by observing his interactions with the system.

Navigation-induced Knowledge Engineering by Example: The NKE paradigm consists of three distinct yet interrelated steps: **(i) Navigation:** NKE

---

[1] Adding the product to a shopping cart could be a good indicator for such a change.

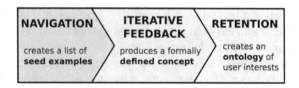

Fig. 1. NKE combines navigational methods with active iterative relevance feedback to create a preliminary ontology

starts by interpreting navigational behavior of users to *infer* an initial (seed) set of positive and negative examples. **(ii) Iterative Feedback:** NKE supports users in *interactively refining* the seed set of examples such that the final set of objects satisfies the users' intent. and **(iii) Retention:** NKE allows users to *retain previously explored sets of objects* by grouping them and saving them for later retrieval. Thus, the idea of NKE is to use clues from navigational behavior of users in a given system to infer a seed set of positive and negative examples that are later refined interactively by users to advance towards their search goal. This set of examples is later used to infer semantic structures in an active machine learning task.

In the following, we will formulate the underlying requirements related to the three steps in greater detail:

(i) Navigation: *The first requirement for NKE is the ability of a system to approximate the informational need of a user and produce positive and/or negative examples.* Many ways of approximating users' informational needs can be envisioned and are deployed in a multitude of traditional recommender systems. One way of approximating users' needs was followed in the DBpedia Navigator, an early prototype by Lehmann et al. [20]. The DBpedia Navigator could be used to browse over Wikipedia/DBpedia articles. Each viewed article was added automatically to the list of positive examples. A user then could review this list and decide to move entries to the list of negative examples, instead. Another well-known recommender system, which is based on user interaction, is the Amazon.com sales web site. Each view of a product is remembered and statistically analyzed to give a wide variety of personalized suggestions[2]: "More Items to Consider", "Customers with Similar Searches Purchased", "Bestsellers Electronics: Point & Shoot Digital Cameras", "The Best Prices on the Most Laptops", "Customers Who Bought Items in Your Recent History Also Bought" are some examples. The most prominent distinction, however, is the clear lack to explicitly give feedback and refine the presented recommendations.

(ii) Iterative Feedback: *The second requirement for NKE is to support the user in actively managing the list of examples to steer the learning process.* In NKE, the user expresses her informational need by creating a list of positive and negative examples. Although the initial list is gathered automatically by a system as an interpretation of navigational behavior, a chance for correction and iterative refinement is given at a later stage. With this requirement, the

---

[2] Taken from the frontpage of `http://amazon.com` accessed on Oct, 13th 2010.

paradigm gives control to the user, who can actively model her search inquiry based on a seed list of examples. Examples selected by users can be seen as a gold standard of labeled data for active machine learning and the learning result can be used to suggest more objects for labeling.

(iii) Retention: *The third requirement for NKE is to enable the user to sufficiently refine and review the learned result, and let her save it for later retrieval.* Retention is a critical part of the NKE paradigm. After the phase of iterative feedback is concluded, the user has to be able to judge whether the learning result matches her needs and is worth saving. To be able to re-use the saved concept, NKE requires users to assign a name to it. The philosophy is straight-forward: A concept, which is saved by one user to ease further navigation, is likely to be useful to other users as well. As we will see later, the saved concepts will form a taxonomy of user interests, which can be directly exploited as navigational suggestions. Also, the created taxonomy can be considered a raw material, which can be facilitated into a full-fledged domain ontology at low cost.

In the following, we explain the concepts involved in NKE in more detail:

Knowledge Source: NKE requires objects that are represented in a structured form and stored within a knowledge base. The following are typical examples:
- *OWL Individuals* in an OWL knowledge base and their RDF properties.
- saved *bookmarks* on Delicious[3] and their tags.
- *products* on Amazon[4] and the product properties.
- *newspaper articles* in a newspaper database and the article attributes like authors, keywords or links to other articles.

Note that the latter three examples can also be modeled in RDF and OWL[5], which we use for our demonstration. The NKE paradigms can be applied to all formalisms fitting the resource-feature scheme.

Supervised Machine Learning Algorithm: As users choose exemplary resources from the knowledge source, the material for applying a supervised machine learning algorithm is prepared. This algorithm can easily be exchanged and optimized according to the data structure of the knowledge source. In our implementation, we use an algorithm (Inductive Logic Programming) that relies on positive and negative examples, but positive-only or negative-only can be sufficient when using other algorithms. Furthermore, the given examples do not need to be binary in any way and could be assigned a weight, instead. The only limitation is that the algorithm needs to produce learned concepts, adhering to the requirements below.

Learned Concepts: The properties of the learned concepts are central to the NKE paradigm. The learned concepts need to serve as a classifier. This classifier can be either binary (retrieving only those resources from the pool that are covered) or assign a weight (e.g. between 0 and 1) to every resource[6].

---

[3] http://www.delicious.com
[4] http://amazon.com
[5] For tags, see [13]. For products, see [10].
[6] If the weight is combined with a threshold, the classifier becomes binary again.

The retrieved set of resources is called $r_{classified}$ or *extension* of the concept. As each learned concept is defined by its extension, they form a partial order by inclusion: Given learned concepts C and D, D is a subconcept of C, iff $r_{classified\ by\ D} \subseteq r_{classified\ by\ C}$. Therefore resources, which are retrieved by a learned concept will also be retrieved by all higher order concepts. The ordering relation is important. As learned concepts can be saved by a user for retention, the ordering relation clearly creates a distinction between user generated data (such as tags, which have no structure per se) and user generated knowledge.

If the classifier is additionally backed by a formalism for an intensional definition, a binary relation can be defined, which should have the same or similar properties as the inclusion relation on the extensions. Naturally, OWL-DL fulfills all the requirements for such a formalism. The subclassOf relation ($\sqsubseteq$) – as it is transitive and reflexive – creates a preorder over OWL class expressions.

Exploratory Search with Iterative Refinement: In our approach, learned concepts can be understood in the following way: As the user explores a knowledge base, she is interested in certain kinds of resources, i.e. she tries to find a set of resources r_{target} that matches her informational need such as *All bookmarks on Java tutorials covering Spring* or *All notebooks with more than 2GB, Ubuntu and costing less than 400 euros*. To express her need, she navigates to resources thereby providing a seed subset of examples $r_0 \subset r_{target}$ in iteration 0. During each iteration i (with i ranging from 0...n), the learning algorithm proposes to the user a new set of resources $r_{classified}$ retrieved via the learned concept. The user then selects more resources from $r_{classified}$ and adds them to r_i creating a new set r_{i+1}. This process can be repeated by the user, until she considers the learned concept a *solution* $lc_{solution}$. The standard measures recall, precision and F-measure apply. The learned concept is correct, if $r_{target} = r_{classified}$.

Two basic assumptions underly our notion of exploratory search: 1. the user either knows all the members of r_{target} or she can quickly evaluate membership with the help of the presented information. 2. Furthermore, the user should be able to make an educated guess about the size of r_{target}. NKE therefore requires an informed user, who can judge whether the search was successful. Although this seems to be a hard requirement for a user, we argue that it can be met quite easily in most cases. Albeit, one limitation of the NKE paradigm is that users who do not know how to evaluate candidate results might be more successful with other methods.

We can also identify several reasons why a NKE-based search might fail: 1. a solution $lc_{solution}$ exists, but the learning algorithm is unable to find it 2. a solution does not exist, because the knowledge source lacks the necessary features and 3. the user selects examples that contradict her informational needs.

Generated Ontology: Saving a learned concept plays a central role in NKE. The design of any NKE system should create strong incentives for users to save solutions once they are found. One such incentive is the ability to retain sets or to view or export a subset of resources $r_{classified}$ for later retrieval. Additionally it is necessary that the user assigns a label upon saving. This way a hierarchy

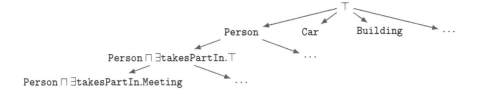

Fig. 2. Illustration of a search tree in OCEL

of terms is created which forms an ontology for the domain. Such an ontology – as it was created by users to query resources – is a useful candidate to support future navigation or other tasks.

3   Proof-of-Concept: An OWL-Based Implementation

In this section, we introduce and present a proof-of-concept implementation (HANNE[7]), where we employ *DL-Learner* [15] to learn class expressions. We first briefly describe DL-Learner before we explain the HANNE prototype.

3.1   HANNE: Technical Background

DL-Learner[15][8] extends Inductive Logic Programming to Description Logics (DLs), OWL and the Semantic Web. It provides a DL/OWL-based machine learning framework to solve supervised learning tasks and support knowledge engineers in constructing knowledge. In this paper, we use the OCEL algorithm implemented in DL-Learner, because its induced classes are short and readable. OWL class expressions form a subsumption hierarchy that is traversed by DL-Learner starting from the top element ($\top$ in DL syntax or *owl:Thing*) with the help of a refinement operator and an algorithm that searches in the space of generated classes. For instance, Figure 2 shows an excerpt of an OCEL search tree starting from the $\top$ concept, where the refinement operator has been applied for the class expressions $\top$, Person etc. The exact details of the construction and traversal of the search tree are beyond the scope of this paper.

When OCEL terminates, it returns the best element in its search tree with respect to a given learning problem. The path leading to such an element is called a refinement chain. The following is an example of such a chain:

$\top \rightsquigarrow$ Person $\rightsquigarrow$ Person $\sqcap$ takesPartinIn. $\top$

$\rightsquigarrow$ Person $\sqcap$ takesPartIn.Meeting

The way the refinement chains are constructed fits the iterative style of the NKE paradigm. DL-Learner supports the use of SPARQL endpoints, and scales

---

[7] http://hanne.aksw.org
[8] http://dl-learner.org

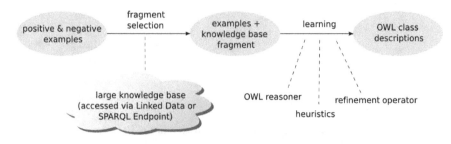

Fig. 3. Process illustration [9]: In a first step, a fragment is selected based on objects from a knowledge source and in a second step the learning process is started on this fragment and the given examples

to very large knowledge bases by using an approach that extracts fragments that are small enough for faciliating real time OWL reasoning[9]. The process is sketched in Figure 3.

3.2   HANNE User Interface

The user interface presented in Figure 4 is a domain-independent implementation of the NKE paradigm, that works on any SPARQL endpoint. For our demonstration, we chose DBpedia as an underlying knowledge base, where the defined goal is to find all 44 U.S. Presidents. Naturally, the list of all U.S. Presidents can be found much faster using e.g. Wikipedia[9]. For the long tail of arbitrary information, such precompiled lists are, however, not easily available. With HANNE, users are able to model such lists according to their information needs as a by-product of navigation.

In our demonstration, the user starts searching for "Bush" to create an initial set of examples. She uses the search components of HANNE marked with 1 in Fig. 4 for that purpose. From the retrieved list of instances, she can select George H.W. Bush and George W. Bush as positive and Kate and Jeb Bush as negative examples. This selection forms the seed set of instances for the second phase - iterative refinement - for which the components marked with 2 (Fig. 4) are used. The user initiates this phase by using the "start learning" button.

The iterative feedback is implemented as follows; using our 4 initial examples, dbo:Presidents is learned. By requesting the instance matches of the concept (button "matching"), the user can iteratively select more instances as either positive or negative examples and thereby refine the concept. The count of instance matches and the accuracy of the concept is displayed to help the user estimate whether the concept satisfies her navigation needs. After selecting 3 more positive examples (George Washington, Eisenhower, Roosevelt) and 2 more negatives (Tschudi and Rabbani) the concept has been narrowed down to (dbo:President and foaf:Person), which only covers 264 instances out of 3 million DBpedia resources. During the iterative feedback process, HANNE displays related concepts on the right side (marked with 3). These related concepts

---

[9] http://en.wikipedia.org/wiki/List_of_Presidents_of_the_United_States

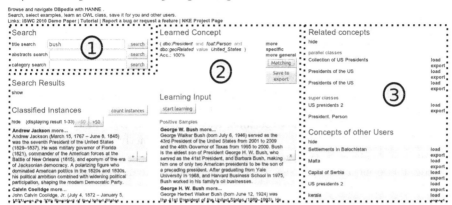

Fig. 4. Screenshot of `http://hanne.aksw.org`: US Presidents in DBpedia

were saved by other users and are either sub, parallel or super classes of the
learned concept. The retrieval of all 44 presidents can be successful in 3 different
ways: 1.) the iterative process is continued until all 44 presidents are added to
the positive list (successful retrieval of the extension) 2.) the learned concept
correctly retrieves all 44 presidents (e.g. `dbo:President and dbo:geoRelated`
`value United_States and dbo:spouse some Thing` retrieves 42 instances) or
3.) a previously saved concept matches the information need (e.g "Collection of
U.S. presidents" on the right side).

Solution 1 has created an *extensional* definition and solution 2 an *intensional*
definition of the search. Both can be saved and labeled by the user to retain it for
later or to export it (either a definition or the SPARQL query to retrieve the instances).

4   Evaluation

Our evaluation scenario is closely tied to the proof of concept implementation
presented in the previous section. The evaluation is based on Wikipedia categories, such as *Ships built in Michigan* or *Battles involving Prussia*, which are
included in DBpedia. The rationale for evaluating HANNE with the Wikipedia
category system are manifold: (1) the categories can be considered a hierarchical structure to more effectively group and browse Wikipedia articles (2) the
categories are maintained manually (which is very tedious and time-consuming)
and (3) they do not enforce a strict is-a relation to their member articles, which
means that the data contains errors from a supervised learning point of view.

From (1) follows that each category corresponds to a potential information
need of at least one user, which took the effort to create the collection. For each
category (e.g. *Wrestlers at the 1938 British Empire Games*[10]), we can formulate

---

[10] `http://en.wikipedia.org/wiki/Category:Wrestlers_at_the_1938_British_`
`Empire_Games`

a search task by preceding the category name with "Find all" . Our goal is to evaluate how satisfactory this search task can be addressed with regular search functionality and NKE based methods. (2) is relevant for evaluating the potential of NKE in aiding users in maintaining such a category hierarchy. Finally, (3) allows to assess NKE in realistic noisy scenarios. The evaluation is split in a quantitative and a qualitative part.

The quantitative part will evaluate how well the underlying machine learning approach is able to find all members of a category when compared with a keyword-based search. For the experiment, we used a SPARQL query to retrieve a list of categories from DBpedia, which contained exactly 100 members that had an infobox as well as an abstract property. This selection based on member count is done to avoid bias towards particular domains of categories. The abstract is required for the search engines and the infobox data is required as input for the underlying machine learning algorithm. We envision the following scenario: A user is intensively researching a topic given by the respective category without being able to use the category itself. The main goal of the user is to find **all** members exhaustively and is willing to invest some effort to reach it.

First, we simulate a keyword-based search. The keywords are initially generated from the category names, which we consider as sets of words. In a second step, we remove all stop words and build the power set over the remaining words W resulting in a set of $Q = \{q_1, \ldots, q_n\}$ queries with $n = 2^{|W|} - 1$, ("−1" as we exclude the empty word). Additionally, we optimized the strings by using the singular form and special treatment of hyphens ($-$) resulting for example in this set of queries: $\{\{Wrestler, 1938, British, Empire, Game\}, \{Wrestler, 1938, British, Empire\}, \{Wrestler, 1938, British, Game\}, \{Wrestler, 1938, Empire, Game\}, \ldots\}$. The search is conjunctive, so usually the queries with many words are very specific but return few results, whereas searching with fewer words finds more but less specific results. For each query set we vary the maximum number of results ($limit$) the user would look at to 20, 100 or 200. Let $results(q)$ be the number of results returned by query $q \in Q$ and $category$ the articles in the considered category. We calculate the precision and recall in the following manner:

$$union(\{q_1, \ldots, q_n\}) = \left| \left(\bigcup_{i=1}^{n} results(q_i) \right) \cap category \right|$$

$$best(\{q_1, \ldots, q_n\}) = \max_{i=1}^{n} |results(q_i) \cap category|$$

$$inter(\{q_1, \ldots, q_n\}) = \left| \bigcap_{i=1}^{n} results(q_i) \cap category \right|$$

$$R_{max} = \frac{union(Q)}{|category|} \quad R_{best} = \frac{best(Q)}{|category|} \quad R_{avg} = \frac{avg(Q)}{|category|} \quad P_{best} = \frac{best(Q)}{|limit|} \quad P_{min} = \frac{inter(Q)}{|limit|}$$

R_{max} is an optimistic recall estimate taking the combination of all queries into account, R_{best} only considers the query returning most results from the considered category and R_{avg} is a pessimistic estimate, which takes only those results into account, which have been returned by each query. Similarly, P_{best} is the

Table 1. Evaluation results for all three approaches (BC = Virtuoso bif:contains, SO = Solr, DL = DL-Learner) with three different limits

Limit	BC 20	SO 20	DL 20	BC 100	SO 100	DL 100	BC 200	SO 200	DL 200
R_{max}	0.03	0.07	0.07	0.12	0.22	0.29	0.17	0.28	0.31
R_{best}	0.03	0.06	0.06	0.11	0.20	0.28	0.15	0.25	0.30
R_{avg}	0.01	0.01	0.04	0.03	0.05	0.19	0.04	0.05	0.21
P_{best}	0.13	0.21	0.30	0.11	0.16	0.28	0.08	0.11	0.15
P_{min}	0.04	0.05	0.20	0.03	0.04	0.19	0.02	0.02	0.10

precision of the best query, i,e. the percentage of returned results for the best query. P_{min} considers the precision of the intersection of all queries.

We used two different query engines: (1) The Virtuoso[11] internal word index based on the titles and abstracts, which we ranked based according to the node in-degree following the search approach in the RelFinder tool [6]. (2) a Lucene/-Solr index, which was build from the titles and abstract and also ranked based on node in-degree.

The keyword-based search is now compared to an iterative refinement based on the DL-Learner algorithm used in HANNE. We assume that the user has basic domain knowledge and can name 5 members of the category, which serve as initial positive examples. In a realistic application such an initial set could have been selected from the navigation history, string search or facet-based browsing. Here, the 5 positive articles are chosen randomly from the 100 members. Then 5 negative examples are randomly chosen from the set of members of parallel categories, i.e. (non-equal) categories with the same direct predecessor in the category graph. The OCEL algorithm, implemented in DL-Learner, is then started and the resulting OWL Class Expression is transformed into a SPARQL query with the same *limit* as above. From the result set up to 5 correctly found articles are added to the positive examples as well as 5 negatives. This process is repeated 5 times. We can calculate precision and recall analogously by replacing the set of a set of keyword-based queries with the set of resulting SPARQL queries, one for each of the 5 iterations.

The results in Table 1 show that DL-Learner as a recommender is competitive with the search approaches we implemented and has higher values in all cases. Detailed results are available at http://aksw.org/Projects/NKE.

In this experiment, we are not aiming to establish that our implementation is better than a key word based search, as we neglected a plethora of optimization opportunities for string search such as more normalization, ngrams, query expansion techniques, relevance feedback, etc. We think of the searches we implemented as a reasonable baseline, which are actually used in many systems (e.g. RelFinder and DBpedia Lookup). While our searches are based on artificial queries, future research could conduct a similiar evaluation with other queries or actual queries obtained from a live system. In addition, the quality of the results for DL-Learner directly depends on the availability of rich semantics and data

---

[11] Virtuoso is the triple store underlying the DBpedia SPARQL endpoint we used bif:contains for accessing its text index.

quality. In our evaluation, DL-Learner did not find any members for a total of 23 categories. For 11 categories, we could not find any negative examples, as there were no parallel categories, thus the learned concept was `owl:Thing`. We expect these limitations to become less relevant with the emergence of more knowledge bases that are rich in semantics.

The qualitative part is concerned with the quality of the learned OWL class expression. The authors as experts in ontology engineering (not domain experts) have manually reviewed the results and we will discuss our findings here. We use the excerpt of results shown in Table 2 for discussing four cases of concepts learned.

Single feature concepts. In case the categorization depended on a single feature, DL-Learner normally learned the appropriate concept. These learned concepts could be added as an intensional definition to the knowledge base and help to (1) automatically categorize new data as well as (2) find missing properties such as `familia` or `subdivisionName`.

Overly specific concepts. In (3) DL-Learner learned an overly specific concept matching only 53 instances. We inspected the data in the endpoint and found that the object values for the property `shipBuilder`, which is highly relevant for this category, were of mixed quality: the property had literal and URIs as objects and the value `Defoe_Shipbuilding_Company` was the one that occurred most frequently (53 occurrences) followed e.g. by `Benton_Harbor,_Michigan` (only 3 occurrences). Note that we used DBpedia 3.6.1 as a basis for our experiments; newer versions might provide higher data quality and completeness. The keyword searches failed completely in this case, as neither "ship", "built" or "Michigan" are selective enough.

Indirect Solution concepts. An indirect solution was found for the category in example (4) as no feature (e.g. `champion value US_Open`) was available to construct the concept. The solution reads like a paraphrase and uses the property `subdivisionName`, which is often used by U.S. cities, which are naturally well curated in the English Wikipedia.

Zero member concepts. The examples (5) and (6) are two of the categories for which the learned concept is retrieving zero members from the considered category. The category `Northland Region` (5) does not have a clear member of relation, but is rather used as a tag to group all related articles[12], thus members are too heterogeneous. For the category `FC Salyut Belgorod players` (6) a pertinent and specific feature could not be found in the data, thus it is virtually impossible to learn a formalization, which is specific enough.

5   Related Work

5.1   Navigation

Several navigation and knowledge exploration methods can be used in combination with the proposed NKE paradigm.

---

[12] http://en.wikipedia.org/wiki/Category:Northland_Region

Table 2. Learned concepts for sample categories with limit 100

Nr	Category	Concept (Manchester OWL Syntax)
1	Pleuronectidae	`familia value Pleuronectidae`
2	Villages in Milicz County	`Village and subdivisionName value Milicz_County`
3	Ships built in Michigan	`Ship and shipBuilder value Defoe_Shipbuilding_Company`
4	United States Open champions (tennis)	`TennisPlayer and (foaf:Person or residence` `    some subdivisionName some PopulatedPlace))`
5	Northland Region	`River and geoRelated value New_Zealand and` `mouth some Thing`
6	FC Salyut Belgorod players	`currentclub some (chairman some Thing and name some` `    position value Defender_%28football%29 )`

For knowledge bases with a sufficiently small schema, techniques like facet-based browsing are commonly used. One of those tools is the Neofonie Browser[13]. For very large schemata, facet-based browsing or browsing based on the class hierarchy can become cumbersome and graphical models are used. One example of that uses graphical models is the RelFinder [6]. A user can enter a number of interesting instances and the tool visualises the relationship between those instances as an RDF graph, which can then be explored by the user. Another navigation method is user specific recommendations. Once a user has viewed certain entities, e.g. products, recommender systems can suggest similar products. Often, this is done based on the behavior of other users, but some systems use background knowledge for recommendations. In the simple case, this can be taxonomical knowledge [28], but has recently also been extended to OWL ontologies [20]. The NKE paradigm is based on the latter idea and translates it to the knowledge engineering use case.

Models of user navigation have been successfully used in a range of related domains. For example, in the domain of tagging systems, navigational models [8] as well as behavioral and psychological theories are exploited to evaluate taxonomic structures [7], to assess the motivation for tagging [25], or to improve the quality of emergent semantics [14] and social classification tasks [29]. While navigational models have been applied to improve or evaluate (unstructured) semantics in these domains, they have not been extensively applied to *structured* knowledge bases. This paper sets out to address this gap.

5.2 Knowledge Engineering

Knowledge engineering aims to incorporate knowledge into computer systems to solve complex tasks. It spans across several disciplines including artificial intelligence, databases, software engineering and data mining. Most traditional Knowledge Engineering methodologies heavily rely on a phase-oriented model built on collaboration of a centralized team of domain experts and ontology engineers[21,22,26]. In particular, Pinto et al. [22] characterize the future settings for evolving ontology building as:

---

[13] http://dbpedia.neofonie.de

Highly distributed: Anyone can contribute more knowledge.

Highly dynamic: Several contributors may be changing knowledge at the same time, with high change rates.

Uncontrolled: There is no control over what information is added, and the quality and reliability of that information. In this case, there will be a lot of noise (positive and negative contributions), and not everybody contributing to the ontology will be focused on the same task or have the same purpose.

In their survey, they argue that future methodologies will need to cope with these properties to be successful and to scale up to the increasing availability of ontologies. In NKE, web users take the role of domain experts and elicitation is done *en passant* during the navigation process. To the best of our knowledge, NKE is currently the only methodology that is not only able to cope with all three properties, but is also designed to exploit them to generate ontologies.

[12] distinguish between the *domain axiomatization* and the *application axiomatization*. Although in NKE, the generated *ontology of user interest* is similar to the mentioned application axiomatization, the DOGMA approach might not be directly applicable to NKE as it uses a domain ontology view to interpret the application model. Ontology matching algorithms could be employed, however, instead of the proposed double articulation to mediate between the application ontologies.

In the following, we briefly discuss work related to Ontology Learning, Knowledge Base Completion and Relational Exploration. Many approaches to Ontology learning rely on Natural Language Processing (NLP) and have the goal of learning ontologies from plain text. Other approaches range from using game playing [24] to Formal Concept Analysis (FCA) and Inductive Logic Programming (ILP) techniques. The line of work which was started in [23] and continued by, for instance [1], investigates the use of formal concept analysis for completing knowledge bases. It is mainly targeted towards less expressive description logics and may not be able to handle noise as well as a machine learning technique. In a similar fashion, [27] proposes to improve knowledge bases through relational exploration and implemented it in the RELExO framework[14].

A different approach to extending ontologies is to learn definitions of classes. For instance, [2] proposes to use the non-standard reasoning tasks of computing most specifics concepts (MSCs) and least common subsumers (LCS) to find such definitions. For light-weight logics, such as $\mathcal{EL}$, the approach appears to be promising. There are also a number of approaches using machine learning techniques to learn definitions and super classes in OWL ontologies. Some of those rely on MSCs as well [4,5,11] while others use so called top down refinement approaches [18,19]. Indeed, the HANNE and Geizhals backends are based on extensions of this work in [19] and [17].

In related research on natural language interfaces, [3] investigate so called intensional answers. For instance, a query "Which states have a capital?" can return the name of all states as an extensional answer or "All states (have a

---

capital)." as an intensional answer. Such answers can sometimes reveal interesting knowledge and they can also be used to detect flaws in a knowledge base.

6 Conclusions and Future Work

The contribution of this paper lies in the presentation of a new paradigm - Navigation-induced Knowledge Engineering by Example - that conceptually integrates user navigation into a coherent framework for knowledge engineering by the masses. We provided a concise definition of NKE, provide a general proof-of-concept prototype demonstrating its technical feasibility, and show its practical applicability in two different application domains. It is the hope of the authors that the presentation of this paradigm ignites and stimulates further work on the development of navigational approaches to knowledge engineering.

Acknowledgements. This work was supported by a grant from the European Union's 7th Framework Programme provided for the project LOD2 (GA no. 257943).

References

1. Baader, F., Ganter, B., Sattler, U., Sertkaya, B.: Completing description logic knowledge bases using formal concept analysis. In: IJCAI 2007. AAAI Press (2007)
2. Baader, F., Sertkaya, B., Turhan, A.-Y.: Computing the least common subsumer w.r.t. a background terminology. J. Applied Logic 5(3), 392–420 (2007)
3. Cimiano, P., Rudolph, S., Hartfiel, H.: Computing intensional answers to questions - an inductive logic programming approach. Journal of Data and Knowledge Engineering, DKE (2009)
4. Cohen, W.W., Hirsh, H.: Learning the CLASSIC description logic: Theoretical and experimental results. In: Doyle, J., Sandewall, E., Torasso, P. (eds.) Proceedings of the 4th International Conference on Principles of Knowledge Representation and Reasoning, pp. 121–133. Morgan Kaufmann (May 1994)
5. Esposito, F., Fanizzi, N., Iannone, L., Palmisano, I., Semeraro, G.: Knowledge-intensive induction of terminologies from metadata. In: McIlraith, S.A., Plexousakis, D., van Harmelen, F. (eds.) ISWC 2004. LNCS, vol. 3298, pp. 441–455. Springer, Heidelberg (2004)
6. Heim, P., Hellmann, S., Lehmann, J., Lohmann, S., Stegemann, T.: RelFinder: Revealing relationships in RDF knowledge bases. In: Chua, T.-S., Kompatsiaris, Y., Mérialdo, B., Haas, W., Thallinger, G., Bailer, W. (eds.) SAMT 2009. LNCS, vol. 5887, pp. 182–187. Springer, Heidelberg (2009)
7. Helic, D., Strohmaier, M., Trattner, C., Muhr, M., Lerman, K.: Pragmatic evaluation of folksonomies. In: 20th International World Wide Web Conference (WWW 2011), Hyderabad, India, March 28-April 1, pp. 417–426. ACM (2011)
8. Helic, D., Trattner, C., Strohmaier, M., Andrews, K.: On the navigability of social tagging systems. In: The 2nd IEEE International Conference on Social Computing (SocialCom 2010), Minneapolis, Minnesota, USA, pp. 161–168 (2010)
9. Hellmann, S., Lehmann, J., Auer, S.: Learning of OWL class descriptions on very large knowledge bases. Int. J. Semantic Web Inf. Syst. 5(2), 25–48 (2009)

10. Hepp, M.: GoodRelations: An ontology for describing products and services offers on the web. In: Gangemi, A., Euzenat, J. (eds.) EKAW 2008. LNCS (LNAI), vol. 5268, pp. 329–346. Springer, Heidelberg (2008)

11. Iannone, L., Palmisano, I., Fanizzi, N.: An algorithm based on counterfactuals for concept learning in the semantic web. Applied Intelligence 26(2), 139–159 (2007)

12. Jarrar, M., Meersman, R.: Formal ontology engineering in the DOGMA approach. In: Meersman, R., Tari, Z. (eds.) CoopIS/DOA/ODBASE 2002. LNCS, vol. 2519, pp. 1238–1254. Springer, Heidelberg (2002)

13. Kim, H.L., Scerri, S., Breslin, J.G., Decker, S., Kim, H.G.: The State of the Art in Tag Ontologies: A Semantic Model for Tagging and Folksonomies. In: Proceedings of the 2008 International Conference on Dublin Core and Metadata Applications, Berlin, Deutschland, pp. 128–137. Dublin Core Metadata Initiative (2008)

14. Koerner, C., Benz, D., Strohmaier, M., Hotho, A., Stumme, G.: Stop thinking, start tagging - tag semantics emerge from collaborative verbosity. In: Proc. of the 19th International World Wide Web Conference (WWW 2010), Raleigh, NC, USA. ACM (April 2010)

15. Lehmann, J.: DL-Learner: learning concepts in description logics. JMLR 2009 (2009)

16. Lehmann, J., Bizer, C., Kobilarov, G., Auer, S., Becker, C., Cyganiak, R., Hellmann, S.: DBpedia - a crystallization point for the web of data. Journal of Web Semantics 7(3), 154–165 (2009)

17. Lehmann, J., Haase, C.: Ideal downward refinement in the $\mathcal{EL}$ description logic. In: De Raedt, L. (ed.) ILP 2009. LNCS, vol. 5989, pp. 73–87. Springer, Heidelberg (2010)

18. Lehmann, J., Hitzler, P.: A refinement operator based learning algorithm for the $\mathcal{ALC}$ description logic. In: Blockeel, H., Ramon, J., Shavlik, J., Tadepalli, P. (eds.) ILP 2007. LNCS (LNAI), vol. 4894, pp. 147–160. Springer, Heidelberg (2008)

19. Lehmann, J., Hitzler, P.: Concept learning in description logics using refinement operators. Machine Learning Journal 78(1-2), 203–250 (2010)

20. Lehmann, J., Knappe, S.: DBpedia navigator. In: Semantic Web Challenge, International Semantic Web Conference 2008 (2008)

21. López, M.M.F.: Overview of Methodologies for Building Ontologies. In: Proceedings of the IJCAI 1999 Workshop on Ontologies and Problem Solving Methods (KRR5), Stockholm, Sweden, August 2 (1999)

22. Pinto, H.S., Martins, J.P.: Ontologies: How can they be built? Knowledge and Information Systems 6(4), 441–464 (2004)

23. Rudolph, S.: Exploring relational structures via $\mathcal{FLE}$. In: Wolff, K.E., Pfeiffer, H.D., Delugach, H.S. (eds.) ICCS 2004. LNCS (LNAI), vol. 3127, pp. 196–212. Springer, Heidelberg (2004)

24. Siorpaes, K., Hepp, M.: OntoGame: Weaving the semantic web by online games. In: Bechhofer, S., Hauswirth, M., Hoffmann, J., Koubarakis, M. (eds.) ESWC 2008. LNCS, vol. 5021, pp. 751–766. Springer, Heidelberg (2008)

25. Strohmaier, M., Koerner, C., Kern, R.: Why do users tag? Detecting users' motivation for tagging in social tagging systems. In: International AAAI Conference on Weblogs and Social Media (ICWSM 2010), Menlo Park, CA, USA. AAAI (2010)

26. Studer, R., Benjamins, R., Fensel, D.: Knowledge engineering: Principles and methods. Data & Knowledge Engineering 25(1-2), 161–198 (1998)

27. Völker, J., Rudolph, S.: Fostering web intelligence by semi-automatic OWL ontology refinement. In: Web Intelligence, pp. 454–460. IEEE (2008)

28. Ziegler, C.-N., Lausen, G., Konstan, J.A.: On exploiting classification taxonomies in recommender systems. AI Commun. 21(2-3), 97–125 (2008)

29. Zubiaga, A., Koerner, C., Strohmaier, M.: Tags vs shelves: from social tagging to social classification. In: Proceedings of the 22nd ACM Conference on Hypertext and Hypermedia, pp. 93–102. ACM (2011)

FacetOntology: Expressive Descriptions of Facets in the Semantic Web

Daniel A. Smith and Nigel R. Shadbolt

Web and Internet Science Research Group,
Electronics and Computer Science,
University of Southampton,
Southampton, UK
{ds,nrs}@ecs.soton.ac.uk

Abstract. The formal structure of the information on the Semantic Web lends itself to faceted browsing, an information retrieval method where users can filter results based on the values of properties ("facets"). Numerous faceted browsers have been created to browse RDF and Linked Data, but these systems use their own ontologies for defining how data is queried to populate their facets. Since the source data is the same format across these systems (specifically, RDF), we can unify the different methods of describing how to query the underlying data, to enable compatibility across systems, and provide an extensible base ontology for future systems. To this end, we present FacetOntology, an ontology that defines how to query data to form a faceted browser, and a number of transformations and filters that can be applied to data before it is shown to users. FacetOntology overcomes limitations in the expressivity of existing work, by enabling the full expressivity of SPARQL when selecting data for facets. By applying a *FacetOntology definition* to data, a set of facets are specified, each with queries and filters to source RDF data, which enables faceted browsing systems to be created using that RDF data.

1 Introduction

Faceted browsing is a form of information retrieval where results can be filtered based on different properties of the data, known as facets. For example a faceted browser used for a library has facets such as *Author, Category, Editor* and *Publisher*. A user can select their *Author* of choice, and see only books by that author, as well as the categories those books fall into, their editors and publishers. Faceted browsing is effective under two contrasting scenarios [1]: (a) when users know what they are looking for and can therefore limit by facets which they know the values for, and (b) when users do not know the values they are looking for, but use the facet listings to determine the possible values for a domain; this is unlike keyword searching. Keyword searching is most effective when users know what they are looking for, but not when they do not know the keywords.

H. Takeda et al. (Eds.): JIST 2012, LNCS 7774, pp. 223–238, 2013.

The graph structure of information on the Semantic Web, where instances have multiple properties according to numerous ontologies and vocabularies is ideal for faceted browsing. In fact, numerous browsers have been developed to support faceted browsing over Semantic Web data. For example, /facet [2], Rhizomer [3], Exhibit [4], and mSpace [5] have been developed to browse RDF and Linked Data. These systems use their own specialised descriptions of how to query data, typically providing a list of predicates and classes, which are used to generate SPARQL queries over a knowledge base. All existing systems are limited in their expressivity, and therefore limit the filters that can be placed against data. We propose that standardising the method to select facets, using a method that enables the full expressivity of SPARQL, would benefit multiple faceted browsers so that a definition of the human readable facets of a domain can be described once, and then multiple faceted browsers can use that information to configure their framework. In addition, faceted definitions are themselves RDF, and can therefore be published, shared amongst collaborators, and extended, to be used as the basis for new definitions.

In this paper, we present FacetOntology[1], which defines a standard set of queries to define facets in RDF. Specifically, it defines clauses that can be used when generating SPARQL queries to select data, to produce a set of facets from source RDF data. Faceted browsing systems are then configured to use the resulting data and facets. In order to apply the transformations and configure the faceted browser frameworks, a "maker" is created for each supported faceted browsing system. The maker uses the *FacetOntology definition*, which creates a standalone faceted browser based on the facets specified in the definition. The maker harvests the raw data from the Semantic Web, and performs queries to extract the relevant data and relationships between the facets.

We demonstrate the capability of FacetOntology using two faceted browser "maker" systems that configure faceted browsers for a particular dataset and domain from a *FacetOntology definition*. Our approach creates an abstract definition of the data cleanup transformations, which means that data querying can be repeated programatically, when source data is updated. This enables faceted browsers to be up-to-date and reflect changes in the source data automatically, rather than represent the state of data from a single point in time.

We situate our work with exemplar faceted browsers using the data from the BBC iPlayer, Francophone Music Criticism, the Rich Tags bibliographic database and Usability UK knowledge base of usability methods. We demonstrate that FacetOntology is effective by applying it to these exemplars using two different faceted browser frameworks. We advance the state-of-the-art in faceted browsing by providing a standard methodology for configuring faceted browsers that enables faceted definitions to be shared and extended, with a richer expressivity than existing approaches.

In Section 2 we overview existing work in this area. Section 3 presents our FacetOntology, an ontology to define facets over multiple datasets, and in

---

[1] FacetOntology OWL representation: `http://danielsmith.eu/resources/facet/#`

Section 4 we present a user interface for creating *FacetOntology definitions*. Then, in Section 5 we discuss our implementations of FacetOntology browser "makers" for Exhibit and mSpace, and exemplar public browsers that use FacetOntology. Finally, in Section 6 we conclude.

2   Related Work

One of the first Semantic Web browsers, Tabulator [6], enabled users to specify URIs of data, and explore that data directly. It harvested data by keeping a local knowledge base within the user's web browser, and populated that store with more data as the user browsed. Whenever the user explored information that linked to additional data, that data was harvested into the store. This approach suffers from scalability problems when more data is downloaded, and performance problems prevent it from browsing large-scale data sources.

In order to combat the scalability problems, more recent approaches populate server-side triplestores. A data harvesting process is responsible for gathering and asserting data into a server-side triplestore, and a client-side browser issues SPARQL queries to that triplestore in order to populate the facets of a faceted browser. One benefit of using a triplestore is that it does not typically directly retrieve RDF documents; that task is offloaded to the process that populates the triplestore, allowing a static set of data to sit indexed on a server, ready to be queried efficiently. Traditionally, using a triplestore would mean foregoing the gathering of additional linked data, however there is ongoing work by triplestore vendors (such as OpenLink Virtuoso) which automatically dereference URIs and crawl linked data, based on queries used.

One such approach that uses a triplestore is mSpace [7], a faceted browsing framework that runs in a web browser. mSpace supports the querying of server-side triplestores using SPARQL, and therefore does not require a user's computer to load data on-demand, allowing the interface to scale-up to larger datasets than Tabulator. The definition of the facets present in a triplestore are defined in the "mSpace Model". The mSpace model defines the RDF classes of the instances in the triplestore, and the predicate relationships that connect them. mSpace uses this information to generate SPARQL queries used to populate the facets in its interface.

The approaches of /facet [2] and Rhizomer [3] also use a triplestore. However, they use the ontologies of the data as the facets, instead of a model. This approach works by querying the store for all RDF classes, and presenting them to the user as possible facets. Following the user's initial choice, all predicates that connect to the chosen class are then available as facets. This benefit of using this approach over mSpace's model-based approach is that when new data is added to the store that uses additional predicates or RDF classes, they appear as facets automatically, without having to specify their relationships in a model. When RDF is structured and filtered for use in /facet and Rhizomer, the results present facets correctly. They provide an intuitive interface for users, and configurable system for administrators. However, in the case where data contains inconsistent information, for example when RDF data is asserted from the open Semantic Web, the methodology is unsuitable, because non-human-readable information may be exposed. This is because the Semantic Web is a general-case open world system, designed for machine-readability, where a single RDF source file can contain information on anything at all, not limited by domain or resource. Without a definition model to filter the data, extraneous information creates additional sparsely populated facets of limited use, and this problem would grow as the amount of "wild data" is loaded into the triplestore.

In addition to the work above, there have also been work that defines vocabularies to describe facets. Firstly, in the Longwell project [8], the list of facets used by its faceted browser was specified in RDF as a list of rdf:Property predicates, according to the Fresnel Facets Ontology[2]. For example a simple bibliographic faceted browser is specified[3] in N3 as:

```
@prefix facets: <http://simile.mit.edu/2006/01/ontologies/fresnel-facets> .
@prefix rdf: <http://www.w3.org/1999/02/22-rdf-syntax#> .
@prefix dc: <http://purl.org/dc/elements/1.1/> .
@prefix ow: <http://www.ontoweb.org/ontology/1#> .

:publicationFacets rdf:type facets:FacetSet ;
  facets:types ( ow:Publication ) ;
  facets:facets (
    dc:type
    dc:publisher
    dc:contributor
    dc:subject
  ) .
```

The benefit to this approach is that it is in RDF, and thus is machine readable, and can reference any predicate. However, it is limited because it can only list individual predicates, rather than more complex relationships.

Taking this approach further in the ClioPatria faceted browser [9], individual widgets (in particular, individual facets) are formally described using RDF against an ontology. Compared to the above approach of Fresnel in Longwell, the Cliopatria approach is richer, and supports richer facets than individual predicates. It also defines a number of specific types of facet, in order to enable specific features. For example, "Autocomplete" facets that allow users to start typing the name of the items, and "Hierarchy" facet that automatically expands a tree, given the predicate that specifies the hierarchical relationship. Their approach also defines a "facetTarget" which specified the RDF class of the type of result item. Thus, each of their facets specify facets that filter those results. However, as in the Longwell example, more complex filtering is not specified, for example querying against a chain of predicates, rather than a single property.

Hence, in this paper we extend the features and expressivity of these vocabularies, so that facets can be specified, but also enable more complex filtering than single predicates.

---

[2] SIMILE Longwell Fresnel Facets Ontology: http://simile.mit.edu/2006/01/ontologies/fresnel-facets

[3] Fresnel Facet example from: http://simile.mit.edu/wiki/Longwell_User_Guide#Facet_Configuration

3   FacetOntology: Expressive Definitions of Faceted Metadata

In this section, we describe our ontology "FacetOntology", which can be used to describe facets and their relationships from multiple RDF sources. It provides a vocabulary for describing a set of facets, and for each facet it defines descriptions of its data source. The descriptions describe RDF classes and predicate relationships, which define links to other facets. These descriptions allow SPARQL queries to be generated, the results of which are used to populate a facet's values. In order to enable the full expressivity of SPARQL to be used when selecting data, we have defined "SPARQL hinting" values to be used in order to further filter data. The description results in a bounded human-readable abstraction of a subset of the RDF data sources, that can be used by a faceted browser.

In order to illustrate how FacetOntology is used, we use a running example with data from the Classical Music domain. Our example dataset is comprised of the details about tracks of music, specifically a piece's title, composer and album. It features the music from the album "Best of Bach" that is composed by the composer "Johann Sebastian Bach", and is serialised in N3, see Figure 1.

```
@prefix : <http://facetontology.example.com/data#> .
@prefix facet: <http://danielsmith.eu/resources/facet/#> .
@prefix rdfs: <http://www.w3.org/2000/01/rdf-schema#> .

:album1 a :Album .
:album1 :track :piece1 .
:album1 :track :piece2 .
:album1 :composer :composer1 .
:album1 rdfs:label "Best of Bach" .

:piece1 a :Piece .
:piece1 rdfs:label "Cantata BWV 1" .
:piece1 :originalName "Wie schon leuchtet der Morgenstern" .

:piece2 a :Piece .
:piece2 rdfs:label "Cantata BWV 2" .
:piece2 :originalName "Ach Gott, vom Himmel sieh darein" .

:composer1 a :Composer .
:composer1 rdfs:label "Johann Sebastian Bach" .
```

Fig. 1. Example N3 definition of the data used by our running example from the Classical Music domain

3.1   Facet Definitions

A *FacetOntology definition* is used to describe facets within a dataset, the descriptions are organised into a *FacetCollection* which is a set of facets used to describe a particular dataset. The items contained in a *FacetCollection* are a *FirstOrderFacet* and one or more *ConnectedFacets*. In more detail:

1. A *FirstOrderFacet*, is described by an RDF class (using the `rdf:type` predicate). For example, the *Piece* facet is comprised of instances of type `:Piece`, as shown in Figure 1.
2. The *ConnectedFacets*, are described by how they connect in the data graph in relation to the first-order facet. This is modelled as a chain of predicates that are used to generate the SPARQL query that gathers the data for the values of a facet. For example, the *Album* facet relates to the *Piece* facet via the `:track` predicate, as shown in Figure 1.

In our classical music example, there are three classes, *Piece*, which represents a single piece of music with its title, *Album*, which represents a grouping of pieces, and *Composer*, which represents the composer of music. These classes are linked using the `:track` predicate and the `:composer` predicate, and all of these classes have various attributes modelled using different predicates. In our FacetOntology definition for this data, the *Piece* class is used to define the *FirstOrderFacet*, with the predicate `:track` being used as the predicate chain to define the *Album* facet. The predicate chain is used in the two following SPARQL queries to harvest data to populate the facet's values in the classical music's *FacetCollection*, where the variables `?label` and `?uri` gather the label and URI, respectively.

1. Query to gather the first-order facet of *Piece*:

```
SELECT ?label ?uri WHERE {
        ?uri rdf:type :Piece .
        ?uri rdfs:label ?label }
```

This query returns a table of URIs and labels that represent all of the items to populate the *Piece* facet's values.

2. Query to gather connected metadata about the *Album* (where the variable `?firstorder` will contain the URI of the *Piece* in the first-order facet:

```
SELECT ?label ?uri ?firstorder WHERE {
        ?firstorder rdf:type :Piece .
        ?uri :track ?firstorder .
        ?uri rdfs:label ?label }
```

This query returns a table of URIs and labels to populate the values of the *Album* facet, as well as the URI of *FirstOrderFacet* items (*Pieces*, in this case) on the album. A faceted interface can use these URIs to make the link between the *FirstOrderFacets* and the *Album* facet, so that when users filter on an album, the pieces that are on it can be shown. The predicate that connects a *Piece* to an *Album* is directional from the *Album* to the *Piece*, we must also indicate that the direction of this predicate is reversed, using the `facet:reverse` directive.

In order to define the *Composer* facet, we must define a predicate chain that first joins to *Album*, and then to *Composer*, as the ontology that defines the classical music data describes composers as having composed albums, and there is no direct link to the individual pieces. As such, the same predicate definition that is used for the *Album* facet (see above) is first defined, and then a predicate composedAlbum is described, in order to complete the predicate chain.

After using FacetOntology to define the facets in our classical music example, it produces the following facet descriptions in RDF, serialised as N3 in Figure 2.

```
@prefix : <http://facetontology.example.com/data#> .
@prefix facet: <http://danielsmith.eu/resources/facet/#> .
@prefix rdfs: <http://www.w3.org/2000/01/rdf-schema#> .
@prefix xsd: <http://www.w3.org/2001/XMLSchema#> .

:mspace a facet:StandardFacetCollection .
:mspace facet:faceturi :item .
:mspace facet:rdfsource "http://facetontology.example.com/data.n3" .

:mspace facet:faceturi :piece .
:mspace facet:faceturi :album .
:mspace facet:faceturi :composer .

:piece a facet:FirstOrderFacet .
:piece a facet:Facet .
:piece facet:class :Piece .
:piece rdfs:label "Piece" .

:album a facet:ConnectedFacet .
:album a facet:Facet .
:album facet:class :Album .
:album rdfs:label "Album" .
:album facet:nextpredicate :album_predicate .

:album_predicate a facet:Predicate .
:album_predicate facet:predicateuri :track .
:album_predicate facet:reverse "True"^^xsd:boolean .

:composer_facet a facet:ConnectedFacet .
:composer_facet a facet:Facet .
:composer_facet facet:class :Composer .
:composer_facet rdfs:label "Composer" .
:composer_facet facet:nextpredicate :composer_predicate .

:composer_predicate a facet:Predicate .
:composer_predicate facet:predicateuri :track .
:composer_predicate facet:reverse "True"^^xsd:boolean .
:composer_predicate facet:nextpredicate :composer_predicate2 .

:composer_predicate2 a facet:Predicate .
:composer_predicate2 facet:predicateuri :composer .
# nb: facet:reverse is not defined for composer_predicate2
```

Fig. 2. Example N3 definition of a FacetOntology for the running example of Classical Music

The above example is the typical use case for FacetOntology, however there is a specialised case, when there is only the definitions from literals from the *FirstOrderFacet* available. This specialised case requires an additional definition so that a SPARQL query takes the structure into account accurately. In order to specify that a facet's values use a predicate other than `rdfs:label` the facet must be marked up as having a `facet:type` of `facet:TypeLiteral`. The predicate to query must then be specified as the `facet:labeluri`. In our classical music example, a user may wish to have a facet for the "Original Name," as specified by the predicate `:originalName` (see Figure 1). If we use the `nextpredicate` definition by specifying `:originalName`, the query engine will try to find an instance at that predicate, which will fail since there is only a literal. Thus, instead we specify that this Facet has a Facet Type (`facet:facettype`), and that it is `facet:TypeLiteral`. We then add a definition that it also has a label URI (`facet:labeluri`) of `:originalName`. This informs the engine that the facet instances are not RDF instances, and are instead literals. This also means that internally, instance URIs are not used to uniquely identify the instances, their string values are used instead. See Figure 3 for an example definition of the `originalName` Facet.

```
@prefix : <http://facetontology.example.com/data#> .
@prefix facet: <http://danielsmith.eu/resources/facet/#> .
@prefix rdfs: <http://www.w3.org/2000/01/rdf-schema#> .

:mspace facet:faceturi :originalname .

:originalname a facet:ConnectedFacet .
:originalname a facet:Facet .
:originalname facet:class :Piece .
:originalname rdfs:label "Original Name" .
:originalname facet:type facet:TypeLiteral .
:originalname facet:labeluri :originalName .
```

Fig. 3. Example N3 definition of a Facet for "Original Name," demonstrating the use of the `TypeLiteral` property, using FacetOntology for the running example

In the next section, we outline all of the data transformation directives that FacetOntology provides so that data can be filtered and modified before inclusion in a faceted browser.

3.2   Data Transformation

One of the areas where existing work is limited is that it assumes that data is under our control, and can be edited freely. However, our approach instead makes the assumption that data is published by others, and make be possibly updated by them. Therefore, our approach is to build in data transformations that modify the data programmatically so that it can be updated repeatedly. Similarly, we support transforming the syntax of data values programmatically so that the relevant aspects of data are used in the facets. For example, removing the time from a time/date value, or applying a regular expression to clean up the data field.

One of the powerful directives in FacetOntology is the `facet:preprocess` directive which allows transformations to be applied to values in a facet, before they are used in the faceted browser. As shown in Table 1, the values of this directive specify the transformation that should be applied. For example, specifying `iso8601:striptime` will remove the time from ISO8601 [10] formatted dates, which is useful if you want to show only dates in a facet, and the times do not matter. We predict that additional directives will be required in the future, in particular with regards to data cleaning (e.g. applying regular expressions), and thus the interpretation of the preprocess value is extensible.

Similarly, we have specified the `facet:regex-PCRE-match` predicate, which expects a perl-compatible regular expressions (PCRE) [11] with a single matched group, used to extract the required information from the facet value. In more detail, Table 1 outlines the predicate directives that can be used against Facets.

Table 1. FacetOntology directives for transformation of data, applied to each facet

Directive	Value	Description
facet:preprocess	iso8601:striptime	Remove the time from an ISO8601 formatted date.
facet:preprocess	iso8601:year	Extract the year from an ISO8601 formatted date.
facet:regex-PCRE-match	PCRE with single match group	Match the regular exression using the first match.

In the next section we discuss additional selection and filtering flexibility via SPARQL query hinting.

3.3 SPARQL Query Hinting

In addition to data transformation, it can be desirable to compute new data from existing value(s). For example to modify units, such as converting kilometres to miles, or to generate a value based on a combination of multiple properties, such as generating a screen resolution from a width and height. It also may be desirable to specify additional filters and patterns above those built into FacetOntology.

Another requirement is to provide additional context for each facet value, rather than only have a single value, for example to show the birth/death date of a composer, or the URL to a thumbnail of the depiction of a value. To this end, we have devised a set of SPARQL-hinting definitions in order to override the default variable bindings, add more complex "AS" algebra to the SELECT binding, and to inject additional patterns into the WHERE clauses. Through this approach, we allow selection of data to be filtered using SPARQL's full expressivity, for example using complex boolean "FILTER"s, and mathematical algebra using multiple data values in "AS" select expressions. Specifically we define the predicate directives in Table 2.

Table 2. FacetOntology directives for hinting how to generate the facet query and which variables to read results from

Directive	Description
`facet:SPARQL-Item-Variable`	Specify the variable to bind to the item URI.
`facet:SPARQL-Context-Variable`	Specify additional variable to select (e.g. to provide extra context on this item).
`facet:SPARQL-Context-Variable-As`	Specify additional binding to get from results (as specified by a SPARQL-Context-Select-As).
`facet:SPARQL-Context-Select-As`	Specify additional "AS" definition to use in SELECT, specify the resulting binding with a SPARQL-Context-Variable-As.
`facet:SPARQL-Context-Variable`	Specify additional variable to select (e.g. to provide extra context on this item).
`facet:SPARQL-Value-Variable`	Specify the variable to bind to the item value.
`facet:SPARQL-Value-Select-As`	Specify an "AS" definition to use for the SELECT.
`facet:SPARQL-Additional-Pattern`	Add an additional pattern into the WHERE.

For example, to calculate a screen resolution by multiplying the width and height, a facet would be defined as follows:

```
:res-facet a facet:ConnectedFacet ;
    facet:class ont:System ;
    facet:type facet:TypeLiteral ;

    facet:SPARQL-Item-Variable "item";
    facet:SPARQL-Context-Variable "res_w";
    facet:SPARQL-Context-Variable "res_h";

    facet:SPARQL-Context-Variable-As "res_name_upper";
    facet:SPARQL-Context-Select-As "UCASE(?res_name) AS ?res_name_upper";

    facet:SPARQL-Value-Variable "val" ;
    facet:SPARQL-Value-Select-As "?res_w * ?res_h AS ?val" ;

    facet:SPARQL-Additional-Pattern "?item res:width ?res_w";
    facet:SPARQL-Additional-Pattern "?item res:height ?res_h";
    facet:SPARQL-Additional-Pattern "?item res:resolution_name ?res_name";

    rdfs:label "Resolution" .
```

results in the following SPARQL query being generated:

```
SELECT ?res_w ?res_h ?item (?res_w * ?res_h AS ?val) (UCASE(?res_name) AS ?
    res_name_upper)
WHERE {
    ?item rdf:type ont:System .
    ?item res:width ?res_w .
    ?item res:height ?res_h .
    ?item res:resolution_name ?res_name }
```

The SPARQL result processor knows from the definition to use the `?item` binding as the item URI and `?val` as the value. Additionally, it will also provide the values of width and height, because `SPARQL-Context-Variable` has been used to enable these bindings, and it will select `?res_name_upper` as an additional context field. Thus, through the SPARQL-hinting mechanism it is possible to provide additional context, and perform functions on the data before it is selected, either through additional patterns (which can contain FILTERs), and by using SPARQL functions on the data, in "AS" expressions.

By extending existing approaches of describing data, one of our aims is to enable FacetOntology to support any possible dataset, and to provide as rich an expressiveness as possible. Through the use of the above SPARQL hinting rules, FacetOntology queries can utilise the full specification of SPARQL functions and operators, and therefore is the full expressivity of SPARQL, which in turn has the expressiveness of relational algebra [12]. We have designed these SPARQL hinting patterns so that they encapsulate at least the full expressiveness of previous facet vocabularies (as discussed in Section 2), and are in fact now more expressive than previous work. In addition, additional context can be included using the SPARQL hinting definitions, to enable a richer faceted browser interface. In the next section we present a user interface to create FacetOntology definitions.

4   A User Interface for Creating FacetOntology Definitions

A FacetOntology definition is required before a facets browser can be created. This process can be performed manually, which is indeed the way that the data definition is performed in the existing approaches. One of our aims is to make it easier for end-users to create faceted browsers, ideally with a minimal of technical knowledge. To this end, we have created a user interface called "Data Picker" to aid in the creation of FacetOntology definitions.

The "Data Picker"[4] tool provides users with an intuitive interface for picking the metadata that they wish to be included in a FacetOntology definition. The tool works by querying a data set for all RDF classes, and a subset of the labels of individuals of that type. These are displayed, and the user can select a class which they want to explore further. All of the predicates that are joined to individuals of that class are then shown, again with a sample of their values. A user can then select which attributes they wish to be marked up using FacetOntology (see Figure 4). While the user is selected facets, they are presented with a preview of the facets, so that they can verify they are what they wanted, as expected. Finally, the users publish their FacetOntology definition, ready to send to a "maker" tool, which configures a faceted browser framework.

The need to aid manual marking up of facets is not a new problem, even within the domain of the Semantic Web. Work by Oren et. al. [13] looked at

---

[4] Data Picker: http://facetportal.danielsmith.eu/picker/

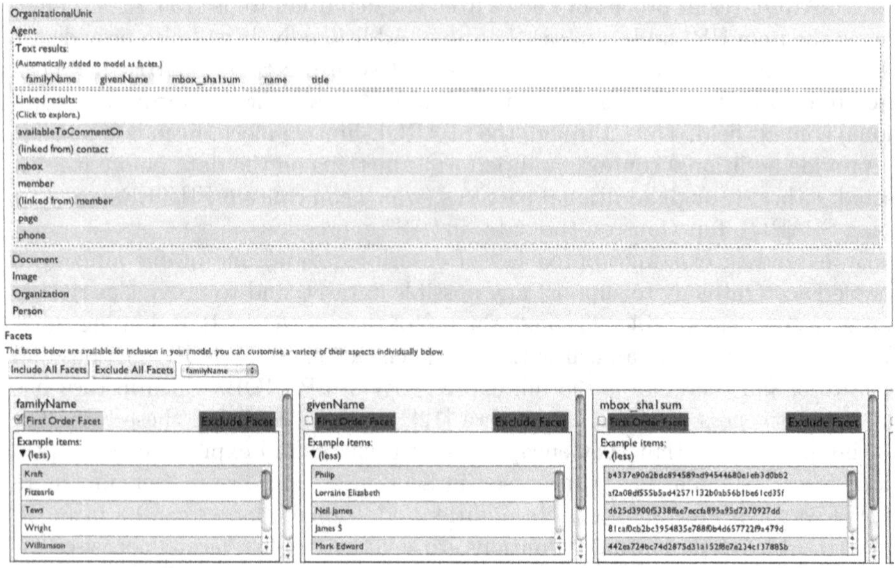

Fig. 4. A screenshot of the data picker running on the University of Southampton SPARQL endpoint. The user has selected the "Agent" class and a number of literals have been previewed as facets ("familyName", "givenName" and "mbox_sha1sum"), and data properties are available to explore, labeled as "linked results" (such as "member", "page", and "phone").

ranking facet quality, as an aid to automatically marking up facets, where their technique was formally proven to show an improvement of quality. Similarly, AKTiveRank [14] presents a technique for ranking ontologies, using structural metrics. Such metrics could be used to inform a facet creation tool, in order to enable an increased level of automation.

In summary, the cost of creating a FacetOntology is linear with the amount of facets in a domain — if a domain has 4 facets, a single First Order facet is defined by its RDF Class, and 3 Connected Facets are defined by how they link to the First Order facet. In a domain with a larger amount of facets, a larger definition is required, which will take more time to create.

5   Implementations of Faceted Browser Makers

To demonstrate and validate our approach, we have created two different faceted browser "makers" that use different faceted browser frameworks to create interfaces over the same data source. An advantage of this approach is that if a developer is more comfortable with a particular faceted browser framework, then they can use the one they wish. Similarly a developer may wish to migrate from one browser or platform to another, and by having the abstract FacetOntology definition, they do not have to reconfigure a new system. This may occur because the data requirements of a faceted browser change, for example a system may work fine with 1000 items using Exhibit, but as that number increase, then migration to a server-backed browser such as mSpace may be appropriate. Using FacetOntology for this purpose would make that approach more straightforward.

The first implementation we discuss using FacetOntology is our FacetOntology Exhibit maker[5], which reads a FacetOntology definition URI, and outputs a configured Exhibit installation. The core components of this process are generating the JSON data file, and then configuring an Exhibit view that uses that data file. Creation of the data file is performed by loading the data sources into a store, and querying them based on the predicates, filters and SPARQL hints defined in the FacetOntology definition. Following this process, an Exhibit HTML file is created based on a template (to allow styling of the resulting Exhibit) which contains each of the facets that have been defined.

The second implementation using FacetOntology is the "mSpace Maker"[6], which produces mSpaces from FacetOntology definitions. mSpace Maker automates the creation of mSpace faceted browsers, by gathering data into an mSpace database, and configuring mSpace to use this database's schema. The process to configure an mSpace differs from that of Exhibit above, because rather than holding all data in a single file, the data is stored in a relational database with a schema that has been optimised for the patterns of querying that mSpace performs. Thus, the mSpace Maker works by creating a table for each facet, and uses foreign keys to link them together. The facets and their database table and column names are written into the mSpace configuration, so that the mSpace server can create SQL queries to join the tables together. Upon completion of the data import, a URL to the mSpace is returned to the user (because mSpace cannot be used locally, unlike Exhibit). In the next section we describe public faceted browsers that we have created that use FacetOntology, and were created using our makers.

5.1   Exemplar Uses

We demonstrate the flexibility of FacetOntology by applying it to different domains, where the use of a faceted browser is different. Each of these browsers demonstrates additional support that FacetOntology supports above and beyond configurations that have been previously attempted. It is through developing these configurations that the core requirements for FacetOntology have been developed.

1. **Francophone Music Criticism**[7]: This faceted browser demonstrates faceted filtering of 19th century music reviews from a digital repository. This browser

---

[5] FacetOntology-Exhibit: `https://github.com/danielsmith-eu/facetontology-exhibit`
[6] mSpace Maker: `http://mspacemaker.danielsmith.eu/`
[7] Francophone Music Criticism: `http://fmc.ecs.soton.ac.uk/`

demonstrates basic FacetOntology usage, over five key facets (Collection, Journal, Year, Author and Title), and results in a set of PDFs that the user can then download from a digital repository.

2. **Rich Tags**[8]: Faceted browsing over multiple digital repositories. This faceted browser runs over a more sophisticated data collection engine which routinely collects data from multiple institutional repositories into a single store. Thus, the FacetOntology definition is applied to all sources of data, resulting in a single browser over the composite data.

3. **BBC iPlayer**[9]: A browser for faceted filtering of television programmes that have been broadcast over the previous 7 days, for streamed viewing online. This browser demonstrates transforming broadcast time data into individual dates for use in a date-picker facet (see Figure 5). The browser provides richer faceted exploration of the data than the BBC site, and includes data from multiple sources (specifically, the Contributor credits data), so that users can filter shows by actor/presenter, which is not possible on the BBC site. In order to demonstrate our Exhibit maker, we have also used the FacetOntology definition of the iPlayer data to create a Exhibit of the same data, as shown in Figure 6.

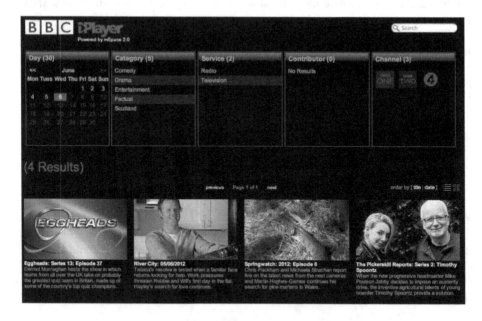

Fig. 5. The BBC iPlayer mSpace faceted browser created using FacetOntology definitions

---

[8] Rich Tags: `http://www.richtags.org/`

[9] BBC iPlayer mSpace (updates daily): `http://iplayer.mspace.fm/`

Fig. 6. The BBC iPlayer Exhibit faceted browser (unstyled) created using Facet-Ontology definitions

4. **UsabilityUK**[10]: A browser of usability methods and JISC-funded projects that use those methods. UsabilityUK demonstrates a faceted browser where users can search for more than one type of record. Specifically, users can search for projects, experts or usability methods. This contrasts with other typical faceted browsers where users search for a single type of record (such as a TV programme, a song, or document).

Our approach supports a range of different requirements in a range of different domains, demonstrating that FacetOntology is fit for purpose as a general approach to semantic description of facets, which benefits from the full expressivity of SPARQL.

6   Conclusion

In this paper, we present FacetOntology, an approach to defining facets for faceted browsers. We applied our approach to BBC iPlayer, digital repositories and a usability techniques knowledge base, showing that our approach is effective and can be applied to many domains. In more detail, we verified that our approach works for different types of browsing across these domains, has the full expressivity of SPARQL, and has been applied to different faceted browsing software.

---

[10] UsabilityUK: http://usabilityuk.org/

For future work, we plan to investigate appropriate methodologies for enabling additions to the FacetOntology data transformations (as discussed in Section 3.2) to be created by the community, when additional transformations are required.

References

[1] Hearst, M., Elliott, A., English, J., Sinha, R., Swearingen, K., Yee, K.: Finding the flow in web site search. Communications of the ACM 45(9), 49 (2002)

[2] Hildebrand, M., van Ossenbruggen, J., Hardman, L.: /facet: A Browser for Heterogeneous Semantic Web Repositories. In: Cruz, I., Decker, S., Allemang, D., Preist, C., Schwabe, D., Mika, P., Uschold, M., Aroyo, L.M. (eds.) ISWC 2006. LNCS, vol. 4273, pp. 272–285. Springer, Heidelberg (2006)

[3] García, R., Gimeno, J.M., Perdrix, F., Gil, R., Oliva, M.: The rhizomer semantic content management system. In: Lytras, M.D., Damiani, E., Tennyson, R.D. (eds.) WSKS 2008. LNCS (LNAI), vol. 5288, pp. 385–394. Springer, Heidelberg (2008)

[4] Huynh, D., Karger, D., Miller, R.: Exhibit: lightweight structured data publishing. In: Proceedings of the 16th International Conference on World Wide Web, pp. 737–746 (2007)

[5] Schraefel, M.C., Wilson, M., Russell, A., Smith, D.A.: mspace: improving information access to multimedia domains with multimodal exploratory search. Commun. ACM 49(4), 47–49 (2006)

[6] Berners-Lee, T., Chen, Y., Chilton, L., Connolly, D., Dhanaraj, R., Hollenbach, J., Lerer, A., Sheets, D.: Tabulator: Exploring and Analyzing linked data on the Semantic Web. In: Proceedings of the 3rd International Semantic Web User Interaction Workshop (2006)

[7] Schraefel, M.C., Smith, D.A., Owens, A., Russell, A., Harris, C., Wilson, M.: The evolving mspace platform: leveraging the semantic web on the trail of the memex. In: HYPERTEXT 2005: Proceedings of the Sixteenth ACM Conference on Hypertext and Hypermedia, pp. 174–183. ACM Press, New York (2005)

[8] SIMILE: Longwell RDF Browser (2003-2005), http://simile.mit.edu/longwell/

[9] Hildebrand, M., Van Ossenbruggen, J.: Configuring semantic web interfaces by data mapping. In: Visual Interfaces to the Social and the Semantic Web (VISSW 2009), vol. 443, p. 96 (2009)

[10] International Organization for Standardization (ISO): ISO 8601:2004 Data elements and interchange formats, Information interchange, Representation of dates and times (2004)

[11] Hazel, P.: PCRE: Perl Compatible Regular Expressions (2005)

[12] Angles, R., Gutierrez, C.: The expressive power of SPARQL. In: Sheth, A.P., Staab, S., Dean, M., Paolucci, M., Maynard, D., Finin, T., Thirunarayan, K. (eds.) ISWC 2008. LNCS, vol. 5318, pp. 114–129. Springer, Heidelberg (2008)

[13] Oren, E., Delbru, R., Decker, S.: Extending Faceted Navigation for RDF Data. In: Cruz, I., Decker, S., Allemang, D., Preist, C., Schwabe, D., Mika, P., Uschold, M., Aroyo, L.M. (eds.) ISWC 2006. LNCS, vol. 4273, pp. 559–572. Springer, Heidelberg (2006)

[14] Alani, H., Brewster, C., Shadbolt, N.: Ranking ontologies with AKTiveRank. In: Cruz, I., Decker, S., Allemang, D., Preist, C., Schwabe, D., Mika, P., Uschold, M., Aroyo, L.M. (eds.) ISWC 2006. LNCS, vol. 4273, pp. 1–15. Springer, Heidelberg (2006)

An Ontological Framework for Decision Support

Marco Rospocher and Luciano Serafini

FBK-irst, Via Sommarive 18 Povo, I-38123, Trento, Italy
{rospocher,serafini}@fbk.eu

Abstract. In the last few years, ontologies have been successfully exploited by Decision Support Systems (DSSs) to support some phases of the decision-making process. In this paper, we propose to employ an ontological representation for *all* the content both processed and produced by a DSS in answering requests. This semantic representation supports the DSS in the whole decision-making process, and it is capable of encoding (i) the request, (ii) the data relevant for it, and (iii) the conclusions/suggestions/decisions produced by the DSS. The advantages of using an ontology-based representation of the main data structure of a DSS are many: (i) it enables the integration of heterogeneous sources of data available in the web, and to be processed by the DSS, (ii) it allows to track, and to expose in a structured form to additional services (e.g., explanation or case reuse services), all the content processed and produced by the DSS for each request, and (iii) it enables to exploit logical reasoning for some of the inference steps of the DSS decision-making process. The proposed approach have been successfully implemented and exploited in a DSS for personalized environmental information, developed in the context of the PESCaDO EU project.

1 Introduction

Decision support systems (DSSs) are information systems that support users and organizations in *decision-making* activities. DSSs have been applied in several diverse application contexts, to help to take decisions in domains like the medical, legal, computer security, and power consumption management ones.

At an abstract level, we can identify three phases in a decision-making process [1]:

1. the formulation of the decision-making problem;
2. the gathering, storing, and fusion of the data relevant for the given problem;
3. the reasoning on the data to take a decision;

To support the implementation of such process, DSSs usually comprise three main modules [2]: the (i) *dialogue* or *user* module, which supports the interaction of the user with the system, to formulate the problem and receive in output the result of the the DSS computation, the (ii) *data* module, which allows to store the data collected and processed by the DSS, and the (iii) *model* module, which implements the decision support strategy.

Being studied, both theoretically and technically, since the late 1960s, research in DSSs field has taken advantage in the last decade of the achievements and results of Semantic Web technologies. In particular, ontologies have recently been adopted in

H. Takeda et al. (Eds.): JIST 2012, LNCS 7774, pp. 239–254, 2013.

DSSs in various application domains [3,4,5,6,7,8,9,10], exploited for several purposes: to support via reasoning some of the decision support phases [6,3,8], to characterize the data manipulated by the DSS [11,3], and to define the tasks and parameters of the various modules of the system [9]. That is, so far ontologies have been adopted by DSSs to support only parts of the decision-making process, mainly to represent the data and to support their processing for taking decisions.

In this paper we propose to exploit an ontology-based knowledge base as the main (enhanced) data structure of a DSS, where *all* the content and data for a specific decision support request, processed and produced by the system, are stored. In details, our approach consists in designing the ontology underlying the knowledge base, i.e. the *T-Box* in Description Logics (DL) terminology, so that it is capable of formally representing all the details of the three decision-maikng phases described above, i.e., (i) the decision support request submitted by the user to the system, (ii) the data that the system processes for the given request, and (iii) the new content and conclusions produced by the DSS from the available data and in view of the given request, possibly together with some details on how the DSS arrived to those conclusions.

Each single request submitted to the system triggers the instantiation of a new *A-Box* of this ontology. The instantiation incrementally occurs in subsequent steps, coherently with the decision-making process phases. Therefore, at the end of the processing of a request by the DSS, the A-Box associated with the request contains a structured and comprehensive description, a *semantic request story-plan*, of the output produced by the DSS, linked to the data and the request that triggered that output.

The ontological representation of the DSS data that we propose is used to support the DSS activities

- as the main, shared, data structure of the DSS;
- as content exchange format between the different modules of the DSS;
- to keep track of all the intermediate data and results produced by the DSS in the course of solving a problem.

The advantages of using a semantic (ontology based) representation of the main data structure of a DSS are many. First, differently from what happened in the past where DSS were closed systems, in the semantic web era most of the knowledge and data useful to support a decision is available (in heterogeneous formats) in the web. As one of the main objectives of ontologies is to define shared domain models, an ontology-based representation of the knowledge in a DSS facilitates the integration of structured knowledge and data available in the web. Second, in the semantic services era we are now, a DSS can be seen as any other web-service and therefore it can be combined with other semantic services. Keeping a semantically rich track of the entire decision process followed by a DSS in order to reach a conclusion/suggestion/decision, and exposing it by adopting for instance the Linking Open Data[1] principles, enables the combination of the DSS with other complex services, such as explanation services (for which, just information about input-output is not enough) or case reuse/adaptation services (which can adapt the entire reasoning chain done by the DSS to slightly different cases). Finally, the third advantages is the fact that some of the inference steps of the DSS can be

---

[1] http://linkeddata.org/

performed via state of the art logical reasoning services, as for instance rule reasoners or ontology reasoners.

The proposed approach has been successfully applied in a running personalized environmental DSS, in the context of the PESCaDO EU project, where its features and advantages have been empirically demonstrated.

The paper is organized as follow. First, in Section 2 we review the state of the art on the usages of ontologies in DSSs, while in Section 3 we briefly describe the context of the PESCaDO project. In Section 4, we present the general Decision Support Knowledge Base (DSKB), providing some directions on how to organize its components, together with details on how we actually implemented them for PESCaDO. In Section 5, we describe the steps followed by a DSS in building a semantic request story-plan, reporting the PESCaDO case as an example of such process. In Section 6, we present some examples of how the semantic representation of request story-plans can be exploited for further purposes (natural language reports generation, query-answering services over a repository of archived story-plans). We also remarks (Section 7) some checks to perform when building a DSKB. Finally, in Section 8 we conclude with some final remarks.

2   Ontologies for Decision Support: State of the Art

In the last decade, ontologies have been extensively used in decision support systems, to support tasks in several application domains: clinical management [3], system audit management [4], network security management [5], justice and legal advice [6,7], waste-water management [8], power consumption management [9], electronic issue management [10].

In [10], an ontology driven case-based reasoning system for electronic issue management is described. The ontology is used to formally represent issue management concepts. [9] presents the development of a system that supports decision-making for tasks aimed to reduce power consumption of oil-and-gas production enterprises. The system relies on two ontologies to support communication between the different modules of the system: one describing the objects of the domain, and one describing the tasks supported by the DSS.

[11] presents a framework for ontology-based DSSs in pervasive computing environment. The ontology is exploited to obtain a shared and common understanding of the knowledge domain of the pervasive computing environment, and an explicit conceptualization that describes the semantics of the data managed by the DSS. [3] presents an ontology-based fuzzy DSS to support neuroradiologists in the diagnosis and monitoring of multiple sclerosis. The ontology is used by the system to encode expert knowledge (i.e., qualitative linguistic labels) and the key data processed by the DSS (e.g., possible lesions). [8] shows how to use a static domain ontology for environmental decision support, to capture, understand and describe the knowledge about the physical, chemical and microbiological environment of a waste-water treatment plant, in order to provide reasoning-support for the decision-making phase. [6] presents an OWL ontology to support the inference process in a legal case-based reasoning systems.

As it emerges from all these works, so far ontologies have been successfully exploited to support *some* of the phases of the decision-making process that we described in Section 1, in particular the second phase (e.g., [11,3]), and the third phase (e.g., [6,3,8]). In this paper, we push further the usage of ontologies in DSSs to support all the phases of the decision-making process. In particular, we propose to use the ontology as the main data structure of the DSS, capable of representing *all* the information processed and produced by the DSS, i.e., the decision support request submitted by the user, the data relevant for that request, and the conclusions triggered by the decision support mechanism implemented in the DSS. Furthermore, our approach is not tailored to a specific domain, but it can be exploited in different application contexts.

3   The PESCaDO Use Case

Citizens are increasingly aware of the influence of environmental and meteorological conditions on the quality of their life. One of the consequences of this awareness is the demand for high quality environmental information and decision support that is tailored (i.e., personalized) to one's specific context and background (e.g., health conditions, travel preferences). Personalized environmental information may need to cover a variety of aspects (e.g., meteorology, air quality, pollen) and take into account a number of specific personal attributes of the user (e.g., health, age, allergies), as well as the intended use of the information. For instance, a pollen allergic person, planning to do some outdoor activities, may be interested in being notified whether the pollen situation in the area may trigger some symptoms, or if the temperature is too hot for doing physical exercise, while a city administrator has to be informed whether the current air quality situation requires some actions to be urgently taken.

The goal of the PESCaDO EU project[2] is to develop a multilingual web-service platform providing personalized environmental information and decision support. This platform takes advantage of the fact that nowadays, the Web already hosts a great range of *environmental nodes* (i.e. web-sites, web-services, open data repositories) that offer data on each of the above aspects, such that, in principle, the required basic data are available. The challenge is thus threefold: first, to discover and orchestrate these environmental nodes; second, to process the obtained data in accordance with the decision support needs of the user; and, third, to communicate the gained information in the user's preferred mode.

For a general overview of the running PESCaDO system[3], and the type of information produced, check the demonstration video[4], or directly play with the on-line demonstrator[5]. Shortly, users submit a decision support request to the system (e.g. *"I want to do some hiking in Nuuksio Park tomorrow: is there any health issues for me?"*), specifying in full details the type of request, the type of activity (if any) they want to perform, their profile, the geographic area and the time period to be covered. Then, the system (i) determines the data relevant for the request, (ii) retrieves the data from

---

[2] http://www.pescado-project.eu

[3] A more comprehensive description of the system workflow can be found in [12].

[4] http://www.youtube.com/watch?v=c1Ym7ys3HCg

[5] Accessible from the project web-site, or directly here http://193.145.50.130/

environmental nodes providing them[6], (iii) processes these data providing conclusions (e.g., warnings, recommendations) according to the needs of the users, and, finally, (iv) generates reports (e.g., text, tables, graphics) to be communicated to the user.

4   The Decision Support Knowledge Base

We propose a reference architecture for designing ontology-based knowledge bases for decision support systems, called Decision Support Knowledge Base (DSKB). It aims at guiding the development of an OWL [13] ontology capable of representing in a connected and comprehensive way all the content relevant for a given decision support request. In details, in our approach each decision support request is associated with an A-Box (i.e., a set of individuals and assertions on them) instantiating the T-Box part of the DSKB (see Figure 1[7]). The DSKB T-Box, to which we refer to as *Decision Support Ontology* (DSO), comprises three main components, namely Problem, Data, and Conclusions. These three components are connected by relations between the corresponding elements. As shown by Figure 1, these relations are[8]: *hasData* and *hasConclusion*, which relates a problem description with the data relevant for it and

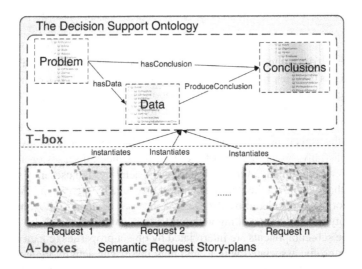

Fig. 1. The Decision Support Knowledge Base

---

[6] More precisely, environmental data are hourly distilled from environmental web-sites and stored in a dedicated repository, so when a decision request is submitted to the system, for efficiency reasons, this database is actually queried for data (instead of directly querying web-sites in real-rime).

[7] The three parts of each request A-Box correspond to the components of the T-Box.

[8] The set of object properties here presented is not exhaustive, and can be further extended depending also on the application context.

the conclusions provided for it by the DSS, and *ProduceConclusion* which connects the data with the conclusions they trigger. As we will remark in Section 4.4, these object properties allow to relate the instances of the different components of the DSO, to assembly a connected semantic request story-plan. Furthermore, these properties are particularly useful for explanation purposes: e.g., the *ProduceConclusion* allows to keep track of what data triggered a certain conclusion of the DSS.

Next, we describe each part of the DSKB, reporting the details of its implementation in the context of the PESCaDO DSS[9].

4.1 Problem Component

The purpose of this component of the DSKB is to formally describe all the aspects of decision support problems that the user can submit to the system. In its simplest form, this component could consist in a taxonomy of the request types supported by the system, enriched by the additional input parameters that are needed by the DSS to provide adequate decision support to the users (e.g., date/time and location of the request). In more advanced situations, a problem description may include also aspects of the users profile (e.g., age, preferences, diseases), or other additional problem features that may affect the decision support provided by the DSS. This component may also be used by the DSS dialogue module to guide (and constrain) users in composing a request.

PESCaDO Problem component The Problem component defined for the PESCaDO system comprises three interrelated parts (see Figure 2): *Request*, *Activity*, and *User*. *Request* describes a taxonomy of request types supported by the system (e.g., "Is there any health issue for me?", "Do environmental conditions require to take some administrative actions?"). *User* enables to describe the profile of the user involved in the request. Examples of the aspects modelled in this component are the user typology (e.g., end-user or administrative user), the age of the user, the gender, diseases or allergies the user may suffer from. Finally, *Activity* describes the activities the user may want to undertake, and that may affect the decision support provided by the system. For instance, different factors are considered by the DSS if the user decides to do some physical outdoor activity rather than travelling with public transportation.

These three sub-components are interrelated by object properties (e.g., a request has to have a user profile associated with it, and may involve an activity the user wants to undertake) and subclass axioms which constrain the possible combination allowed. The PESCaDO user interface module has been developed to dynamically read this constraints from the ontology, to guide the users in formulating their decision support problems. For instance, the subclass restriction "hasRequestUser only AdministrativeUser" on class "CheckAirQualityLimits" states that a request for checking the air quality limits can be submitted only by administrative users. Similarly, the restriction "hasRequestActivity some (AttendingOpenAirEvent or PhysicalOutdoorActivity or Travelling)" on

---

[9] The DSO of the PESCaDO DSKB consists of 210 classes, 99 object properties, 42 datatype properties, and 641 individuals, and it is available at: http://www.pescado-project.eu/ontology.php

Fig. 2. Excerpt of the PESCaDO Problem component

class "AnyHealthIssue" enables to propose to the users only some activities, those of type "AttendingOpenAirEvent" or "PhysicalOutdoorActivity" or "Travelling", forcing them to select one of those. Further parameters are also defined to allow the specification of all the necessary details to compose a complete problem description: for instance, the time period and geographical region considered in the request.

4.2 Data Component

The purpose of this component of the DSKB is to formally describe the data accessed and manipulated by the DSS. For instance, in the case of an environmental DSS, the Data component could describe physical phenomena observations like temperature, humidity, or wind speed, while in the case of a financial DSS, it may represent stock market rates or currency exchanges.

To some extent, this component play the role of the *domain ontology* of the DSS application. Differently from the other two components of the DSO which are more application-oriented, an ontology to be used as Data component may be already available in the web, and thus reused in the DSKB. By adopting an ontological representation of the data processed by the DSS, we favour the integration of (structured) data provided by heterogeneous sources, like web-sites or nodes of the Linking Open Data cloud. For instance, this approach enables to easily exploit for decision making purposes the data exposed by smart city initiatives, like the SmartSantander project[10].

All the aspects of the data that may affect the conclusions taken by the DSS should be described in this component: e.g., validity, provenance, trust, uncertainty. For instance, the Data component may incorporate standardization efforts like the Open Provenance Model [14].

---

[10] http://www.smartsantander.eu/

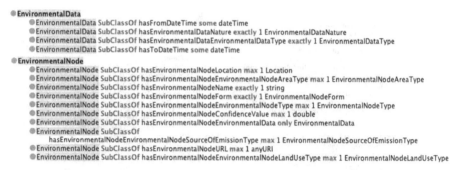

Fig. 3. Excerpt of the PESCaDO Data component

PESCaDO Data component The Data component in PESCaDO describes the environmental data used by the system to provide decision support: e.g., meteorological data (e.g., temperature, wind speed), pollen data, and air quality data (e.g., NO2, PM10, air quality index). Environmental related data like traffic and road conditions are also represented. All the necessary details to comprehensively describe observed, forecast, or historical data are described, including values (both quantitative and qualitative values are supported), the time period covered by the data, and the type of the data (e.g., instantaneous, average, minimum, maximum). Concerning the values, the mapping between qualitative values and quantitative ones is also encoded in the ontology: for instance, a *moderate* quantity of birch pollen in the air correspond to a concentration between 10 and 100 grains per meter cube of air. Detailed information on the environmental node providing the data is also representable, like its type (e.g., measurement station, web-site, web-service), geographical location, and confidence value.

The development of this component of the PESCaDO DSKB has involved (i) the reuse of (part of) some already available environmental domain ontologies (e.g., SWEET[11]), (ii) the application of techniques for automatic ontology extension [15] (e.g., to define the pollen sub-domain), and (iii) the contribution of environmental domain experts. An excerpt of the Data component of the PESCaDO DSKB, specifically the characterization of environmental data and nodes, is illustrated in Figure 3.

The ontological representation of the data processed by the PESCaDO DSS is used to integrate for decision-making purposes the input data coming from heterogeneous sources, and obtained with different techniques, like by querying environmental web-services or by distilling data from text and images offered by environmental web-sites.

4.3 Conclusions Component

The purpose of this component of the DSKB is to formally describe the output produced by the DSS by processing the problem description and the data available. Examples of the content to be produced are warnings/suggestions/instructions, as well as the results of further processing of the data (e.g., data aggregations, data analysis results). Details on the confidence of the system about this content may also be represented, for instance

---

[11] http://sweet.jpl.nasa.gov/ontology/

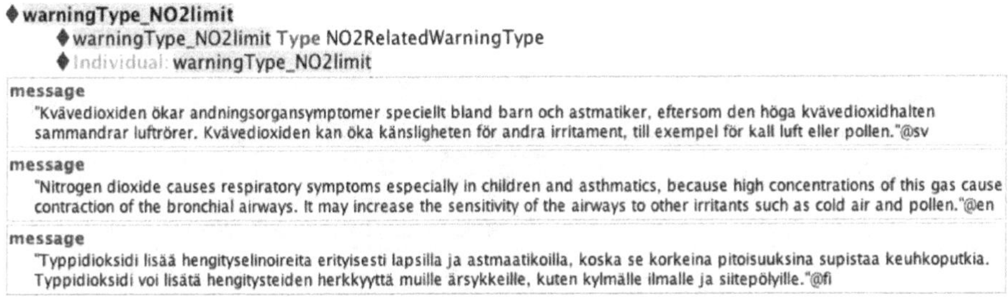

◆ **warningType_NO2limit**
　　◆ warningType_NO2limit Type NO2RelatedWarningType
　　◆ Individual: **warningType_NO2limit**

message
　"Kvävedioxiden ökar andningsorgansymptomer speciellt bland barn och astmatiker, eftersom den höga kvävedioxidhalten sammandrar luftrörer. Kvävedioxiden kan öka känsligheten för andra irritament, till exempel för kall luft eller pollen."@sv

message
　"Nitrogen dioxide causes respiratory symptoms especially in children and asthmatics, because high concentrations of this gas cause contraction of the bronchial airways. It may increase the sensitivity of the airways to other irritants such as cold air and pollen."@en

message
　"Typpidioksidi lisää hengityselinoireita erityisesti lapsilla ja astmaatikoilla, koska se korkeina pitoisuuksina supistaa keuhkoputkia. Typpidioksidi voi lisätä hengitysteiden herkkyyttä muille ärsykkeille, kuten kylmälle ilmalle ja siitepölyille."@fi

Fig. 4. Excerpt of the PESCaDO Conclusions component

supporting the possibility to assign a weight. Furthermore, if needed, this component may also allow to represent the feedback left by the users about their degree of satisfaction of the content produced by the DSS for the submitted request. We also recall that the "ProduceConclusion" object property links the conclusions produced by the DSS to the data that triggered those conclusions, an information that may be exploited for explanation purposes.

PESCaDO Conclusions component The Conclusions component in PESCaDO allows to describe conclusions like warnings, recommendations, and suggestions that may be triggered by environmental conditions, or exceedances of air pollutants limit values that may be detected from the data. An example of warning type encoded in the PESCaDO DSKB is described in Figure 4, together with the associated warning message to be reported to the users (available in all the three languages supported in PESCaDO, English, Finnish, and Swedish). A warning issued by the PESCaDO system for a given decision support request, is represented as a new instance in the A-Box associated with the request, having an object property asserting the type of warning, and a datatype property asserting the relevance (a decimal value in [0..1]) of the warning for the current problem request.

4.4　Semantic Request Story-Plans

The three components of the DSO provide a schema to represent the main aspects of a decision support request. Therefore, for any given decision support request submitted to the system, the actual content about these three aspects of the given request can be formalized as a set of individuals, and assertions on them, instantiating the T-Box part of the DSKB. This results in a connected set of triples, what we called a semantic request story-plan. A semantic request story-plan is an RDF graph covering all the aspects of any decision support request: its formulation, the data relevant for it, and the conclusions generated by the DSS from the data.

An A-Box of the PESCaDO DSKB Figure 5 shows an excerpt of an A-Box instantiating the proposed schema. The three blocks in Figure 5 represents the three components of the T-Box of which the individuals, subjects of the assertions, instantiate some

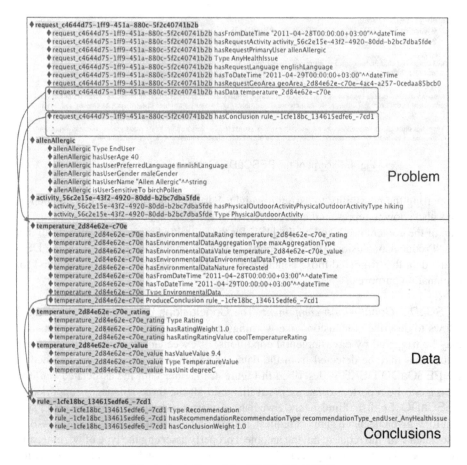

Fig. 5. An A-Box of the PESCaDO DSKB

classes. Note the connections between the individuals of different components, high-lighted by the red boxes and corresponding arrows.

Next, we show how a semantic request story-plan is composed by the DSS while processing a request.

5   Incrementally Building Semantic Request Story-Plans

Coherently with the main phases of a decision-making process (see Section 1), semantic request story-plans are incrementally built by DSSs in three consequent phases.

Phase1: Instantiation of the problem. In the first phase, the problem part of the DSO is instantiated. This occurs when the user submits the request to the DSS via the dialogue module. That is, a module of the system processes the input selections and parameters provided by the user, and generates the instances and assertions characterizing the user decision support request. A consistency check of the input instances with the schema defined by the DSO can be performed via reasoning, to verify that the user request is compliant with the problem supported by the DSS.

Phase2: Instantiation of the data. In the second phase, the actual data which will be used by the DSS to provide decision support, and generate the final conclusions, are instantiated in the A-Box, and connected to the instances describing the problem being processed. First, the DSS determines which data are relevant for the user decision support request submitted. For this purpose, different strategies and techniques can be exploited. In PESCaDO, we encoded in the DSO some mappings between the three parts (request, user, activity) of the Problem component, and the types of environmental data formalized in the Data component, to represent that certain environmental data are relevant for some problem aspects. These mappings, defined together with the domain experts involved in the project, are formalized as OWL hasValue restrictions on the classes of the Problem component. For instance, a restriction of the form "hasRelevantAspect **value** Rain" on the class characterizing the users sensitive to some pollen, states that data about precipitation should be retrieved and taken into consideration when providing decision support for this typology of user. By modelling the mapping this way, the environmental data types for which to retrieve data about, can be automatically determined via DL-reasoning, simply checking the new assertions inferred by the reasoner to the request, user, and activity individuals forming the current user decision request.

Once the data to be used are determined, the module of the system in charge of retrieving these data from the data sources can instantiate them in the DSKB according to the schema defined in the Data component. The connection of the ontology individuals formalizing the data with the individuals formalizing the request under processing is also instantiated (see the "hasData" assertions - first red box - Figure 5).

Phase3: Instantiation of the conclusions. In the third phase, the conclusions triggered by the data according to the user decision support request are instantiated in the A-Box. The way conclusions are computed depends on the techniques for decision support implemented in the DSS. For instance, in PESCaDO we implemented a module responsible for computing conclusions which combines some complementary techniques, based on DL-reasoning and rule-based reasoning. More in details, a two-layers reasoning infrastructure is currently in place: the first layer exploits the HermiT reasoner [16] for the OWL DL reasoning services. The second layer is stacked on top of the previous layer, and implements the Jena [17] RETE rule engine, which performs the rule-based reasoning computation. Next we report an example of rule for triggering the introduction of a recommendation to pollen-sensitive users in case of abundant pollen levels:

```
[ruleAbundantPollen:
    (?request rdf:type pescadoProblem:AnyHealthIssue)
    (?request pescadoProblem:hasUser ?user)
    (?user pescadoProblem:isSensitiveTo ?pollen)
    (?pollen rdf:type pescadoData:PollenDataType)
    (?request pescado:hasGeoArea ?geoArea)
    (?request pescado:hasData ?data)
    (?data pescadoData:hasEnvironmentalDataType ?pollen)
```

```
(?data pescadoData:hasAggregationType pescadoData:max)
(?data pescadoData:hasRating ?rating)
(?rating pescadoData:hasRatingValue pescadoData:abundantPollen)
makeTemp(?rec)
->
(?rec rdf:type pescadoConclusions:Recommendation)
(?rec pescadoConclusions:hasRecommendationType
     pescadoConclusions:recommendation_abundantPollen)
(?rec pescadoConclusions:hasWeight '1.0'^^xsd:double)
(?request pescado:hasConclusion ?rec)
(?data pescado:ProduceConclusion ?rec)]
```

Once the conclusions are instantiated in the DSKB according to the schema defined in the Conclusions component, they are also connected to the individuals formalizing the request under processing (see the "hasConclusion" assertions - second red box - Figure 5), and to the data triggering them (see the "ProduceConclusion" assertions - third red box - Figure 5).

Each semantic request story-plan is maintained by a DSS at least for the lifetime of the processing of the request by the system. Then, the DSS can dispose of the story-plan, or it can archive it in a dedicated cases repository for other purposes (see Section 6). Note that, especially for web-based DSS like PESCaDO, where simultaneous requests may be submitted by users to the system, several semantic request story-plans may be there at the same time in the DSKB. To manage them in parallel, the DSS can adopt an *ontology pooling* mechanism [18]: multiple in-memory ontologies (aka *pools*) are available in the system, and each decision support request submitted to the DSS is assigned exclusively to one of these pools. This solution, adopted in the PESCaDO DSS, allows to keep the size of the ontology used in each pool relatively small (on average the PESCaDO system is working with A-Boxes containing approximatively 20,000 triples), allowing to efficiently exploit the ontology also for some DL-reasoning tasks, like the ones we previously described in this section.

6   Exploitation of Semantic Request Story-Plans

In this section, we present a couple of examples where the semantic representation of a request story-plan can be further exploited to offer additional enhanced services.

6.1   Natural Language Generation of DSS Reports

At the end of a DSS computation, the A-box associated with a decision support request contains a complete "semantic" snapshot of all the information processed and produced by the DSS for the given request: it contains a complete description of the user request, the data relevant for the request and that were used for the decision support computation, and the conclusions and inferred content produced by the DSS together with the information on what triggered those conclusion. All this information can be used to automatically generate a text, summarizing and explaining in natural language the most relevant information to be reported to the user. In the context of an environmental DSS, like in PESCaDO, this automatically generated text, which may complement information provided by the system in graphical or tabular form, is especially appreciated by laymen, or even media corporations which may directly spread it through their communication channels.

In PESCaDO, an approach for multilingual personalized information generation from dynamically instantiated ontologies is adopted [19]. Two modules are involved in the information generation: the text planning module and the linguistic generation module. In particular, the text planning module consists of a content selection and a discourse structuring phase, both performed on a dynamically instantiated ontology (e.g., the T-Box + an A-Box of our DSKB) extended by an additional ontology module capable of representing content selection schemas and elementary discourse units. The output of the text planning module is thus an instantiated ontology enriched with information on the content selected, and the way the text should be organized. This constitutes the input of the linguistic generation module, which produces the text in the three languages supported by the system. Next we report an excerpt of the kind of output produced by the PESCaDO system by exploiting the semantic request story-plan.

Situation in the selected area between 08h00 and 20h00 of 07/05/2012. The ozone warning threshold value ($240g/m^3$) was exceeded between 13h00 and 14h00 ($247g/m^3$), the ozone information threshold value ($180g/m^3$) between 12h00 and 13h00 ($208g/m^3$) and between 14h00 and 15h00 ($202g/m^3$). The minimum temperature was $2°C$ and the maximum temperature $17°C$. The wind was weak (S). There is no data available for carbon monoxide, rain and humidity.

Ozone warning: ozone irritates eyes and the mucous membranes of nose and throat. It may also exacerbate allergy symptoms caused by pollen. Persons with respiratory diseases may experience increased coughing and shortness of breath and their functional capacity may weaken. Sensitive groups, like children, asthmatics of all ages and elderly persons suffering from coronary heart disease or chronic obstructive pulmonary disease, may experience symptoms. [...]

6.2   Semantic Archive of Request Story-Plans

The semantic request story-plans produced by the DSS could be archived in a semantic repository (e.g., a triple store like Sesame [20] or Virtuoso [21]), whose schema is defined by the DSO of the DSKB. This allows to build an incrementally growing archive of all the decision support requests handled by the DSS, together with the data used to process each request and the conclusions generated, as well as some feedback of the user on the decision support provided by the system.

This semantic archive of request story-plans can be exploited for several purposes, enabled by the possibility to semantically access/query its content. For instance, the archive could be used to fine-tune the decision support strategies implemented in the DSS by querying and inspecting the requests not positively rated by the users. Similarly, in the case of DSSs implementing case-based reasoning strategies, the positively rated requests could be used to strengthen the selection of cases used for taking decisions.

Furthermore, thanks to the precise semantic provided by the DSO, this archive could be exposed to the world, adopting the Linked Open Data philosophy. Therefore, the content of the archive could be further exploited by other applications or web-services, like for instance a case reuse/adaptation service which can adapt to slightly different cases the decision-making process done by the DSS, the main phases of which are tracked in each semantic request story-plan.

Relevant statistics could also be produced by semantically querying the archive. For instance, in the context of PESCaDO, one may be interested in checking how frequent is the occurrence of warnings, triggered by environmental conditions, reported to sensitive users, or which geographic areas are more frequently part of the decision support requests submitted by users, or which type of requests are more frequently submitted by each of the various typologies of users supported by the system.

7   On Engineering a DSKB

As any other engineering artefact, the DSKB has to undergo some checks to verify its adequacy for the DSS it supports. Next, we remark some of the important aspects to be taken into consideration.

In addition to checking the formal consistency of the DSO part of the DSKB with state of the art OWL reasoners (e.g., Hermit), and to verifying its correct instantiation with the usage in the DSS, it is crucial to assess the completeness and appropriateness of the content represented in the DSO with respect to the requirements of the application context in which the ontology is used. This requires to check that:

- the types of decision support requests to be supported by the system are formally representable in the Problem component;
- the data relevant for the decision support to be provided are characterized in the Data component;
- the types of decision support conclusions and explanations to be generated by the DSS are formalized in the Conclusions component;

In PESCaDO, the adherence to these guidelines of the three components of the PESCaDO DSO has been evaluated separately, and with different strategies. The Problem component has been evaluated by checking that it is able to represent all and only the possible types of decision support requests that the PESCaDO DSS is supposed to handle. This task was eased by the availability of a detailed description of the typology of these requests in the PESCaDO Use Cases Specification document (Deliverable D8.5[12]). The check showed that all the types of requests defined in the use cases can be represented by the Problem component of the PESCaDO DSO. For the other two components (Data and Conclusions) of the PESCaDO DSO, we asked some environmental expert users (4 people) to judge the appropriateness and completeness of their content, by filling some closed questionnaires. Instead of asking them to directly inspect the content of the ontology, we presented a set of representative decision support requests, the type of environmental data that the system determined to be relevant (and, thus, that the ontology is able to represent) for each of these requests, as well as the type of conclusions that the system is able to provide for them. For the Data component, the results showed an appropriateness of 94% and a completeness of 92% of the content, while for the Conclusions component we obtained an appropriateness of 90% and a completeness of 87%. Therefore, despite some minor adjustments to be made, the PESCaDO DSO proved to be adequate with respect to the proposed guidelines.

---

[12] http://www.pescado-project.eu/Pages/Pdfs-pages/D8.5.pdf

8   Conclusions

In this paper we proposed to employ an ontology-based knowledge base as the main data structure in DSSs. In our approach, to each decision support request submitted to the DSS corresponds a semantic request story-plan in the knowledge base, which describes in a structured way (i) the request itself, (ii) the data relevant for the request, and (iii) the conclusions/suggestions/decisions generated by the system by processing the data. We described the possible usages and advantages offered by the proposed approach, demonstrating them in a concrete implementation for an environmental DSS, developed in the context of the PESCaDO EU project. In details, we showed that a semantic representation of the content processed and produced by the PESCaDO DSS enables (i) to integrate heterogeneous sources of data available in the web (e.g., web sites, web services), (ii) to track, and to expose in a structured form to additional services (e.g., a natural language report generation service), all the content processed and produced by the DSS for each request, and (iii) to exploit logical reasoning for several of the inference steps of the DSS decision-making process.

Acknowledgements. The work described in this paper has been partially funded by the European Commission under the contract number FP7-248594, PESCaDO (Personalized Environmental Service Configuration and Delivery Orchestration) project. The authors would like to thank all the partners involved in the project, and Chiara Di Francescomarino for her support in designing the evaluation of the PESCaDO DSO.

References

1. Laskey, K.B.: Decision Making and Decision Support (2006), http://ite.gmu.edu/~klaskey/SYST542/DSS_Unit1.pdf
2. Marakas, G.M.: Decision support systems in the twenty-first century. Prentice-Hall, Inc., Upper Saddle River (1999)
3. Esposito, M., De Pietro, G.: An ontology-based fuzzy decision support system for multiple sclerosis. Eng. Appl. Artif. Intell. 24(8), 1340–1354 (2011)
4. Ishizu, S., Gehrmann, A., Minegishi, J., Nagai, Y.: Ontology-driven decision support systems for management system audit. In: Proc. of the 52nd Annual Meeting of the ISSS (2008)
5. Choraś, M., Kozik, R., Flizikowski, A., Renk, R., Hołubowicz, W.: Ontology-based decision support for security management in heterogeneous networks. In: Huang, D.-S., Jo, K.-H., Lee, H.-H., Kang, H.-J., Bevilacqua, V. (eds.) ICIC 2009. LNCS (LNAI), vol. 5755, pp. 920–927. Springer, Heidelberg (2009)
6. Wyner, A.: An ontology in owl for legal case-based reasoning. Artif. Intell. Law 16(4), 361–387 (2008)
7. Casanovas, P., Casellas, N., Vallbé, J.-J.: An ontology-based decision support system for judges. In: Proceedings of the 2009 Conference on Law, Ontologies and the Semantic Web: Channelling the Legal Information Flood, pp. 165–175. IOS Press (2009)
8. Ceccaroni, L., Cortés, U., Sànchez-Marrè, M.: OntoWEDSS: augmenting environmental decision-support systems with ontologies. Environmental Modelling & Software 19(9), 785–797 (2004)
9. Zagorulko, Y., Zagorulko, G.: Ontology-based approach to development of the decision support system for oil-and-gas production enterprise. In: Proceedings of the 2010 Conference on New Trends in Software Methodologies, Tools and Techniques: Proceedings of the 9th SoMeT 2010, pp. 457–466. IOS Press, Amsterdam (2010)

10. Pangjitt, T., Sunetnanta, T.: A model of ontology driven case-based reasoning for electronic issue management systems. In: International Joint Conference on Computer Science and Software Engineering (2011)

11. Saremi, A., Esmaeili, M., Habibi, J., Ghaffari, A.: O2dss: A framework for ontology-based decision support systems in pervasive computing environment. In: Proceedings of the 2008 Second Asia International Conference on Modelling & Simulation (AMS 2008), pp. 41–45. IEEE Computer Society, Washington, DC (2008)

12. Wanner, L., Vrochidis, S., Tonelli, S., Moßgraber, J., Bosch, H., Karppinen, A., Myllynen, M., Rospocher, M., Bouayad-Agha, N., Bügel, U., Casamayor, G., Ertl, T., Kompatsiaris, I., Koskentalo, T., Mille, S., Moumtzidou, A., Pianta, E., Saggion, H., Serafini, L., Tarvainen, V.: Building an environmental information system for personalized content delivery. In: Hřebíček, J., Schimak, G., Denzer, R. (eds.) ISESS 2011. IFIP AICT, vol. 359, pp. 169–176. Springer, Heidelberg (2011)

13. Smith, M.K., Welty, C., McGuinness, D.L.: Owl web ontology language guide. W3C Recommendation (2004)

14. Moreau, L., Clifford, B., Freire, J., Futrelle, J., Gil, Y., Groth, P.T., Kwasnikowska, N., Miles, S., Missier, P., Myers, J., Plale, B., Simmhan, Y., Stephan, E.G., Van den Bussche, J.: The open provenance model core specification (v1.1). Future Generation Comp. Syst. 27(6), 743–756 (2011)

15. Tonelli, S., Rospocher, M., Pianta, E., Serafini, L.: Boosting collaborative ontology building with key-concept extraction. In: Proceedings of the Fifth IEEE International Conference on Semantic Computing, Stanford, CA, USA (2011)

16. Shearer, R., Motik, B., Horrocks, I.: HermiT: A Highly-Efficient OWL Reasoner. In: Ruttenberg, A., Sattler, U., Dolbear, C. (eds.) Proc. of the 5th Int. Workshop on OWL: Experiences and Directions (OWLED 2008 EU), Karlsruhe, October 26-27 (2008)

17. hp: Jena - A Semantic Web Framework for Java (2002),
    http://jena.sourceforge.net/index.html

18. Moßgraber, J., Rospocher, M.: Ontology management in a service-oriented architecture. In: 11th International Workshop on Web Semantics and Information Processing, WebS 2012 (2012)

19. Bouayad-Agha, N., Casamayor, G., Mille, S., Rospocher, M., Saggion, H., Serafini, L., Wanner, L.: From ontology to NL: Generation of multilingual user-oriented environmental reports. In: Bouma, G., Ittoo, A., Métais, E., Wortmann, H. (eds.) NLDB 2012. LNCS, vol. 7337, pp. 216–221. Springer, Heidelberg (2012)

20. Broekstra, J., Kampman, A., van Harmelen, F.: Sesame: A generic architecture for storing and querying rdf and rdf schema. In: Horrocks, I., Hendler, J. (eds.) ISWC 2002. LNCS, vol. 2342, pp. 54–68. Springer, Heidelberg (2002)

21. Erling, O., Mikhailov, I.: Virtuoso: Rdf support in a native rdbms. In: Semantic Web Information Management, pp. 501–519. Springer (2009)

Development of the Method for the Appropriate Selection of the Successor by Applying Metadata to the Standardization Reports and Members

Isaac Okada[1], Minoru Saito[1], Yoshiaki Oida[1], Hiroyuki Yamato[2],
Kazuo Hiekata[2], and Shinya Miura[3]

[1] System Engineering Knowledge Improvement div.,
System Engineering Technology Unit, Fujitsu Limited., Tokyo, Japan
`{isaac-okada,saito.minoru,oida.yoshiaki}@jp.fujitsu.com`
[2] Graduate School of Frontier Sciences, The University of Tokyo, Chiba, Japan
`{yamato,hiekata}@k.u-tokyo.ac.jp`
[3] Faculty of Engineering, The University of Tokyo, Tokyo, Japan
`miura@is.k.u-tokyo.ac.jp`

Abstract. In businesses and organizations, it is difficult to find the successor for various activities by considering a person's knowledge and actual experience. In this study, we find the successor to a member of a standardization activity. By assigning metadata to profiles and annual activity reports of members engaged in standardization activities, the relationship between the profiles and the annual activity reports is described as an RDF graph and visualized with nodes and links. This paper has two objectives. Objective-1 is the development and evaluation of a method to design the best combination of search queries to discover an appropriate successor. Objective-2 is the proposal and evaluation of an easy and understandable visualization method of the successor search results obtained in objective-1. The proposed procedure nominates candidates for the successor effectively and the results are visualized in the case study.

Keywords: metadata, RDF, semantic technology.

1   Introduction

Many companies have a huge and growing amount of resources in databases (DBs). In general, each resource is managed individually in accordance with the resource type (e.g., persons, goods, documents), but the relationships among such resources are not managed.

In official standardization activities performed by Fujitsu Limited for ISO, IEC, IEEE and etc., the "Name List DB" listing the members engaged in official standardization activities and the "Annual Activity Report DB" made by the standardization activity members have been individually accumulated and managed.

Some studies have covered the linkage among different data sources. Kashima [1] addressed link prediction by analyzing a network structure (i.e., link mining), and Matsumura et al. [2] addressed information utilization by linking museum information

H. Takeda et al. (Eds.): JIST 2012, LNCS 7774, pp. 255–266, 2013.

and local community information. Also, "VIVO" (http://vivoweb.org) visualizes relationships of researchers. Our research differs from these studies in that we aim to obtain effective results from the actual data in companies by focusing on limited business needs.

When choosing a successor for a standardization activity, the selected successor typically has knowledge, experiences, and a human network similar to or common to the predecessor. If we are able to support this member selection process through a rational and effective way by organizing the relationships among DB data, we can enhance the successor's appropriateness, validity, and adequacy in an objective manner. In this paper, we have two objectives. Objective-1 is the development and evaluation of a method to design the best combination of search queries to discover an appropriate successor. Objective-2 is the proposal and evaluation of an easy and understandable visualization method of the successor search results obtained in objective-1.

2   Proposed Methods

2.1   Overview

To find effective methods to create superior search queries, resources are linked to each other using metadata technology and the relationship is visualized (Figure 1).

The proposed methods are based on the assumption that the "visualization" of linkages among resource elements will facilitate our searching and identifying similar relationships from the resources. Specifically, the proposed methods assign metadata to the two resources, "Name List DB" for the official standardization activities (persons) and "Activity Report DB" (documents), and "visualize" the relationship by which the same metadata link various DBs.

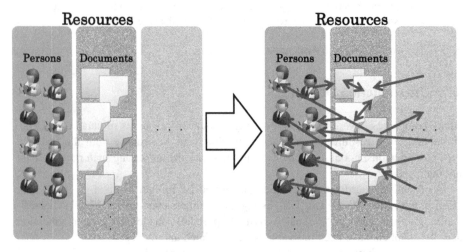

Fig. 1. Linking among resources using the metadata technology

2.2   Verification of Targeted Sample Data

This verification targeted the sample data of the following two DBs.
(1) "Name List DB" for members engaged in the official standardization activity (persons):
DB Content: Name, Department, Standardization activity field, Organization, and other information
(2) "Annual Official Standardization Activity Report DB" (documents):
DB Content: Documents (Word/PDF; Writer Name, Organization, and other information)
Conventionally, both the physical location and the management of these DBs have been independently controlled, without coordination between them.
(Note: In this research, the real names of persons were converted into aliases.)

2.3   Metadata Design

Metadata were designed by using "FOAF" (The Friend of a Friend project: http://www.foaf-project.org) for the Name List (persons), and "Dublin Core" for the Annual Activity Report (documents). Metadata unique to this system were additionally designed as "fkw". The unique metadata included "fkw:belongsTo" (standardization organizations to which the members belong) and "fkw:similarTo" (similar documents based on keywords) (Table 1).

Table 1. List of designed metadata

Meta-data for persons
(from the Name List of the members engaging in the standardization activity)

Meta-data field (Property)	Details	Domain (rdfs:domain)	Range (rdfs:range)	Example of meta-data value
foaf:familyName	Last name	foaf:Person	rdfs:Literal	Fuji
foaf:firstName	First name	foaf:Person	rdfs:Literal	Taro
fkw:belongsTo	Standardization organization to which the member belongs	foaf:Person	fkw:StdOrganization	<http://intap.org>
fkw:stdOrgOfficialPosition	Post assigned by the standardization organization	foaf:Person	rdfs:Literal	Member
fkw:company	Name of Company to which the member belongs to	foaf:Person	rdfs:Literal	FUJITSU LIMITED
fkw:country	Nationality	foaf:Person	rdfs:Literal	Japan

Meta-data for documents (from the Standardization Activity Reports)

Meta-data field (Property)	Details	Domain (rdfs:domain)	Range (rdfs:range)	Example of meta-data value
fkw:similarTo	Similar document based on keywords	fkw:Document	fkw:Document	<http://id103.doc>
dcterms:creator	Writer	fkw:Document	foaf:Person	<http://id96.per>
dcterms:subject	Theme of the standardization activity	fkw:Document	fkw:StdOrganization	<http://intap.org>
fkw:fileName	File name	fkw:Document	rdfs:Literal	ActRepo.pdf
fkw:publishedyear	Name of Company to which the member belongs to	fkw:Document	rdfs:Literal	FUJITSU LIMITED

Meta-data for standardization activity organizations
(from the Name List of the members engaging in the standardization activity)

Meta-data field (Property)	Details	Domain (rdfs:domain)	Range (rdfs:range)	Example of meta-data value
fkw:orgName	Organization name	fkw:StdOrganization	rdfs:Literal	INTAP

2.4   Automatic Assignment of Metadata

In designing metadata, explicit information in the data, such as the names of persons engaged in a standardization activity, is automatically set as the values of the metadata.

For the Annual Standardization Activity Report DB (documents), the proposed method is based on the assumption that an efficient metadata application can further consolidate the relationship between the Name List and the Activity Report, and so the two types of information were added to the data through the following two methodologies.

(1) Evaluation of Similarities among Activity Reports

Writers of documents with many similarities can have knowledge and experiences similar to each other. Therefore, a keyword extraction method using statistical information of word co-occurrence [3] was combined with the "inverse document frequency."

By this combined method, listed below, 10 keywords were extracted from each activity report as a set. Then, similarities among these sets were evaluated by Simpson's Coefficient, which was used for the evaluation of the similarity between sets. If the value calculated by the coefficient was greater than the threshold value 0.5, the relevant reports were determined to be similar to each other.

[Step 1] Extract up to 10 keywords from each document.

1. Calculate the expected co-occurrence frequency of each term in the document [3]. The expected co-occurrence frequency of term w and term g is shown as the product of n_w and p_g.

n_w: The total number of terms in a sentence where the term w appears

p_g: $\dfrac{\text{The total number of terms in a sentence where the term } g \text{ appears}}{\text{The total number of terms in the whole document}}$

2. Calculate the χ^2 value which indicates the bias of the term co-occurrence [3].

$$\chi^2(w) = \sum_{g \in G} \frac{(freq(w,g) - n_w p_g)^2}{n_w p_g} \tag{1}$$

where $freq(w,g)$ is the co-occurrence frequency of term w and term g. The inverse document frequency (IDF) is calculated by the following equation.

3. Calculate the IDF

$$IDF(t) = 1 + \ln\left(\frac{N_{all}}{N_t}\right) \tag{2}$$

Where

N_{all}: The total number of documents

N_t: The number of documents where the term t appears

4. Calculate IDF value $\times \chi^2$ for each term, and select 10 terms having the highest results from the top as keywords.

[Step 2] Calculate the similarity among documents in accordance with keywords common to the documents.

1. Calculate the similarity by the Simpson's Coefficient

$$\text{Simpson's Coefficient} = \frac{|X \cap Y|}{\min(|X|, |Y|)} \tag{3}$$

where

    X : Set of keywords of document 1

    Y : Set of keywords of document 2

2. As documents with results more than a certain threshold value are considered to be similar to each other, each document name is set to the other's metadata field fkw: similarTo.

(2) Setting of Standardization Organization Name on Which an Activity Report Focuses.

The Name List was compiled as a dictionary of Standardization Organization Names and the top three organization names were set as metadata on the activity report data to facilitate finding a standardization activity or similar activities on which the activity report focused.

Once metadata were applied to the Name List DB, the DB was referred to as the "Profile DB".

2.5   Metadata Description in RDF

By describing the applied metadata in the Resource Description Framework (RDF) (Figure 2), it was possible to plot the members engaging in standardization activities and the annual standardization activity reports in RDF graphs to visualize the inter-DB relationships (Figure 3).

NameSpace

foaf: <http://xmlns.com/foaf/0.1/>(Metadata relevant to persons)
dcterms: <http://purl.org/dc/terms/>(Metadata relevant to Web content)
fkw: <http://know.who.org/2011/06/stdorg#>(Metadata specific to this system)

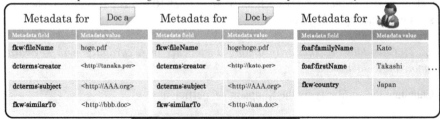

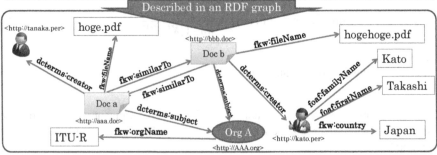

Fig. 2. Applying metadata and RDF description

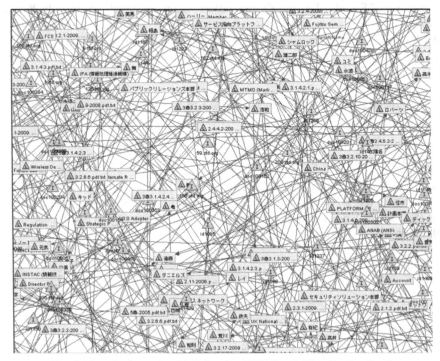

Fig. 3. Visualization using RDF Gravity (RDF Graph Visualization Tool)

By executing SPARQL queries on these RDF graphs (i.e. a group of on-tologized elements in both DBs), the same graph patterns were easily extracted.

2.6 Assumption of Successor Selection Logic for the Replacement of Members Engaging in a Standardization Activity

When the replacement of a member engaging in a standardization activity is needed because of reassignment or other reasons, customarily a colleague of the predecessor is selected as the successor. Accordingly, the successor has the knowledge and a human network similar to those of the predecessor, and thus such selection is considered as an appropriate decision.

The proposed method is based on the assumption that the conventional successor selection logic consists of five queries, as shown below (Table 2), on the basis of interviews from the relevant persons. The queries are described in SPARQL.

Table 2. Queries for researching persons who have the similar engineering expertise

Query	Details
Q1	Belonging to the same standardization organization as the predecessor.
Q2	Belonging to a standardization organization on which the predecessor wrote the activity report as a theme.
Q2'	Writing (or has written) an activity report on a standardization organization to which the predecessor belongs.
Q3-1	Writing (or has written) a report similar* to the report written by the predecessor.
Q3-2	Writing (or has written) a report similar** to the report written by the predecessor.

* This "similar to" means that the reports have been written on the same subject (dcterms:subject).
** This "similar to" means that the reports are linked with (fkw:similarTo).

2.7 Selection Procedure of Best Query Combination

The following procedure was adopted to select the best combination of queries.

1. Select the correct cases for training data.
2. Search by the assumed successor selection logic queries and combinations of these queries.
3. Calculate the information content (I_{ij}) of query combinations as follows:

$$
I_{ij} =
\begin{cases}
- \log_2 \dfrac{N_{ij}}{N_{all}} & (N_{ij} \text{ includes a successor}) \\
0 & (N_{ij} \text{ includes no successor})
\end{cases}
\tag{4}
$$

where

N_{ij}: The number of successor candidates as matching results

N_{all}: The number of all successor candidates

4. Select the best query combinations that maximize I_i .

$$
I_i = \sum_j I_{ij}
\tag{5}
$$

2.8 Visualization

The visualization method used by this system is as follows. Visualization of the resource link by the network helps the user's understanding of the relationship between resources. First, resources such as standards organizations and members are set as nodes. Second, the metadata link the resources. Then, the network is generated and visualized. A metadata value about the name of the resource is shown as a node label.

3   Case Studies for Verifying the Availability of Linked Data

This system was verified by whether it could find a successor in a similar manner to the conventional method for actual replacement cases.

Specifically, the following metadata were used to select the cases in which members engaging in standardization activities were replaced by the same process described in Section 2:

(1) The 5 queries mentioned above,

(2) Data of approximately 550 persons in the "Profile DB" (persons) for members engaging in the Official Standardization Activity,

(3) Reports of the past 5 years in the "Annual Standardization Activity Report DB" (documents).

Approximately 200 official standardization organizations were extracted. The system was verified as to whether it could reproduce the same results as conventional selections.

3.1   Selection of Correct Cases for Training Data

Table 3 shows the past 5 years' actual replacements of members for standardization activity reports.

Table 3. Five persons who took over themes for standardization activity reports

No.	Predecessor	Successor	Report theme	Year
1	Ishikawa	Kinoshita	OMA	2005 → 2006
2	Ishikawa	Murayama, Hata	ITU·R	2005 → 2006
3	Kinoshita	Ozaki	OMA	2006 → 2007
4	Kikuchi	Yokosuka	INTAP	2007 → 2008
5	Hata	Endo	ITU·R	2008 → 2009

3.2 Searches by the Assumed Successor Selection Logic Queries

After conducting searches by the abovementioned 5 queries through the whole candidate DB (approximately 550 persons), the following was determined:

(1) How many persons were refined from the total candidates,
(2) Whether the actual successor really existed within the refined candidates.

3.3 Results

The 5 queries in 15 different combinations for the data of the targeted 550 persons were executed. Table 4 shows the results of the selection from the candidate DB to refine the number of candidates to 10 or less. If a case matches "Q1" and "Q2" separately, it is not certain that the case matches the combination of "Q1" and "Q2". The information content in Table 5 was calculated using equation (4).

Table 4. Query combinations for the 5 predecessors and results refined from approximately 550 candidates

No	Predecessor	Q1	Q2	Q2'	Q3-1	Q3-2	Q1 ∩ Q2	Q1 ∩ Q2'	Q1 ∩ Q3-1	Q1 ∩ Q3-2	Q2 ∩ Q2'	Q2 ∩ Q3-1	Q2 ∩ Q3-2	Q2' ∩ Q3-1	Q2' ∩ Q3-2	Q3-1 ∩ Q3-2
1	Ishikawa	46	44	19	37	12	27	4	3	5	3	6	5	17	6	9
2	Ishikawa	46	44	19	37	12	27	4	3	5	3	6	5	17	6	9
3	Kinoshita	98	81	24	38	11	48	6	3	6	3	8	6	18	6	9
4	Kikuchi	166	138	18	55	21	71	7	1	13	1	5	3	8	6	13
5	Hata	72	32	19	30	7	10	5	1	5	1	6	3	6	5	7

Table 5. Infromation content

No	Predecessor	Q1	Q2	Q2'	Q3-1	Q3-2	Q1 ∩ Q2	Q1 ∩ Q2'	Q1 ∩ Q3-1	Q1 ∩ Q3-2	Q2 ∩ Q2'	Q2 ∩ Q3-1	Q2 ∩ Q3-2	Q2' ∩ Q3-1	Q2' ∩ Q3-2	Q3-1 ∩ Q3-2
1	Ishikawa	3.58	3.64	4.85	3.89	5.52	4.35	7.10	7.52	6.78	7.52	6.52	6.78	5.01	6.52	5.93
2	Ishikawa	0.00	3.64	0.00	3.89	5.52	0.00	0.00	0.00	0.00	0.00	0.00	0.00	0.00	0.00	5.93
3	Kinoshita	2.49	2.76	4.52	3.85	0.00	3.52	0.00	0.00	6.52	0.00	6.10	6.52	4.93	0.00	0.00
4	Kikuchi	0.00	0.00	4.93	3.32	4.71	0.00	0.00	0.00	0.00	0.00	0.00	0.00	6.10	6.52	0.00
5	Hata	2.93	4.10	4.85	4.19	0.00	0.00	0.00	0.00	0.00	0.00	6.52	0.00	6.52	0.00	0.00

Design method: Select a query combination i so that the equation (5) gives the greatest value for I_i. Evaluate the design method using a cross-validation technique.

Conduct cross-validations for the 5 correct data sets by using 4 of them as training data and one of them as test data. The cross-validation results are given in the table below. The values marked in yellow are the maximum in each validation.

Table 6. Selection of best query combinations (5 cases)

Training data	Q1	Q2	Q2'	Q3-1	Q3-2	Q1∩Q2	Q1∩Q2'	Q1∩Q3-1	Q1∩Q3-2	Q2∩Q2'	Q2∩Q3-1	Q2∩Q3-2	Q2'∩Q3-1	Q2'∩Q3-2	Q3-1∩Q3-2
1,2,3,4	1.52	2.51	3.57	3.74	3.93	1.97	1.78	1.88	3.32	1.88	3.15	3.32	*4.01*	3.26	2.97
1,2,3,5	2.25	3.54	3.56	3.96	2.76	1.97	1.78	1.88	3.32	1.88	*4.78*	3.32	4.11	1.63	2.97
1,2,4,5	1.63	2.85	3.66	3.82	3.93	1.09	1.78	1.88	1.69	1.88	3.26	1.69	*4.41*	3.26	2.97
1,3,4,5	2.25	2.63	4.79	3.81	2.56	1.97	1.78	1.88	3.32	1.88	4.78	3.32	*5.64*	3.26	1.48
2,3,4,5	1.35	2.63	3.57	3.81	2.56	0.88	0.00	0.00	1.63	0.00	3.15	1.63	*4.39*	1.63	1.48

Table 7. Evalation of selection results

	Q2∩Q3-1	Q2'∩Q3-2	Q2'∩Q3-1
training data:1,2,3,4			6.52
training data:1,2,3,5	0		
training data:1,2,4,5			4.93
training data:1,3,4,5			0.00
training data:2,3,4,5			5.01

The average information amount obtained from the results of the selected queries was 3.29. The result is better than 2.67, which was the average of the information amount of all the queries.

3.4   Visualization

Furthermore, visualizing the data relationships also made it possible to find other successor candidates besides the actual successors.

The example shown in Figure 4 indicates that "Endo" and "Ozaki", who have similar relationships, can also be candidates for successors for the former "Ishikawa". "Kinoshita" and "Hata" were actually selected.

As shown in Table 3, "Ozaki" was selected as a successor to "Kinoshita" at a later date and "Endo" was also selected as a successor to "Hata", which shows that the possibilities shown in Figure 4 are reasonable and appropriate.

Thus, visualization of the relationship between profiles (persons) and activity reports (documents) made it possible to propose other options for member nomination.

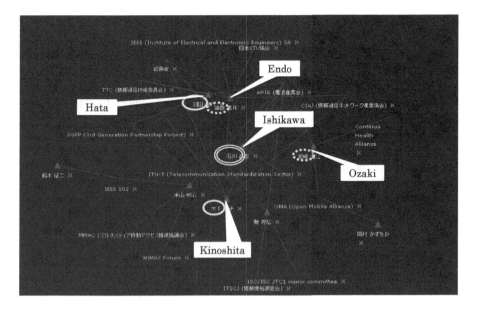

Fig. 4. Possibilities of the other successors

4   Discussion

Though the proposed design procedure gives better results for the successors than does random selection from the candidates in the lists, the accuracy is not good enough. However, the selection results are justified more clearly than the current practice because the results are based on clear rules and queries described in the RDF metadata and the SPARQL queries. The criteria can be clarified in this method and this is one of the contributions of this research.

The proposed design procedure is evaluated for a very small number of case studies and data, so the findings of this research must be confirmed continuously in the real field. Data on human resources are very difficult to manage in a company and compliance and privacy problems might be considered for further research. Though the findings of the case studies might be revealed by other approaches, the number of metadata fields is getting bigger and so this approach works more effectively when employed for search processes in the workplace.

As for the visualization, this manuscript proposes a method for rendering the relationships between resources, such as humans and documents. All the information in the graphs can be shown to the persons in charge to discover new findings. The relationships described in metadata, such as similarity, could be illustrated using varying font sizes and other visualization techniques.

5   Conclusion

A procedure for the selection of queries to find an appropriate successor to an activity member is proposed in this paper. The proposed procedure nominates candidates for the successor and the list of nominees is more accurate than that obtained by random selection. The proposed procedure effectively limits the number of qualified candidates and thus fulfills objective-1.

For objective-2, a visualization method for selection of the successor is proposed and the results of the visualization are shown. The links explain the similarities and commonalities of the data. Discovering an appropriate successor from the visualization results is also discussed in the paper. Visualization is one of the crucial approaches for extracting the background of human affairs. To promote more appropriate and efficient searches for a successor, visualizing data links is essential.

6   Future Research

In order to promote more efficient information search by the procedure for the selection of the queries to find an appropriate successor is proposed in the paper, it is essential to popularize this approach on a broad range of public data.

References

1. Kashima, H.: Survey of Network Structure Prediction Methods. Journal of the Japanese Society for Artificial Intelligence 22(3), 344–351 (2007)
2. Matsumura, F., Kobayashi, I., Kamura, T., Kato, F., Takahashi, T., Ueda, H., Ohmukai, I., Takeda, H.: Collaboration between Linked Open Data of Museum Information and Regional Information. Information Processing Society of Japan, Computers and the Humanities Symposium, "JINMONKON 2011" Journal 2011(8), 403–408 (2011)
3. Matsuo, Y., Ishizuka, M.: Keyword Extraction from a Document Using Word Co-occurrence Statistical Information. Journal of Japanese Society for Artificial Intelligence 17(3), 217–227 (2002)

A Native Approach to Semantics
and Inference in Fine-Grained Documents

Arash Atashpendar, Laurent Kirsch, Jean Botev, and Steffen Rothkugel

Faculty of Science, Technology and Communication
University of Luxembourg
Luxembourg
arash.atashpendar.001@student.uni.lu,
{laurent.kirsch,jean.botev,steffen.rothkugel}@uni.lu

Abstract. This paper proposes a novel approach for enhancing document excerpts with semantic structures that are treated as first-class citizens, i.e., integrated at the system level. Providing support for semantics at the system level is in contrast with existing solutions that implement semantics as an add-on using intermediate descriptors. A framework and a toolset inspired by the Semantic Web Stack have been integrated into the *Snippet System*, an operating system environment that provides support for fine-grained representation and management of documents and the relationships that exist between arbitrary excerpts. The high granularity, combined with native support for semantics, is leveraged to alleviate some of the existing personal information management problems in terms of content retrieval and document engineering. The resulting framework offers some inherent advantages in terms of refined resource description, content and knowledge reuse along with self-contained documents that retain their metadata without depending on intermediate entities.

Keywords: System-integrated Semantics, Personal Information Management, Document Engineering, Visualization.

1 Introduction

Personal information management facilities, as pointed out in [1,2], need to be enhanced with semantics in order to be able to cope with the ever-increasing growth of information. This idea, geared towards the usage of semantics, has been suggested using a wide variety of techniques. Some of the most common solutions include using search by association instead of keywords and directories [2], relying on a personal ontology for expressing the user's domain of interest in terms of objects and relations [1] or enhancing desktop environments with semantics, an idea called the *Semantic Desktop* [3]. However, a native and system-integrated approach to semantics offers advantages that are not available or more cumbersome to achieve in systems that implement semantics as an add-on, instead of a native system component.

H. Takeda et al. (Eds.): JIST 2012, LNCS 7774, pp. 267–277, 2013.
© Springer-Verlag Berlin Heidelberg 2013

As discussed in [4], capturing relationships between document excerpts managed by the same or different authoring tools [5] is of critical importance when creating, retrieving, but also when reading documents. While creating or adapting documents, authors can utilize these relationships to access and assess the context of a specific excerpt. Readers, in turn, tend to read not only linearly, but mostly laterally [6] by using multiple, complementary sources. Accessing and visualizing information on such relationships – added during the authoring process – allows the reader to better grasp the context of specific document excerpts, also preventing wrong associations.

The *Snippet System* [7,8] is an operating system environment that aims at providing novel document management facilities. Most importantly, it replaces the traditional concept of hierarchical folders with so-called *Relations*. These Relations define predicates indicating relationships between individual document excerpts that can be conveniently leveraged by users while authoring or browsing information [9]. A connected document excerpt can have any granularity and may, for example, consist of a single bullet point or an entire slide in case of a presentation, an individual shape or a layer in case of a vector graphics object, or a word, sentence or paragraph in case of a text document. This is enabled by the Snippet System's integrated document model that forms the basis for all of its higher-level document management facilities. Instead of mapping documents onto single files, the model maps them onto graphs of more fine-grained entities, so-called *Snippets* [7]. An inherent advantage of this approach is that individual Snippet graphs can share certain Snippets, thereby synchronizing the corresponding data between the involved graphs. This, in turn, allows for the creation of dedicated Snippet graphs representing Relations. A Relation's graph interconnects subgraphs of Snippets representing excerpts of documents, which are shared with the graphs representing the original documents.

Due to the generic design of Snippets [7], semantic metadata is integrated into Relations' Snippet graphs by using exactly the same data structures as for the representation of a document's actual data. Furthermore, in combination with the fine-grained nature of Snippets, this is leveraged for defining more refined knowledge bases and concept hierarchies that involve document excerpts of arbitrary granularity. On the other hand, existing approaches are typically based on coarse-grained files and therefore require semantic metadata to be separated from the main data, which results in a lower level of expressiveness. This limitation has been somewhat alleviated with the advent of web-based technologies such as RDFa[1] and Microformats[2] that embed rich metadata within web documents. The application of these technologies has been subject to a wide variety of techniques such as embedding semantics in the form of invisible elements into web content [10] or defining annotations over different ranges of text within a web page [11]. The work presented in this paper takes these approaches a step further by providing native support for semantics and integrating metadata at the system level.

---

[1] http://www.w3.org/TR/xhtml-rdfa-primer/

[2] http://microformats.org/

As already argued in [3] and [12], one can take the know-how of the Semantic Web and benefit from the maturity of its standards in order to tackle personal information management problems by applying them to the desktop environment. However, existing attempts at enhancing personal information with semantics rely on indirect and non-native approaches such as trying to glue semantics [3] to existing desktop applications or using free text annotations [12] to integrate knowledge in documents.

The remainder of this paper describes how the Snippet System's support for Relations is enhanced with a semantic framework that is capable of expressing relationships in terms of higher-level constructs such as concept hierarchies and ontologies. The framework relies on a set of well-established formalisms and technologies for representing semantics that are based on a subset of the Semantic Web[3] Stack. Furthermore, a toolset is built on top of this framework, which includes, among other things, a native ontology editor, an ontology browser, a visualization tool and a semantic search engine module that is powered by a triple-matching query interpreter, an inference engine and a Snippet crawler. Section 2 describes the architecture of the framework and the toolset that has been built on top of it, followed by an elaborate example in Section 3 that highlights the advantages of the proposed solution along with specific use-cases in terms of document engineering. References to specific functions and operations of other similar projects given throughout the text replace a dedicated section for related work. Finally, a discussion surrounding a set of conclusions derived from the presented work along with suggestions for future work are given in Section 4.

2 Semantic Framework

The architecture and the toolset discussed in this section, have been implemented as a proof of concept in the form of a prototype running on Mac OS X using Core Data and Cocoa.

2.1 Architecture

A fundamental property of the Snippet System lies in its usage of a generic document model. The same data structures, namely Snippets, are used for representing documents, excerpts and Relations. Thus, semantic relationships as well as the involved excerpts' actual data are equally matched in terms of Snippet representations. As a consequence, semantics are first-class citizens of the document model and no dedicated intermediate layers linking semantic structures to user data are required.

An illustration of the architecture is given in Figure 1. In order to provide support for an RDF-based[4] data model, the object graph of Relations' Snippets

---

[3] http://www.w3.org/standards/semanticweb/
[4] http://www.w3.org/rdf/

represents a labeled, directed multigraph that also makes it possible to distinguish the roles of involved excerpts, i.e., it is clear whether an excerpt is a subject or an object in a given subject-predicate-object triple. Moreover, predicates are realized by Relations through a set of pointers to related excerpts, which correspond to resources and concepts in the Semantic Web Stack. The inherent architectural compatibility of the system with this stack offers the potential to make use of available structured knowledge bases and to query external resources [12] through the use of semantic harvesters to promote knowledge reuse.

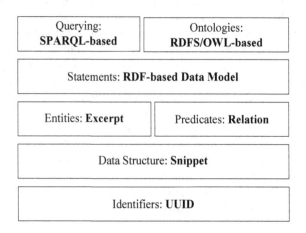

Fig. 1. Framework architecture inspired by the Semantic Web Stack

Furthermore, a native type system added to the framework allows for fixing the number of excerpts that a predicate can point to according to its inherent semantics in terms of collection types. The type system is also used to represent primitive data types as well as meta relationships between ontology concepts and roles in terms of inheritance.

The Snippet System's native support for creating relationships through the use of bidirectional pointers to connect excerpts to each other, coupled with its capability to define any piece of data as an excerpt and thus an end-point in its connections, leads to a uniform representation of relationships across the entire framework, be it between individuals/instances or ontology concepts.

Moreover, similar to the way everything is considered to be a resource in the Semantic Web expressed in terms of entities and roles that can be uniquely identified, excerpts and Relations are also treated as resources that get assigned UUIDs by the Snippet System to provide support for unicity.

Finally, the triple store can benefit from the native indexing capabilities of the Snippet System as its statements are stored as Snippets. Apart from indexing excerpts and roles, a partially implemented cross-indexing mechanism is also in place to allow for retrieving triples using different permutations.

2.2   Toolset

A collection of tools has been developed on top of the discussed architecture. Its main components include the following:

- An **ontology editor** that allows the user to build personal ontologies.
- An **ontology browser** used for creating relationships between document excerpts that also makes use of a consistency enforcer component.
- A **semantic search** module powered by inference containing a query interpreter and a Snippet crawler.
- A **visualization** module for displaying semantic graphs of documents and ontologies.

The framework is also equipped with a native vocabulary inspired by RDF, RDFS[5] and OWL[6] that can be used to create ontologies. Ontologies defined with this vocabulary in the ontology editor are used by the ontology browser to establish relationships between document excerpts. These semantic relationships can then be queried by using the inference-enabled semantic search module, which internally uses a query interpreter and a crawler to perform its tasks. The query interpreter supports queries consisting of triple patterns, conjunctions and disjunctions with a syntax similar to that of SPARQL[7]. Furthermore, semantic navigation allows for browsing documents, inspecting their excerpts and opening related ones. Ontologies and metadata can additionally be visualized to obtain graphs of documents and ontology schemas based on different criteria. Such a graph can for example depict excerpts pertaining to a specific ontology or simply the semantic relationships of a specific document's excerpts, similar to the way tools such as IsaViz[8] and OntoViz[9] visualize ontologies and RDF models. This functionality is described in more detail in Section 3.2.

Moreover, as suggested in [13], binding retrieval to inference constitutes an important step towards improving both semantic search engines as well as inference engines. Similarly, an inference engine has been added to the semantic framework that is leveraged by the presented semantic toolset.

Finally, a consistency enforcer module has been integrated into the ontology browser to prevent the user from introducing inconsistent statements into the triple store. For instance, upon creation of a relationship, it can check the type of an excerpt against the domain or range of the selected predicate and disallow the operation in case of type mismatch.

The discussed functionalities are further illustrated in Section 3 with a concrete use-case example. Furthermore, it should be noted that since the main focus of this research is to demonstrate the effectiveness of native, system-integrated semantics in the realm of desktops and personal information management, the developed prototype only provides a partial implementation of the operators that

---

[5] http://www.w3.org/TR/rdf-schema/
[6] http://www.w3.org/TR/owl2-overview/
[7] http://www.w3.org/TR/rdf-sparql-query/
[8] http://www.w3.org/2001/11/IsaViz/
[9] http://protegewiki.stanford.edu/wiki/OntoViz

are available in the OWL family. Therefore, the expressiveness of the framework does not fit in any specific Description Logic[10].

3   Semantic Document Engineering

Semantic documents, as noted in [14], constitute an important step towards improving classification and retrieval of documents through the use of custom domain models. They can also enable higher-level operations such as domain knowledge extraction from large documents [15] and analysis of knowledge in documents for providing various visualization techniques [16].

The toolset presented in this paper leverages the high degree of granularity provided by the *Snippet System*'s underlying document model, by using the integrated semantic framework discussed in the previous section. This goal is achieved by means of creating semantic relationships between arbitrary document excerpts in order to allow for both semantic inspection of thus created documents as well semantic retrieval of content while creating new documents.

Be it a professor looking for related content in previous exams and lecture notes to prepare an exercise sheet, or a student trying to study for an exam and going through several documents to do so, both types of users – document authors as well as readers – can benefit from semantic inspection and retrieval of related content. The remainder of this section is dedicated to the description of a concrete example involving a physics professor who tries to prepare an exercise sheet by leveraging the facilities provided by a semantic, fine-grained document engineering environment to simplify the task.

The degree of precision and the extent to which users can refine their queries in terms of expressiveness depend on the richness of the ontologies that have been used to define the semantic relationships that get queried by the user. Therefore, the following example is based on the assumption that the used ontologies are sufficiently expressive and have been adapted to the presented use-case.

Before starting the process of assembling related content that can be used in the exercise sheet, a general assessment of the students' performance in different domains can be made based on their results in previous exams. The professor can do so by retrieving all exam problems that students have failed to solve in previous exams. This kind of query can be further refined using separate restrictions on specific topics in which students are more prone to failure according to the professor's personal experience. Once a rough assessment of these areas has been made, the actual process of querying related content for the preparation of the exercise sheet can be carried out with more precision.

3.1   Semantic Content Retrieval

The professor can prepare a list of queries and define restrictions in terms of operators such as conjunction and disjunction to retrieve a set of excerpts that

---

[10] http://dl.kr.org/

pertain to the topics in which students are to be interrogated based on their previous performances. The exercise sheet can be prepared using two distinct approaches. Either exercises fulfilling a set of conditions can be directly retrieved and embedded into the involved exercise sheet as they are, or lecture materials related to certain topics such as exercise sheets, notes, presentations or diagrams can be retrieved through the built-in semantic explorer of each document for creating new exercises from scratch, based on existing materials.

For example, the following requirements can be formulated as a batch of queries that is then fed to the semantic search engine to retrieve all matching excerpts: Questions that have Newtonian Mechanics as their topic plus multiple choice examples covering Quantum Electrodynamics illustrated by diagrams in lecture notes.

A partial representation of the semantic graph corresponding to matching excerpts is illustrated in Figure 2. This illustration also highlights the focal point of this research, namely the natural integration of semantic structures with user data. Concrete document excerpts directly participate in the semantic graph as uniquely identified individuals in the same way that other purely ontological concepts such as "Multiple Choice Question" and "Newtonian Mechanics" are represented in the system.

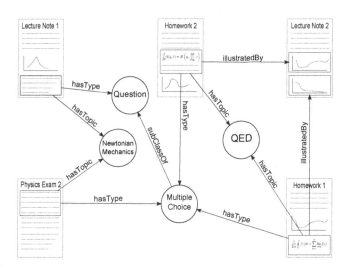

Fig. 2. Partial graph of matching excerpts and their integration with semantic entities

A close look at the illustration in Figure 2 reveals how inference and classification through semantics can alleviate content retrieval. The subset of excerpts on Quantum Electrodynamics only includes questions of type Multiple Choice, as explicitly indicated by the query, whereas those corresponding to the first part of the query, namely questions covering Newtonian Mechanics contain multiple choice questions as well as other question types. The retrieved result set contains multiple choice instances in addition to other relevant questions, because multiple choice questions have been defined as a subclass of the concept of *Question*. Therefore the system infers that they are a specialized question type and returns them as matching instances. Furthermore, even though there are no instances related to each other using the predicate "covers", the system still manages to retrieve matching excerpts because it can infer, using an ontology that relates the two predicates "covers" and "hasTopic" as such, that a question having "hasTopic" as predicate also qualifies as a match.

Thanks to the high degree of granularity provided by the Snippet System, virtually any piece of data in a document can be defined as an excerpt. This feature can be taken a step further to, for example, assign semantic statements to individual curves in a diagram. Provided that diagrams and curves are enriched with such descriptions, the professor can then leverage this functionality to retrieve all diagrams illustrating a specific phenomenon that contain at least a curve describing an asymptotic function.

In addition to running queries to retrieve relevant content, the user can also simply inspect specific documents and use tools such as semantic excerpt navigation or semantic graph visualization to explore content and search for excerpts of interest.

For instance, the content of a document containing a large set of diagrams and formulas can be explored by using both its *Excerpt View* as well as its *Relation View*. Upon selection of an excerpt, the Relation View displays a list of all predicates that describe relationships between the selected excerpt and other excerpts. These can, in turn, either represent other similarly defined entities corresponding to concrete document excerpts or purely descriptive entities such as ontology concepts contained in specific ontologies. For example, the professor can select an excerpt that defines a particular curve in a function graph in order to view all its relationships using the Relation View. Once a specific Relation of the curve is selected, a dedicated Excerpt View displays all excerpts that are related to the selected curve through the predicate in question. These related excerpts can then be invoked to either view them in the context of their containing document or to view their respective Relations. This process can be continued to systematically navigate through the semantic graph of related document excerpts.

3.2   Semantic Content Visualization

The visualization of ontologies and metadata in terms of clusters and subgraphs can reveal implicit information embedded in documents that would otherwise be difficult to extract [17]. Labeled, directed graphs provide a natural representation of RDF data and the visualization module adopts a similar approach. In addition to individual documents and ontologies, users can visualize document excerpts and their relationships along with the ontologies that describe the involved excerpts. This feature displays both schemas as well as instances in terms of subgraphs that delineate the boundaries of involved documents and ontologies. A partial cluster graph visualization of a physics exam and its content's relationships to other documents' excerpts along with the ontologies that describe these excerpts is given in Figure 3.

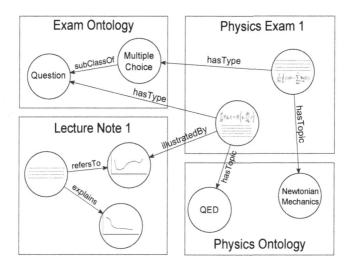

Fig. 3. Cross-boundary cluster graph visualization of content

In addition to creating graphs for individual documents, the visualization module offers two other modes, namely an ontology-centric mode along with a non-restrictive mode that visualizes the entire collection of user documents. Using the ontology-centric mode, documents and user content can be visualized in terms of specific ontologies and their corresponding concepts and predicates. This allows the user to filter and select document excerpts whose relationships belong to a specific ontology to perform instance comparison and schema comparison [18]. In both modes, cluster graphs are used in order to highlight cross-boundary relationships in terms of labeled subgraphs.

4   Conclusion and Future Work

This paper presented a novel approach for alleviating some of the existing personal information management issues by enhancing individual document excerpts with semantics. A semantic framework, developed as a proof-of-concept prototype, has been integrated into the *Snippet System*, an operating system environment with an integrated document model that supports a fine-grained representation of documents and their relationships. The essence of this model's design is that it allows for representing both the actual contents of documents, as well as semantic metadata using exactly the same, generic data structures. This entails that there is only one model for representing both semantics and documents, which eliminates the need for creating bridging mechanisms such as API adapters to connect two separate models, as is the case for most existing approaches.

Moreover, a set of semantic tools has been designed and implemented on top of this semantic framework, most importantly a semantic query interpreter that leverages existing semantic relationships to allow for more advanced queries for retrieving very specific, fine-grained excerpts of documents. Furthermore, base assertions along with ontology definitions are made available to an inference engine that can reason and derive triples and relationships between excerpts in response to user queries. The reasoner module relies on property and concept entailments, e.g., sub/super properties and sub/super classes, to infer triples that are not explicitly stored in its knowledge base. Additionally, a visualization tool has been developed to present the semantic relationships of a specific excerpt, document, ontology or even a set of documents and ontologies in a user-friendly, easy-to-understand fashion, thereby allowing the user to better grasp the context of the involved resources. The visualization module adopts a cluster graph-based approach to display the semantic content of documents as well as ontology schemas in order to delineate boundaries between related document excerpts and ontologies that denote their meaning.

Future work is dedicated to extending the current tool set with a template-based content creation module that allows for creating new documents from a set of excerpts that have been retrieved by the semantic query interpreter as a result of some user query. This feature, in turn, requires an ontology ranking mechanism [19] in order to automatically select the most relevant results. Furthermore, the visualization module can be extended to provide facilities for user interaction, to enable dynamic content browsing and navigation via interactive semantic graphs. Finally, the inference engine's reasoner can be extended to automatically insert new triples back into the triple store according to user-defined inference rules, as well as implicit information that can be inferred from context and user activity.

References

1. Katifori, V., Poggi, A., Scannapieco, M., Catarci, T., Ioannidis, Y.: Ontopim: How to rely on a personal ontology for personal information management. In: Proc. of the 1st Workshop on the Semantic Desktop (2005)
2. Cai, Y., Dong, X.L., Halevy, A., Liu, J.M., Madhavan, J.: Personal information management with semex. In: SIGMOD 2005: Proceedings of the 2005 ACM SIGMOD International Conference on Management of Data, pp. 921–923. ACM Press (2005)
3. Sauermann, L.: The semantic desktop - a basis for personal knowledge management. In: Proceedings of the I-KNOW, pp. 294–301 (2005)
4. Landsdale, M.W.: The psychology of personal information management. Applied Ergonomics 19, 55–66 (1988)
5. Boardman, R., Sasse, M.A.: "Stuff goes into the computer and doesn't come out": a cross-tool study of personal information management. In: Proceedings of the SIGCHI Conference on Human Factors in Computing Systems, CHI 2004, pp. 583–590. ACM, New York (2004)
6. Waks, S.: Lateral Thinking and Technology in Education. Journal of Science Education and Technology 6(4), 245–255 (1997)
7. Kirsch, L., Botev, J., Rothkugel, S.: The Snippet System - Reusing and Connecting Documents. In: Proceedings of the 7th International Conference on Digital Information Management, ICDIM 2012. IEEE (2012)

8. Kirsch, L., Esch, M., Rothkugel, S.: The Snippet System – Fine-Granular Management of Documents and Their Relationships. In: Proceedings of the 6th IASTED International Conference on Human-Computer Interaction (2011)

9. Kirsch, L., Botev, J., Rothkugel, S.: An Extensible Tool Set for Creating and Connecting Reusable Learning Resources. In: Proceedings of World Conference on Educational Media, Hypermedia and Telecommunications, EdMedia 2012, pp. 1434–1442. AACE (2012)

10. Hepp, M., García, R., Radinger, A.: Rdf2rdfa: Turning rdf into snippets for copy-and-paste. In: 8th International Semantic Web Conference (ISWC 2009) (October 2009)

11. Navarro-Galindo, J.L., Jiménez, J.S.: Flexible range semantic annotations based on RDFa. In: MacKinnon, L.M. (ed.) BNCOD 2010. LNCS, vol. 6121, pp. 122–126. Springer, Heidelberg (2012)

12. Lanfranchi, V., Ciravegna, F., Petrelli, D.: Semantic web-based document: Editing and browsing in aktiveDoc. In: Gómez-Pérez, A., Euzenat, J. (eds.) ESWC 2005. LNCS, vol. 3532, pp. 623–632. Springer, Heidelberg (2005)

13. Mayfield, J., Finin, T.: Information retrieval on the semantic web: Integrating inference and retrieval. In: Proceedings of the SIGIR Workshop on the Semantic Web (2003)

14. Brasethvik, T., Gulla, J.A.: A conceptual modeling approach to semantic document retrieval. In: Pidduck, A.B., Mylopoulos, J., Woo, C.C., Ozsu, M.T. (eds.) CAiSE 2002. LNCS, vol. 2348, pp. 167–182. Springer, Heidelberg (2002)

15. Rayson, P., Garside, R., Sawyer, P.: Assisting requirements engineering with semantic document analysis. In: Proceedings of RIAO 2000 (Recherche d'Inforrnafions Assisfie par Ordinateur, Computer-Assisted Information Retrieval) International Conference, College de, pp. 12–14 (2000)

16. Becks, A., Sklorz, S., Jarke, M.: A modular approach for exploring the semantic structure of technical document collections. In: Proceedings of AVI 2000, Int. Working Conference on Advanced Visual Interfaces, pp. 298–301 (2000)

17. Mutton, P., Golbeck, J.: Visualization of semantic metadata and ontologies. In: Proceedings of the Seventh International Conference on Information Visualization, IV 2003, p. 300. IEEE Computer Society, Washington, DC (2003)

18. Telea, R., Frasincar, F., Jan Houben, G.: Visualization of rdf(s)-based information. In: Proceedings of the 7th International Conference on Information Visualization. Press (2003) 1093–9547

19. Finin, T., Peng, Y., Scott, R., Joel, C., Joshi, S.A., Reddivari, P., Pan, R., Doshi, V., Ding, L.: Swoogle: A search and metadata engine for the semantic web. In: Proceedings of the Thirteenth ACM Conference on Information and Knowledge Management, pp. 652–659. ACM Press (2004)

A German Natural Language Interface
for Semantic Search

Irina Deines and Dirk Krechel

University of Applied Sciences RheinMain, Germany
{deines,krechel}@ecmlab.de

Abstract. Semantic data is the key for an efficient information retrieval. It relies on a well-defined structure and enables automated processing. Therefore,
more and more ontologies are specified, extended and interlinked. By now, only
the query language SPARQL provides a precise access to semantic data. Since
most common users are overstrained in formulating queries, which satisfy the
structure of semantic data, more search-interface approaches emerge aiming at
good usability and correct answers. We implemented a Natural Language Interface (NLI), that answers questions formulated in German natural language. In
order to query the domain ontology, the user query is translated into SPARQL
first. Since domain-ontology resources are required for the SPARQL-query formulation, this paper introduces an approach for the identification of resources in
user query. We show a path-based identification of semantically similar resources
and a similarity measure. After running 100 test questions, our system achieves a
precision and recall of 66%.

1 Introduction

Semantic data is modelled in an ontology, which '*is an explicit and formal specification
of a conceptualization*' [1]. In order to describe ontologies W3C proposes the resource-
oriented model called Resource Description Framework (RDF) [2]. More and more
RDF ontologies emerge across the web and are getting interlinked in Linked-Open-
Data project [3]. RDF data can be accessed via SPARQL [4]. However, for a common
user a query formulation is complicated and difficult. It requires at least basic skills
in SPARQL syntax and knowledge of the ontology structure. More user-friendly is a
keyword-based full-text search like Google. However, in cases of complex information
needs, keyword-based search is not able to achieve good precision values [5], [6], [7].
An intuitive interface for complex query formulation is provided by Natural Language
Interfaces (NLI), which are able to process queries formulated in natural language. In
order to find the answer, NLIs need to generate a SPARQL query out of the user
query. Thus, NLIs are aimed at a correct interpretation of the user query to achieve
good precision values.

The NLI system described in this paper is able to answer German user queries addressing RDF data labelled in German. The system was designed to process questions
with a structure and complexity similar to the English training and test questions of the
Workshop Question Answering over Linked Data (QALD-1) [8]. In QALD question
answering approaches for Linked Data were evaluated on the two datasets DBpedia [9]

H. Takeda et al. (Eds.): JIST 2012, LNCS 7774, pp. 278–289, 2013.
© Springer-Verlag Berlin Heidelberg 2013

3.6 and MusicBrainz. The NLI system in this work was evaluated on an extract of DB-pedia 3.7.

Aiming for high precision and recall, this NLI system concentrates on a semantic component which mainly contributes to a correct generation of the SPARQL query. In the scenario of figure 1 a user formulates the question "Wo ist Bismarck gestorben?" (Where did Bismarck die?). According to the domain ontology, the expected answer is the resource :Friedrichsruh . Thus, the user query has to be transformed into the following SPARQL query:
SELECT DISTINCT ?var WHERE{:Otto_von_Bismarck :deathPlace ?var}

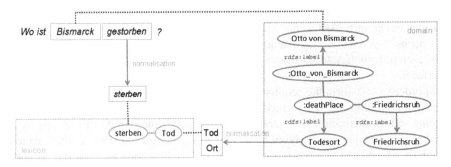

> SELECT DISTINCT ?var WHERE{:Otto_von_Bismarck :deathPlace ?var}

Fig. 1. Identification of resources in question "Wo ist Bismarck gestorben?"

When translating the user query into SPARQL the identification of resources is a crucial factor. Only if all resources are identified in user query a correct SPARQL query can be generated and a right answer is returned. Thereby text-based identification and a semantic-based identification of resources are distinguished. Based on text-based similarity between the query extract "Bismarck" and the German label "Otto von Bismarck", the corresponding resource :Otto_von_Bismarck is recognized easily. However, no text-based similarity exists between "gestorben" (died) and "Todsort", which is the label of the required resource :deathPlace. A semantic-based similarity indicates a relation between those two terms. Thus, if a phrase or term in user query is semantically related to a resource's label, a special treatment is applied, see figure 1. This paper introduces an algorithm for a semantic-based identification of resources. The algorithm first normalizes the question's term "gestorben" and the resource's label "Todsort". The result is the infinitive form "sterben" (die) and the noun component "Tod" (death), which is a part of the noun compound "Todsort". Then, a German lexicon is used to resolve the semantic relation between "sterben" and "Tod". In the end, an implicit path emerges between the normalised user-query term "sterben" and the label "Todsort". After making those paths explicit, semantically similar resources can be identified by following paths.

This paper proceeds with a discussion of other question answering approaches for ontologies. In section 3 the particular components of our German NLI system are introduced, focusing on the identification of semantically similar resources. A path-based algorithm for the identification of resources is explained and a similarity measure determined. Then, in section 4 the evaluation results are presented. The last section 5 derives a conclusion from the evaluation results and shows several intentions of our future work.

2   Related Work

English search interfaces combined with query execution on ontology-based data were investigated in Kaufmann [5], Heim [6] and Damljanovic [7]. In summary, natural-language, graph-based and query-autocompletion-based interfaces are introduced as a trade-off between a user-friendly search solution and efficient querying in semantic data via SPARQL. These user interfaces focus on an intuitive use, as they demand neither any experience in SPARQL nor knowledge about the ontology structure. Moreover, they aspire a precise hit list. Natural-language interfaces were accepted most by the user, even though they suffer from worse recall and precision than the other two query interfaces.

In order to transform the original query into SPARQL, NLI systems usually apply natural-language-processing methods, such as tokenization, stemming, part-of-speech tagging and lemmatisation to the labels of the domain ontology and to the user query. In general, the user query is analysed by a constituency parser too. This resulted natural-language information is sufficient for a text-based identification of resources, as in the identification of :Otto_von_Bismarck in scenario of figure 1. Semantically similar resources can only be recognized by means of semantic relations. Therefore the English NLI systems FREyA [7], Querix [10] and PANTO [11] expand their knowledge base by synonyms from WordNet [12]. They store a resource's natural-language information and its synonyms in a gazetteer, which is either a relational database or a hash map. Since these data stores are simply structured, the identification of resources is very fast. On the other hand, complex structured information such as hierarchies, e.g. the conceptual relations hypernymy respectively hyponymy, cannot be archived. Therefore no other semantic relations besides synonymy are considered in the identification process of these solutions.

The NLI system described in this paper uses the high flexible RDF to structure a resource's natural-language information and its semantic relations. Lexical entries as well as lexical and conceptual relations from the German lexicon GermaNet [13] are modelled in RDF and linked to the domain ontology. As a result, a resource cannot only be identified by its synonyms, but also by any other semantic relation that connects it to the user-query extract. Therefore, the semantically similar resource :deathPlace in figure 1 can be encountered, even though "Sterben" and "Tod" are connected by a hypernymy relation.

3   A German Natural Language Interface for Semantic Search

In this section the components of the German NLI system are introduced first. Then, in detail the identification of semantically similar ontology resources is discussed. The semantic-based identification relies on paths between user-query extract and its semantically similar resource. Those paths are created in 3.2. The identification algorithm and the calculation of similarity between user-query extract and semantically similar resource is described in 3.3.

3.1   System Design

According to figure 2, the system consists of two components, namely the translation process and the knowledge base. The translations process is a major part of the system workflow, as illustrated in figure 2(a). It transforms the user query into SPARQL concentrating on the identification of resources since they are essential in the graph pattern of the SPARQL query. The text-based search for resources as well as the semantic-based search is executed in German labels which are linked to the resources of the domain ontology. For that purpose a normalisation needs to be applied on the user query and the labels first, as illustrated by the system workflow in figure 2(a). The natural-language-processing methods such as parsing, lemmatising, part-of-speech tagging and noun-compound splitting constitute the normalisation and are accomplished by the constituency parser Stanford Parser [14], the part-of-speech tagger TreeTagger [15] and the noun-compound splitter BananaSplit [16]. Then, in case of a semantic-based identification a lexicon is used to resolve the path between normalised user-query extract and semantically similar resource.

After text-based and semantic-based identification of resources, the translation process concludes by generating the SPARQL query. Thereby, an answer type is determined and triples for the graph pattern are built. Like in many other natural-language

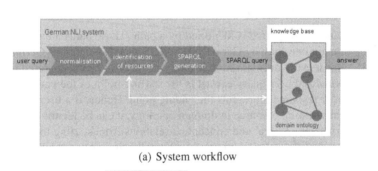

(a) System workflow

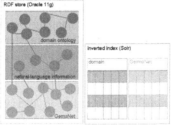

(b) Knowledge-base structure after the initialization phase

Fig. 2. System design

interfaces [7,11,17], the answer-type determination in this work occurs dictionary- and rule-based. Finally, the SPARQL query is sent to the domain ontology and the returned answer is passed to the user.

In order to save execution time, several steps of the translation process are moved to the initialization phase of the NLI system. In that phase the ontology is extended by paths needed for a semantic-based identification of resources. I.e. natural-language information resulted by the normalisation of a resource's label is linked to the domain ontology as well as the German lexicon GermaNet, such as needed for an identification of recourse :deathPlace in figure 1. This approach is described in section 3.2 in detail. The entire ontology is stored in the RDF store of Oracle 11g [18], which at the end of initialization phase consists of the domain ontology with its natural-language information and the linking to GermaNet, see figure 2(b).

For a further reduction of response time, the expanded domain ontology is copied to the inverted index of Apache Solr [19]. As illustrated in figure 2(b), the knowledge base contains after the initialization two different information sources, namely RDF store and inverted index. Since querying via SPARQL in a huge amount of data causes long response times, the index-based search serves as a time saving alternative. On the other hand, index-based search compared to SPARQL suffers from bad precision and recall values [20]. Thus, complex information need is answered by the RDF store precisely. For simple search queries a quick response can be achieved by the inverted index. Depending on complexity of the emerging information need the translation process chooses one of the two stores.

3.2   Expansion of Domain Ontology

A semantically similar resource is identified by a path. Those paths are created by expanding the domain ontology by GermaNet. Since GermaNet contains Germans common words, almost every normalised user-query extract can be encountered in the ontology. The normalised user-query extract is the starting point of the path while its semantically similar resource represents the ending point. Hence, if a user-query extract has a semantically similar resource in domain ontology, it can be identified by walking the path of GermaNet's lexical and conceptual relations from starting to ending point. The more paths can be created, the more semantically similar ontology resources can be recognized.

Since the linking of domain ontology and GermaNet can only be achieved when both share the same representation format, mapping GermaNet to RDF needs to be accomplished first. Thereby, lexical information is transferred into the lemon-LexInfo model [21] which is represented in OWL. The structure of lemon-LexInfo enables a mapping of lexical entries as well as common lexical and conceptual relations. As an example figure 3 shows an extract of the resulted lemon-LexInfo model for GermaNet. The synsets {Sterben,Abgang} (dying) and {Tod,Exitus} (death) are modelled by the sense property and their hypernymy relation is mapped to a hypernymy property. For the part-of relation between synset {Tod,Exitus} and {Todesursache} (cause of death) the property meronymy was used.

lemon-LexInfo is domain-independent and allows linking of lexical entries and domain-ontology resources. Before linking of domain ontology and GermaNet ontology

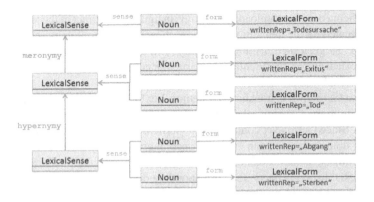

Fig. 3. GermaNet entry "Sterben" (dying) modelled in lemon-LexInfo

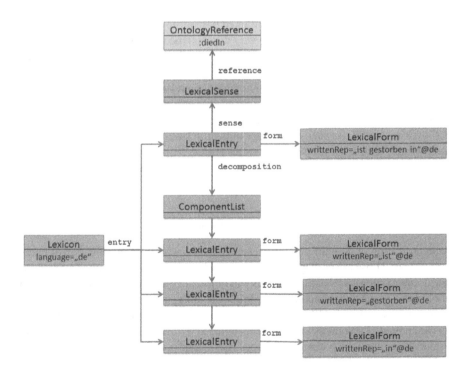

Fig. 4. Sample resource :diedIn labelled as "ist gestorben in" is modelled in lemon-LexInfo

takes place, a domain lexicon of domain-ontology schema is created. For this purpose the generation service[1] [21] of lemon-LexInfo is used. It auto-generates a lexicon in lemon-LexInfo schema out of any uploaded ontology. In figure 4 an extract of the lemon-LexInfo ontology created by the generation service is shown. For the ontology resource diedIn labelled as "ist gestorben in" a lexical entry with its entire canonical name is created and three lexical entries representing the particular components of its tokenized form are generated as well.

Now, that GermaNet is transferred to lemon-LexInfo and the domain ontology is expanded by its appropriate lexical entries, linking of these two lemon-LexInfo ontologies is possible. Since GermaNet contains most German's common words, most lexical entries of the domain lexicon can either be encountered in it by their canonical name or by their lemma. For each domain-lexicon entry that is available in GermaNet, a connection is established between the corresponding domain resource and the GermaNet entry. If a domain-lexicon entry's lemma can be found in GermaNet, its linking to the GermaNet entry is accomplished by a pertonymy relation. Special cases are noun compounds, which usually are only included in a lexicon if their meaning is not decucible by their noun components. If a domain-lexicon entry is a noun compound which cannot be found in GermaNet, then its noun components are linked to the corresponding GermaNet entries.

After the linking process almost every domain-lexicon entry, the ending point of the path, is connected to GermaNet. Few unlinked domain-lexicon entries may be encountered, basically caused by two reasons. Very domain-specific lexical entries, which are mostly named entities, are not available in lexicons. Besides, results generated by the morphological analysis may be incorrect. In that case, either no lemma is returned or a noun compound could not be split or a wrong lemma respectively wrong noun components are found.

In the end, the entire ontology contains about 900.000 triples. Reasoning leads to about 4.6 Mio. triples. Basically, those reasoning rules were applied in Oracle which are widely spread, such as the properties rdfs:subClassOf , rdfs:subPropertyOf , rdfs:range and rdfs:domain. Now, more SPARQL queries can be answered correctly or answered at all.

3.3 Identification of Semantically Similar Resources

The identification of domain-ontology resources starts with the determination of potential resources. Those are user-query extracts for which may exist an equivalent resource in the domain ontology. Usually, subjects and objects in ontologies are labelled by nominal phrases, nouns and named entities. Predicates in general are named after nominal phrases, nouns and verbs [7,11]. Thus, nominal phrases, nouns, named entities and verbs are potential resources and can be recognized by part-of-speech tags. In the sample query of figure 5 the potential resources "Bismarck" and "gestorben" are marked by purple.

Based on potential resources, text-based and semantic-based identification of domain-ontology resources takes place. A text-based identification is accomplished by querying

---

[1] http://monnetproject.deri.ie/lemonsource

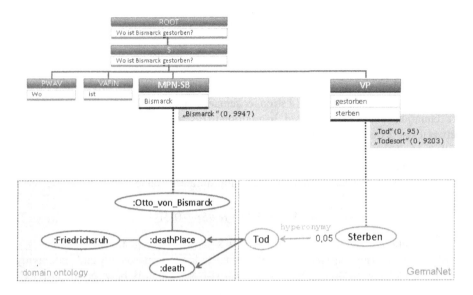

Fig. 5. Text-based identification of resource :Otto_von_Bismarck and path-based identification of semantically similar resource :deathPlace

the potential resource in domain index. The similarity value between potential resource and found domain ontology resource is calculated by a string metric, see figure 5. The identification of semantically similar resources is based on paths created in section 3.2. For the calculation of the similarity value between potential resource and recognized similar resource a specific semantic similarity measure is defined. A path-based algorithm similar to Leacock & Chodorow, Wu & Palmer, Hirst & St-Onge and Banerjee & Pedersen, which are discussed in [22], is chosen to determine the semantic similarity value. Starting point of the path is the lexical-entry equivalent of the potential resource as illustrated in figure 5 for the potential resource "gestroben".

If a similar resource exists in domain ontology, it represents the ending point of the path. Since in initialization phase of section 3.2 links between domain-ontology resources and lexical entries were established, paths from lexical entries to their similar resources are available in the RDF store. Mostly, only those domain-ontology resources could not be linked to the lexicon, which are very specific. Nevertheless, those specific resources usually don't have a semantic equivalent and can be encountered by their name in text-based identification.

The path includes relations which were transferred from GermaNet. Specific distances are assigned to each relation to distinguish the semantic similarity a relation expresses. For instance, for synonymy, holonymy, meronymy and pertainymy the minimum distance value of 0.0 is chosen. Hence, those relations express a semantic similarity of 1.0 between two lexical entries. Since more distance lies in between lexical entries related by hypernymy/hyponymy, a distance value of 0,05 is chosen. Equation (1) calculates the path-based semantic similarity value. The variable n is the number of relations in path and x_k the distance value of the kth relation. Every additional hypernymy or hyponymy relation in path causes a decrease of resulting similarity value.

$$semSim = \prod_{k=1}^{n}(1 - x_k) \tag{1}$$

Using equation (1) to calculate the similarity between "gestorben" and "Todesort" in figure 5 yields the same value of 0.95 as for the path between "gestorben" and "Tod". Since obviously "gestorben" has a stronger semantic similarity to "Tod" than to "Todesort" equation (1) has to be adapted. In addition to distance values, text-based similarity between last lexical entry in path and similar ontology resource needs to be considered in the calculation. Since the lexical entry "Tod" obviously has a higher text similarity to the ontology resource "Tod" than to the ontology resource "Todesort", ontology resource "Todesort" achieves a higher semantic similarity value.

$$semSyn = 0.99 + textSim \cdot 0.01 \tag{2}$$

The calculation of semantic similarity is now split into two equations. Equation (2) is used in case the path consists only of synonymy, holonymy, meronymy and pertainymy relations. Variable $textSim$, representing the text-base similarity between the last lexical entry in path and the similar resource, is ranged between 0.99 and 1.0. In case of paths which contain hypernymy or hyponymy relations equation (3) is defined. Like in equation (2) $textSim$ stands for the text similarity of last lexical entry and similar resource. Variable k contains the number of hypernymy and hyponymy relations in path. Before a decrease of similarity value by the distance values of hypernymy and hyponymy relations is calculated, $textSim$ is ranged between 0.95 and 1.0. The calculation of the semantic similarity value for similar resource "Todesort" from figure 5 yields $0.9203 = (0.95 + 0.375 \cdot 0.05) \cdot 0.95^1$.

$$semHyp = (0.95 + textSim \cdot 0.05) \cdot 0.95^{k-1} \tag{3}$$

4 Evaluation

A good query interface requests a convenient usability as well as correct answers. The correctness of a search interface is expressed by precision and recall [23]. The definition of these measures has slightly to be adapted, though. Recall is the number of questions correctly answered by the search interface, divided by the total number of questions. Precision measures the number of questions correctly answered divided by the number of questions which were answered at all [24].

250 resources which basically specify persons were chosen from the ontology schema of DBpedia 3.7 to evaluate the German NLI system. Resources without a German label were translated semi-automatically. In addition, German labelled resources about 1000 persons were loaded from a SPARQL endpoint [2]. The NLI system was asked 100 questions addressing this data set. 66 questions were answered correctly, 34 either partial or incorrectly. Since our NLI system always returns an answer, recall and precision achieve 66%.

---

[2] http://de.dbpedia.org/onlinezugriff

According to figure 6, the evaluation results show clearly a coherence between the identification of resources and a correct answer. In 58 of 66 correctly answered questions all resources required for a proper SPARQL generation were identified. On the other hand, in 32 of 34 wrongly answered questions only a partial or none identification of resources was accomplished. Hence, a complete identification of resources contributes essentially to a correct translation of user query into SPARQL.

In a total of 40 questions an incomplete resource recognition was detected. The evaluation results reveal, that in 27 of these 40 questions an identification of resources fails because of missing relations in GermaNet. The lack of especially pertonymy relations, such as between the lexical entries "schreiben" (write) and "Schriftsteller" (writer) or between "heiraten" (marry) and "Ehegatte" (spouse), prevent a path construction which is needed for a semantic-based identification. Hence, questions like "Welche Bücher hat Stephen King geschrieben?" (Which books did Stephen King write?) or "Mit wie vielen Männern war Marilyn Monroe verheiratet?" (How many men Marilyn Monroe was married to?) cannot be answered because the resources :writer labelled as "Schriftsteller" and :spouse labelled as "Ehegatte" cannot be found.

Other reasons for incorrect translations of user query are special expressions like the verb "kommen" (to be from) in "Woher kommt X?" (Where is X from?) or the pronoun "derselben" (the same) in "Werden Futurama und Simpsons von derselben Person produziert?" (Are Futurama and Simpsons produced by the same person?). Also transitive verbs like "beeinflussen" (influence) are translated incorrectly, such as both questions "Wen hat X beeinflusst?" (Who did X influence?) and "Wer hat X beeinflusst?" (Who was X influenced by?) produce the same answer. Here, ambiguities arise because of missing verb subcategorization frames in expanded domain ontology and omitted determination of dependencies [25] in user query.

The tests also revealed, that for 7 correctly answered questions only because of reasoning a complete identification of semantically similar resources could be achieved.

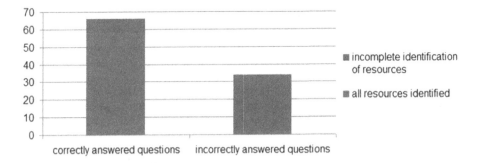

Fig. 6. Evaluation results with regard to the identification of resources

5   Conclusion and Future Work

A German NLI system with a path-based approach for the identification of semantically similar resources was introduced. In the evaluation a precision and recall value of 66% was measured. In 58 of 66 correctly answered questions every semantically similar resource was identified by those paths created during the expansion of the domain ontology in section 3.2. A path starts at the normalised form of the user-query extract and ends at the semantically similar resource. Starting and ending point are connected by the lexical and conceptual relations of GermaNet. Unidentified semantically similar resources were mainly responsible for a wrong answer, since not available in the subsequent formulation of the SPARQL query. Most of these resources could not be recognized because of absent paths, i.e. lack of relations in GermaNet.

A few questions weren't answered correctly because of a wrong interpretation of transitive verbs. This drawback can be resolved in future work. Aiming at higher precision and recall values, an improvement of the translation process is intended. Dependency parsing and subcategorization will be necessary to avoid incorrect translations caused by transitive verbs. For this purpose an integration of the dependency parser MaltParser [26] into the natural-language analysis of the user query is aspired. Moreover, during the expansion of the domain ontology subcategorization frames of GermaNet's verbs have to be determined and included.

References

1. Antoniou, G., van Harmelen, F.: A Semantic Web Primer, 2nd edn. Cooperative Information Systems. The MIT Press (2008)
2. W3C World Wide Web Consortium: Resource Description Framework (RDF). Website, http://www.w3.org/RDF/ (visited on September 14, 2010)
3. Bizer, C., Heath, T., Berners-Lee, T.: Linked Data – The Story So Far. International Journal on Semantic Web and Information Systems 5(3), 1–22 (2009)
4. Prud'hommeaux, E., Seaborne, A.: SPARQL Query Language for RDF. Website, http://www.w3.org/TR/rdf-sparql-query/ (visited on September 14, 2010)
5. Kaufmann, E., Bernstein, A.: How Useful Are Natural Language Interfaces to the Semantic Web for Casual End-Users? In: Aberer, K., Choi, K.-S., Noy, N., Allemang, D., Lee, K.-I., Nixon, L.J.B., Golbeck, J., Mika, P., Maynard, D., Mizoguchi, R., Schreiber, G., Cudré-Mauroux, P. (eds.) ISWC/ASWC 2007. LNCS, vol. 4825, pp. 281–294. Springer, Heidelberg (2007)
6. Heim, P., Ertl, T., Ziegler, J.: Facet Graphs: Complex Semantic Querying Made Easy. In: Aroyo, L., Antoniou, G., Hyvönen, E., ten Teije, A., Stuckenschmidt, H., Cabral, L., Tudorache, T. (eds.) ESWC 2010, Part I. LNCS, vol. 6088, pp. 288–302. Springer, Heidelberg (2010)
7. Damljanovic, D., Agatonovic, M., Cunningham, H.: Natural Language Interfaces to Ontologies: Combining Syntactic Analysis and Ontology-Based Lookup through the User Interaction. In: Aroyo, L., Antoniou, G., Hyvönen, E., ten Teije, A., Stuckenschmidt, H., Cabral, L., Tudorache, T. (eds.) ESWC 2010, Part I. LNCS, vol. 6088, pp. 106–120. Springer, Heidelberg (2010)
8. Semantic Computing Group, CITEC, Bielefeld University and the Knowledge Media Institute (KMi), Open University: Proceedings of 1st Workshop on Question Answering over Linked Data (QALD-1), Collocated with the 8th Extended Semantic Web Conference (ESWC 2011), Heraklion, Greece (June 2011)

9. Lehmann, J., Bizer, C., Kobilarov, G., Auer, S., Becker, C., Cyganiak, R., Hellmann, S.: DBpedia - a crystallization point for the web of data. Journal of Web Semantics 7(3), 154–165 (2009)

10. Kaufmann, E., Bernstein, B., Zumstein, R.: Querix: A natural language interface to query ontologies based on clarification dialogs. In: In: 5th ISWC, pp. 980–981. Springer (2006)

11. Wang, C., Xiong, M., Zhou, Q., Yu, Y.: PANTO: A Portable Natural Language Interface to Ontologies. In: Franconi, E., Kifer, M., May, W. (eds.) ESWC 2007. LNCS, vol. 4519, pp. 473–487. Springer, Heidelberg (2007)

12. Princeton University: Princeton University 'About WordNet'. Website, http://wordnet.princeton.edu (visited on November 7, 2011)

13. University of Tübingen: GermaNet - A German Wordnet. Website, http://www.sfs.uni-tuebingen.de/GermaNet/index.shtml (visited on November 7, 2011)

14. Rafferty, A.N., Manning, C.D.: Parsing three German treebanks: lexicalized and unlexicalized baselines. In: Proceedings of the Workshop on Parsing German, PaGe 2008, pp. 40–46. Association for Computational Linguistics (2008)

15. Schmid, H.: Improvements in Part-of-Speech Tagging with an Application to German. In: Proceedings of the ACL SIGDAT-Workshop, pp. 47–50 (1995)

16. Ott, N.: BananaSplit - Dictionary-based Compound Splitter for German Platform. Website, http://niels.drni.de/s9y/pages/bananasplit.html (visited on February 6, 2012)

17. Moldovan, D.I., Harabagiu, S.M., Pasca, M., Mihalcea, R., Girju, R., Goodrum, R., Rus, V.: The Structure and Performance of an Open-Domain Question Answering System. In: ACL. ACL (2000)

18. Wu, Z., Das, S., Annamalai, M., Murray, C., Beauregard, B.: Oracle Semantic Technologies Inference Best Practices with RDFS/OWL. Oracle (2008)

19. Kuc, R.: Apache Solr 3.1 Cookbook. Packt Publishing (2011)

20. Delbru, R., Toupikov, N., Catasta, M., Tummarello, G.: A Node Indexing Scheme for Web Entity Retrieval. In: Aroyo, L., Antoniou, G., Hyvönen, E., ten Teije, A., Stuckenschmidt, H., Cabral, L., Tudorache, T. (eds.) ESWC 2010, Part II. LNCS, vol. 6089, pp. 240–256. Springer, Heidelberg (2010)

21. McCrae, J., Spohr, D., Cimiano, P.: Linking Lexical Resources and Ontologies on the Semantic Web with Lemon. In: Antoniou, G., Grobelnik, M., Simperl, E., Parsia, B., Plexousakis, D., De Leenheer, P., Pan, J. (eds.) ESWC 2011, Part I. LNCS, vol. 6643, pp. 245–259. Springer, Heidelberg (2011)

22. Pedersen, T., Patwardhan, S., Michelizzi, J.: WordNet::Similarity: Measuring the Relatedness of Concepts. In: Proceedings of the Demonstration Papers at HLT-NAACL (2004)

23. Manning, C.D., Raghavan, P., Schütze, H.: Introduction to Information Retrieval. Cambridge University Press (2008)

24. Cimiano, P., Haase, P., Heizmann, J.: Porting natural language interfaces between domains: an experimental user study with the orakel system. In: Proceedings of the 12th International Conference on Intelligent User Interfaces (2007)

25. Kübler, S., Prokic, J., Groningen, R.: Why is German dependency parsing more reliable than constituent parsing? In: Proceedings of the Fifth Workshop on Treebanks and Linguistic Theories, TLT (2006)

26. Nivre, J., Hall, J., Nilsson, J., Chanev, A., Eryigit, G., Kübler, S., Marinov, S., Marsi, E.: Maltparser: A language-independent system for data-driven dependency parsing. Natural Language Engineering 13(2), 95–135 (2007)

Topica – Profiling Locations through Social Streams

A.E. Cano[1,2], Aba-Sah Dadzie[1], Grégoire Burel[2], and Fabio Ciravegna[1]

[1] OAK Group, Dept. of Computer Science, The University of Sheffield, UK
firstinitial.lastname@dcs.shef.ac.uk
[2] Knowledge Media Institute, The Open University, Milton Keynes, UK
firstinitial.lastname@open.ac.uk

Abstract. This paper presents work in interlinking social stream information with geographical spaces through the use of Linked Data technologies. The paper focuses on filtering, enriching, structuring and interlinking microposts of localised (i.e. geo-tagged) social streams (a.k.a localised forums) to profile geographical areas (e.g., cities, countries). For this purpose, we enriched social streams extracted from Twitter[1], Facebook [2] and TripAdvisor[3] and structured them into well-known vocabularies and data models, such as SIOC and SKOS. To integrate this information into a location profile we introduce the linkedPOI ontology. The linkedPOI ontology captures and leverages DBpedia categories to derive concepts which profile a geographic space.

We exemplify the use of social stream-based location profiling by means of a travel mashup case study. We introduce the *Topica Portal*, which allows users to browse geographical spaces by topic. We highlight potential impact for the future of semantic travel mashup systems.

Keywords: Points of Interest, Social Streams, DBpedia, Travel Mashups.

1 Introduction and Motivation

The development of Social Web platforms that enable real-time posting of information suggests the use of humans as sensors. Citizen sensor networks regard the Social Web as a network of interconnected, participatory citizens who actively observe, report, collect, analyse and disseminate information about events and activities via text, audio and/or video messages in almost real-time [16,17].

Recent studies in user profiling have proposed the use of social activity streams for modelling users' interests, activities and behaviour [1,7,20]. These studies explore users' comments over windows of time in order to reveal hidden features; this aids user profiling in real-time. Although people entities have started to be modelled in real-time, little has been done in modelling location entities.

Geo-tagged social streams gather information from millions of users who leave trails in locations (i.e. "check-ins") in the form of microposts. These *footprints* provide a

---

[1] Twitter, http://twitter.com
[2] Facebook, http://facebook.com
[3] TripAdvisor, http://tripadvisor.com

H. Takeda et al. (Eds.): JIST 2012, LNCS 7774, pp. 290–305, 2013.

unique opportunity to explore the way in which users perceive and engage with a location. In this paper we explore the representation of a location based on the data streams generated from or based on it.

The contributions of this paper are as follows:

○ *we investigate whether the supplement of situational knowledge extracted from social streams can be used to infer higher level contextual information, which can be used to induce a transient representation of a location.*

○ *we propose the **GeoLattice Awareness Stream** model, which is a graph-based formalisation for expressing the profiling of geographical areas based on social awareness streams.*

○ *we introduce the **LinkedPOI** ontology, which provides a semantic representation of this graph-based model, and highlights the dynamic features of a location.*

○ *we demonstrate the applicability of this approach by means of the semantic travel mashup **Topica**.*

2   Related Work

Researchers in social interaction and environmental psychology have documented the way in which mobile phone users tend to provide information about location when they are asked about their current activity [14,21]. Schegloff [15] noted that during a conversation, attention is drawn to: 1) 'where-we-know-we-are'; 2) 'who-we-know-we-are'; 3) 'what-we-are-doing-at-this-point-in-conversation'; from which a '*this* situation' can be translated in some '*this* conversation, at *this* place, with *these* members, at *this* point in its course'. This contextual knowledge is used to infer users' situational features, including a person's level of availability or interruptibility.

The role of geography and (physical) location in online social networks is attracted increasing attention. Experimental work done in location awareness has shown that location sharing services (LSS) (e.g. Foursquare) are used to express not only users' whereabouts but also their moods, lifestyle and events [3]. E.g., Cheng *et al.* [8] modelled the spatial distribution of words in Twitter's user-generated content in order to predict a user's location. Following a top-down approach they propose a probabilistic framework for estimating a Twitter user's city-level location based on the content of the user's tweets even in the absence of any explicit geospatial cues. Their approach is content-based and can automatically identify words in tweets with a strong geo-scope. They however do not, as carried out in the work described, provide topical categorisation of a given geo-scope. Cheng *et al.* further study mobility patterns of users in location sharing services (LSS) [9]; they correlate social status, geographic and economic factors with mobility and perform a sentiment-based analysis of posts to derive unobserved context between people and locations. These approaches are focused on profiling users based on their whereabouts. In contrast this paper proposes the profiling of locations based on social media comments generated from users.

Computing-based applications to support traveling increasingly make use of online resources that encourage end user (traveller) contributions by making use of Semantic Web (SW) and Web 2.0 technologies [10,11]. Well-known examples include *TripIt*[4],

---

DBpedia Mobile[5] and *Revyu.com*[6]. Other online services that make no (explicit[7]) use of SW services still remain popular with travellers – including *TripAdvisor*[8] and *Journeywoman*[9]. A notable advantage in the use of such resources, especially those that take advantage of SW technology to enrich user-contributed information, is the varied and rich contextual information contributed from multiple perspectives, by end users with actual experience of the locations and services they describe.

Hao *et al.* [11] demonstrate the use of information extracted from travelogues to obtain information specifically geared at the traveller, by other travellers. *DBpedia Mobile* [4] is a location-aware SW client, which, based on the user's GPS or IP (Internet Protocol) information, renders a map with nearby locations extracted from the underlying DBpedia dataset. Icons are used to provide information about different entity types in the overview, and the application queries the Linked Data (LD) cloud[10] for additional information on the user's focus. Other applications, such as *Stevie* [5], allow users to share and browse temporal information about points of interest (POIs) – events – on a map, based on the location broadcast by end users' GPS. Stevie annotates the information shared using an ontology, and by linking to corresponding entities in DBpedia.

MetaCarta[11], while not specifically targeted at traveling, may be used to obtain news and breaking events, as well as location-specific information about POIs and local events at a traveller's destination. MetaCarta uses location-specific information collected from both traditional and online news media to provide "geographic intelligence solutions". By linking this to a geographical knowledge base and custom gazetteer, MetaCarta is able to provide in-depth context- and location-aware information, with options to personalise the information presented to an end user.

Tintarev *et al.* [19] demonstrate the added benefit in personalising recommendations of popular POIs for tourists. Hornecker *et al.* [13], like Tintarev *et al.*, recognise the benefit in using personal information to guide the exploration of new areas. To reduce information load and to allow serendipitous discovery, however, Hornecker *et al.* only alert the user to nearby POIs that match their preferences or that are similar to previously visited POIs. We note, as do Sheth *et al.* [17], the specific challenges in retrieving semantically enriched information in dynamically evolving situations, such as commonly occurs in social media.

Recent work has investigated the integration of spatial information into the Linked Data cloud. One example is the `LinkedGeoData` effort [2,18], which adds spatial information to the Web of Data by providing structure to the collaboratively collected *OpenStreetMap*[12] dataset, by following Linked Data principles. The work in [2,18] reports the formulation of an RDF mapping and a set of mapping rules for

---

[5] DBpedia Mobile, `http://mes-semantics.com/DBpediaMobile`

[6] Revyu, `http://revyu.com`

[7] We differentiate "explicit use of SW technology and services" by excluding web sites and services that simply provide, e.g., links to a Twitter account or feed, flickr tags and photos, or include a Facebook like button.

[8] TripAdvisor, `http://www.tripadvisor.com`

[9] Journeywoman, `http://www.journeywoman.com`

[10] Linking Open Data Cloud, `thedatahub.org/group/lodcloud`.

[11] MetaCarta, `http://www.metacarta.com`

[12] OpenStreetMap, `http://openstreetmap.org`

converting OpenStreetMap data into RDF. The lightweight ontologies derived from the data types appearing in OpenStreetMap include subclasses such as nodes, ways, tags and relations. Another effort is the `csxPOI` ontology [5] which models the collaborative creation and editing of POIs; it consists of a set of ontologies modelling `users`, `POIs` and `collaboration`. These ontologies however do not model relations among POIs, such as the `contained-within` and `contains` one-to-many relationships established by the POI W3C working group [22].

3   Profiling Locations through Localised Forums

In this paper we will refer indistinctly to location and point of interest (POI). The W3C POI Working Group[13] defines a point of interest (POI) as a human construct which describes information about locations. According to their definition, a POI is described by a set of coordinates as well as by other features (e.g. an identifier, opening and closing hours). In this sense, a POI can refer to a restaurant as well as to a city or any regional area delimited by a set of coordinates.

Our definition of a location is in line with Hightower's [12], according to which a place is considered to be an evolving set of both communal and personal labels for potentially overlapping geometric volumes. Hightower highlights that a meaningful place can capture the venue's demographic, environmental, historic, personal or commercial significance. This definition brings attention to the role played by a location's forums in the evolution of the representation of this place. In order to capture this community-based, evolving representation of a location, we propose the `GeoLattice Awareness Stream` formalisation.

3.1   GeoLattice Awareness Stream

In terms of location-awareness, a Point of Interest (POI) has been so far modelled as a set of static data (including, e.g. name, address, geo-coordinates) and classified according to the type of services it provides. Nonetheless, there are diverse latent (or hidden) features which can describe volatile and temporal aspects of a location. For instance, under normal conditions *London, United Kingdom* would be classified as a city labeled as, among others: *Urban, Tourism, Fashion*. However, during the August 2011 London riots, the collective opinion gathered through social activity streams (predominantly Twitter) regarding London started profiling it with the following tags: *looting, unrest, police*. The latter tags clearly provide a temporal reclassification of this venue, labelling it, among others, as *Political, Uprising, Violence*.

The `GeoLattice Awareness Stream` model is introduced in this subsection as a formalisation for describing a location based on the comments generated from its space at a specified time. The key tenet in this approach is to enrich a POI by associating transient categories emerging from social streams generated from or about a location.

---

[13] W3C POI Working Group, `http://www.w3.org/2010/POI`

Definition 1 *A GeoLattice Awareness Stream is a sequence of tuples*

$$S := (Poi_{q1}, M_{q2}, R_{q3}, Y, ft), \text{ where}$$

- Poi, M, R *are finite sets whose elements are* PointsofInterest, Messages *and* Resources;
- *Each of these sets is qualified by* q1, q2 *and* q3 *respectively;*
  - *The qualifier* q1 *for a Point of Interest (poi) includes e.g. name, geographical-bounding area, and geo-coordinates.*
  - *The qualifier* q2 *for a message m considers, e.g. the message's source (e.g. Facebook, Twitter) and its geo-coordinates.*
  - *The qualifier* q3 *for a resource r considers:* R_{cat} *(category),* R_k *(keywords),* R_h *(hashtags).*
- Y *is the ternary relation* $Y \subseteq$ Poi $\times$ M $\times$ R *representing a hypergraph with ternary edges. The hypergraph of a GeoLattice Awareness Stream* Y *is defined as a tripartite graph* $H(Y) = \langle V, E \rangle$ *where the vertices are* $V =$ Poi $\cup$ M $\cup$ R, *and the edges are:*
  $E = \{\{poi, m, r\} \mid (poi, m, r) \in Y\}.$
- f_t *is a function that assigns a temporal marker to each Y;* $f_t : Y \to T$.

Given a GeoLattice awareness stream S, a POI awareness stream is defined as the sequence of tuples from S where:

$$S(Poi') = (Poi, M, R, Y', ft), \text{ and}$$
$$Y' = \{(poi, m, r) \mid poi \in Poi' \lor \exists poi' \in Poi', \tilde{m} \in M, r \in R : (poi', \tilde{m}, r) \in Y\}$$

A POI Awareness Stream is therefore the aggregation of all messages which are related to a certain set of points of interest poi $\in$ Poi' and all resources and further points of interest related to these messages.

Based on this formalisation, the following subsection introduces the LinkedPOI ontology, which is a knowledge representation of a location profiled by information extracted from social streams generated *in situ* (a.k.a localised forums).

3.2   Integration of Geographical Spaces into the Linked Data Cloud

The W3C *Point of Interest working group* is an effort that aims to provide a data model and XML-formatted syntax that represents information about Points of Interest on the World Wide Web. However, there is as yet no standard ontological representation of a POI. The following subsection proposes the LinkedPOI Ontology, which models POIs based on the social streams generated within them.

3.3   LinkedPOI Ontology - Representing POIs through Localised Forums

The LinkedPOI (lp) ontology, presented in Fig. 1, considers the representation of a point of interest as a collection of localised forums, which act as channels or discussion areas on which posts are made. A `lp:Post` contains qualified information including e.g. links, typed entities, hashtags; which inherently annotate the content of a post. The representation of an `lp:POI` is therefore the aggregation of concepts derived from the annotations of the posts generated from it.

The LinkedPOI ontology follows a semantic pattern similar to the SIOC ontology, where a forum is a subclass of a container which is an aggregation of items (e.g. Posts). A `lp:POI` enriches the semantics of a `sioc:Forum` by bridging the gap between online community forums and physical space through the definition of "localised forum". A "localised forum" can be understood as a geographically annotated collection of channels or discussion areas from which posts are made.

LinkedPOI makes use of the AO[14], SIOC[15], WGS84[16], and SKOS[17] ontologies:

- the `Annotation Ontology` (`AO`) – is a vocabulary designed to reuse domain ontologies (entities, annotations or semantic tags) to annotate any kind of document;
- `Semantically Interlinked Online Communities` (`SIOC`) – enables the integration of online community information;
- the `World Geodetic System` (`WGS84`) – is a basic RDF vocabulary which provides a namespace for representing information about spatially-located things, using WGS84 as a reference datum, including latitude and longitude;
- `Simple Knowledge Organization System` (`SKOS`) – develops specifications to support the use of systems such as thesauri and classification schemes.

Core Components. The core structure of the LinkedPOI ontology is presented in Fig. 1:

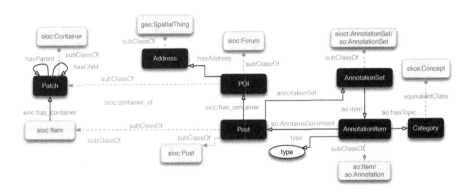

Fig. 1. `LinkedPOI` ontology

---

[14] AO, `http://code.google.com/p/annotation-ontology/wiki/Homepage`
[15] SIOC Ontology, `http://sioc-project.org/`
[16] Geo (WGS84 lat/long) Vocabulary, `http://www.w3.org/2003/01/geo/`
[17] SKOS, `http://www.w3.org/2004/02/skos/`

Patch – a high-level concept that represents a geographical location bounded by a set of geo-coordinates. It is used to group `Patch`s together through the `sioc:has_container` and `sioc:container_of` properties. This class also defines a recursive hierarchy between `Patch`es by defining the parent and child relationships `lp:hasParent`, `lp:hasChild`. E.g. the `Patch` representing the city of Palo Alto is a child of the `Patch` representing the state of California.

POI – a point of interest can be thought of as a "localised forum" . A `POI` is a subclass of `sioc:Forum` and `lp:Patch`. While a `sioc:Forum` is linked to an online space independent of the physical space, a `POI` enriches the concept of a forum by assigning a geographical annotation to it. A `POI` is a localised channel from which posts are streamed; therefore an `lp:POI` acts as a container of `lp:Post`. A `POI` may be assigned an address using the `lp:hasAddress` relation.

Address – represents generic features of a point of interest such as city, street, postcode. An address can be assigned to a `POI` which represents e.g., a restaurant, or hotel. However, when the `POI` represents a bigger regional space such as city, the `Address` will not be defined.

Post – geographically annotated, text-based content posted by a `sioc:User-Account` to an online forum (e.g. Facebook and Twitter). A `Post` is a subclass of `sioc:Post`, inheriting relations such as `sioc:reply` and `sioc:has_creator`, each of which links a post with a user. An `lp:Post` is related to an `lp:POI` through the `sioc:has_container` relation. A `Post` can be related to a set of annotation items through the `lp:annotationSet` relation.

AnnotationSet – a collection of qualified annotations which characterise the content of a `Post`. An `AnnotationSet` is a subclass of `sioct:AnnotationSet` and `ao:AnnotationSet`. It relates to a `lp:Post` through the `lp:annotationSet` and relates to an annotation item through the `ao:item` relation.

AnnotationItem – a subclass of `ao:Annotation` that describes an annotation qualified by type and topic. Types of annotations appearing in microposts include, e.g., links, hashtags and typed entities. An `lp:AnnotationItem` is related to a topic through the `ao:hasTopic` relation, and is related to a `Post` through the `ao:AnnotatesDocument` relation.

Category – a `skos:Concept` equivalent class. It is used to provide a conceptual representation of a piece of information. The definition of a category hierarchy is application dependent, and it represents a finite set of conceptual representations in which localised forums can be classified. A category is linked to a `lp:Post` through the `lp:AnnotationItem` class and can be linked as well through the `lp:WeightedConcept` class when using the `LinkedPOI` Weighting module. A `Category` can be linked to an `lp:Patch` by using the weighting module through the `WeightedConcept` class.

The representation of a point of interest changes according to the change in the concepts involved in the posts being streamed from this location. This dynamic representation of a location makes evident the need to describe the proportion in which a concept is representative of a location in a given point in time. In order to cope with this temporal characterisation of a `POI`, the `LinkedPOI Weighted Module` is introduced (see Fig. 2).

WeightedConcept – a pattern that links an item to a `Category` based on a numeric weight. The calculation of this weight is application dependent. An interpretation of the `WeightedConcept` is the proportion of the conceptual representation of an item that a category represents for a specified point in time. The properties of this class also include `createOn` and `lastUpdatedOn`, of type `Timestamp`, which assigns a temporal annotation to a weighted concept relation. It can be linked to an `lp:Patch` and an `lp:Post` through the `lpw:hasWeightedConcept` relation. While the weighted concept of a `lp:Post` is fixed in time (since the content of a post does not change), the weighted concept of a `lp:Patch` changes as new posts related to this category appear.

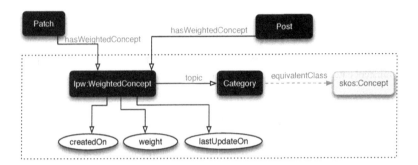

Fig. 2. LinkedPOI Weight Module

3.4   Transient Semantic Classification of a POI

Problem Statement Comments extracted from localised forums can be described as semi-public, natural-language, geo-annotated messages produced by different users and characterised by their brevity. Given these characteristics and the variation in the vocabulary appearing on a POI's stream's comments, finding relevant categories that can accurately qualify a `Post` is a challenging task.

Definition 2 *A temporal classification of a* `POI` *is defined as the aggregation of* R_{cat} *category resources' qualifying messages contained in a specific window of time, denoted by* $[t_s, t_e]$.
An $S(\texttt{Poi'})[t_s, t_e]$ *is defined as* $S(\texttt{Poi'})$ *where* $ft : Y \to T, t_s \leq ft \leq t_e$.

Given the above definition, the task consists of obtaining category resources R_{cat} which can classify a *poi* within a window of time $[t_s, t_e]$. The following section introduces a strategy for deriving these categories.

Entity-Based Discovery of Transient Categories. We propose an approach that uses the categorisation of posts' resources generated from an `lp:POI` (localised forums) ($S(Poi')$) taken in windows of time ($[t_s, t_e]$), to induce a categorisation function. Fig. 3 presents an overview of our approach.

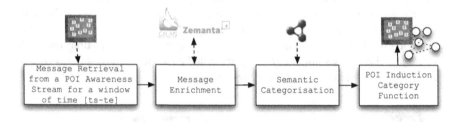

Fig. 3. Category Induction Pipeline: Messages are retrieved from localised forums (`lp:POI`). DBpedia categories are derived for each enriched message. These sets of categories are used to induce a transient categorisation of a Point of Interest.

Enrichment Given a message from a POI stream $S(Poi')$, lightweight *message enrichment* is performed using *Zemanta*[18], and *OpenCalais*[19]. These services perform entity-extraction on the input message, identifying resources which can be qualified as: R_o (entities recognised as an organisation), R_p (entities recognised as a person), R_l (entities recognised as a location) and R_{li} (links to resources). These services also provide DBpedia concepts relevant to each message. Consider the example in Fig. 4, where the extracted entities and DBpedia concepts for a Twitter message are shown.

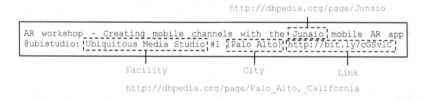

Fig. 4. Message Enriched with Zemanta and OpenCalais services. These services return entity labels as well as DBpedia concepts related to the message.

Semantic Categorisation. In order to semantically categorise a POI stream's message (m), we query DBpedia for all concepts relevant to the extracted entity-based resources. These concepts are then aggregated to those already suggested by the message enrichment services. Given a resource (r), DBpedia (D) is queried for *categories* and *broader categories* using the following construct:

---

[18] Zemanta, `http://www.zemanta.com/`

[19] OpenCalais, `http://www.opencalais.com/`

$$\mathrm{R}_{cat}(r) = \{x_{cat} \cup x_{broaderCat} | < r, \text{dcterms:subject}, x_{cat} >$$
$$\wedge < x_{cat}, \text{skos:broader}, x_{broaderCat} >\in D\} \qquad (1)$$

For each resource (r) a collection of categories (dcterms:subject) and parent categories (skos:broader) of r are extracted from DBpedia. Table 1 presents an example of categories and broader categories derived from DBpedia for the resource Palo_Alto contained in the example of Fig. 4. These categories become a resource category R_{cat} of the POI stream ($S(Poi')$); which can also be considered as lp: annotationItems of a lp:POI.

Table 1. Categories and broader categories derived for the entities extracted from the comment in Fig. 4

Entity	Category
Palo_Alto (of type City)	dcterms:subject *Palo_Alto, California*
	skos:broader Populated_places_in_Santa_Clara
	skos:broader University_towns_in_the_United_States
Junaio (of type Thing)	dcterms:subject *Augmented_reality*
	skos:broader *Mixed_reality*

The following section presents a use case in which POIs are profiled using localised forums, based on the approach presented.

4   The Topica Travel Mashup – A Use Case

The factor that characterises all travellers is that they often possess inaccurate, misinformed, incomplete, outdated, little or no information about the places they are visiting [6] and/or information on how to get there. This is in large part because travellers are immersed in an environment that is in many cases new or barely known to them. Even for the (seasoned) traveller returning to a previously visited location, an element of uncertainty, due to (potentially) outdated knowledge, exists. This highlights the need for reliable, current, easy to interpret information. This section highlights the potential for travel mashups to bridge this information gap, by identifying travel information needs elicited by different travel scenarios.

The use of citizen sensing for travel applications has resulted in a new generation of applications which demand real-time processing of very large amounts of data. One of the characteristics that make citizen sensing extremely useful for situational awareness in travel mashups is the different dimensions reported about a particular event occurring at a specified geographical location. Multiple perspectives of a single event can be derived by taking into account the information contained in the spatio-temporal metadata captured by the device from which a piece of content is posted.

In this section, we present the use of social stream aggregation as a data source that conveys meaningful, collective information for modelling dynamic characteristics of a POI. Although a POI is typically represented as a set of static data (e.g. name, address, geo-coordinates), there are many latent (or hidden) features which can describe volatile and temporal aspects of it. E.g., an Italian restaurant may be well known for good, hand-made ravioli, or for the (transient) two-for-one offer available during the current month. *Topica*[20] is a mash-up that uses localised forums to enhance collective information about a POI, by leveraging structured data extracted from the LD cloud.

4.1 Application Scenario

Consider the following scenario:

> *Alice, who grew up in Rethymnon, Crete, has moved back to take up a new job in the city, after having been away for almost 10 years. Alice wishes to explore what is new in Rethymnon. She starts with the city center and its immediate surroundings, to find out if the venues she often frequented in her late teens match her (evolved) interests. Recognising also that some of the places she remembers may no longer exist, Alice decides to make use of* Topica *to discover venues in her (physical) neighbourhood. The* Topica *app, which she is able to access from a (desktop) web browser and also her new smartphone, allows her to browse current topical information about nearby POIs, based on (collective) information contributed by visitors to these POIs and also from more official sources (such as the tourist board, event organisers and venue managers).*

Although Alice is not, strictly, a traveller, her information needs in this scenario match those of a traveller seeking information about a selected location during the *the actual journey* stage of travel. We demonstrate the approach taken in *Topica* to realise this scenario, i.e., to support topical recommendation of POIs based on POIs' latent features. *Topica* analyses publicly available social stream information filtered by location, in this case, the city of Rethymnon on the island of Crete, in Greece.

Topica models space based on the topical information buzzing in social media streams. This information is aggregated to profile regional areas enabling *Topica*'s users to expand a topical search from a focus to include increasingly wider radii of regional interest. From a SW point of view, *Topica* identifies the topics that model a POI by extracting DBpedia categories from (key) *entities* (e.g. `Location`, `Organisation`, `People`, `Places`) and *keywords* obtained from the POIs' related social awareness streams – in this case, an aggregation of Facebook, Twitter, and Tripadvisor comments regarding the POI.

From Alice's point of view, *Topica* allows the retrieval of topical information of interest to her, from the collective information obtained by aggregating comments contributed by users in the locality. She is able to select from these topics that match her interests. She chooses to focus on music (*Jazz, World Music*) and food (*Italian*). Using these filters *Topica* retrieves POIs by topic as well as the related current, valid concepts and categories that feature the POI.

Alice decides to try out one of the Italian restaurants suggested by *Topica* with a couple of friends. During the meal she posts a comment on *Facebook Places* about the restaurant, and comments on her main meal (ravioli with a creamy lobster sauce)

---

[20] *Topica,* http://nebula.dcs.shef.ac.uk/sparks/topica

and the live Jazz band playing. This information is in turn consumed by *Topica*, which re-profiles the regional area (the POI) accordingly, to include, in addition to Italian-_Cuisine – what prompted Alice to select this restaurant – also Sea_Food and Jazz_Music.

4.2   *Topica* Application Architecture

Topica facilitates the retrieval of POIs by characterising them with latent topical features extracted from the collective perception of a POI over a period of time. As demonstrated in section 4.1, *Topica* exploits localised forums to enrich POIs with structured data from the LD Cloud and provide a filter-based visualisation of POIs. The fundamental structure of Topica consists of three layers: i) the Information Sources Layer; ii) the Generation Layer; and iii) the Presentation Layer, which build on each other (Fig. 5).

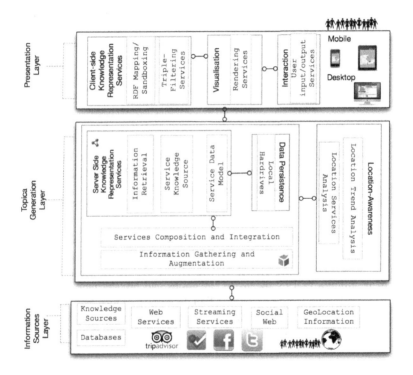

Fig. 5. *Topica* architectural design

Information Sources Layer. A POI is represented as a geographically bounded area from which social streams coming from Twitter, Facebook and TripAdvisor are filtered and associated with the POI.

Generation Layer. This layer consists of the following:

1. *Mashup Logic*: This stage can be subdivided into three main inter-dependent sections:

   (a) *Information Gathering and Augmentation*: When data is retrieved from the Information Sources layer, further related information is extracted from external knowledge sources, including OpenCalais, Zemanta, and DBpedia Spotlight. This information includes entities and related categories and broader categories as defined in Subsection 3.4. The data collected is used by *Topica* to generate a list of potential DBpedia resources. In order to model the topics mentioned in a comment, *Topica* uses this resource list to query DBpedia. For each resource DBpedia's categories and broader categories are extracted. E.g., for the resource `Ravioli`, these categories include: `Pasta` and `Italian_Cuisine`; and broader categories for `Italian_Cuisine` includes: `Mediterranean_Cuisine` and `Italian_Culture`. The set of categories collected from a POI's related comments are weighted following a tf-idf (term frequency-inverse document frequency) function. The information augmentation stage communicates with the Knowledge Representation services (see 1b) which keep a local, up to date knowledge repository containing the information handled in the mashup.

   (b) *Server-side Knowledge Representation Services*: This stage handles services related to data modelling, retrieval and storage of information, providing an interface to the data persistence stage and to the presentation layer. The knowledge representation of a POI is achieved by means of the `LinkedPOI` ontology (refer subsection 3.3).

   (c) *Data Persistence*: Topica makes use of a virtuoso server which exposes the knowledge representation of POIs.

2. *Location Awareness*: While a regional space can be described statically based on a name and a set of geo-coordinates, a dynamic representation of the location is achieved through location profiling. This stage involves the process of profiling a POI by means of categories and broader categories which are weighted using the `LinkedPOI` Weighted module (see Fig. 2). Processes in this stage include, e.g., location trend and topical analysis. The `LinkedPOI` ontology takes advantage of `sioc:Container` to model POIs as elements of a bounded area (`linkedPOI:Patch`). This allows SPARQL querying for concepts featuring a bounded area.

Presentation Layer. *Topica* makes use of the *Prism Framework*[21] to enable synchronised semantic filtering of POIs. Prism is a JavaScript-based semantic framework that allows the creation of synchronised semantic filters using a seed query. Prism relies on *OpenStreetMap* and *Google maps*[22]. It computes a SPARQL query against a set of filter parameters in order to select a subset of the objects returned by the seed query. *Topica* uses the following Prism filters:

---

[21] Prism, `http://evhart.online.fr/prism`

[22] Google maps, `http://code.google.com/apis/maps/index.html`

1) The *Location lens* is used to filter POIs according to their location, using geo-coordinate information. This activity is supported through the use of a map widget.

2) The *Tag lens* enables the selection of POIs according to their associated tags. This involves the generation of a geo-tag filter to allow the retrieval of POIs labeled by a given tag.

3) The *Search lens* provides a text filter that operates on the POIs' messages. This filter extracts POIs based on comments in emerging social streams that contain the search text.

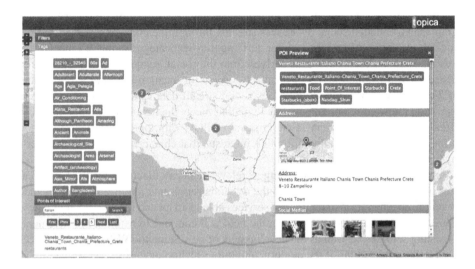

Fig. 6. Overview of the *Topica* user interface, showing sub-windows with additional detail for the POIs selected

By employing the Prism framework *Topica* is able to support the user to retrieve information stored in the LD cloud using intuitive visualisation (see Fig. 6). The visual representation provides a consistent and synchronised view of POIs according to the filter parameters provided by the user. When the user is satisfied with the values returned by the filters, they can access a description of a particular POI by clicking on a map object from the POI list. The description, presented in a pop-up visualisation, includes the name and address of the POI, its tags and topics and the different social messages associated with this POI.

4.3   *Topica* Mashup - Summary of Key Features

Topica improves on existing POI retrieval applications in that: (1) *Topica* does not require end users to explicitly contribute information about events – the mashup extracts these from users' social interaction data; (2) *Topica* enriches POIs by adding (semantic) DBpedia resources extracted from comments related to them. By using additional

SW services such as OpenCalais and Zemanta, we provide even richer annotation and widen recall during information retrieval.

Topica currently uses fixed snapshots in time for each of its location datasets (Dec 2010 – Jan 2011) – the restricted dataset was to allow a focus on the design and the development of the backing technology. We aim however to move to dynamic batch update, to ensure that the fully working tool allows the end user to retrieve the most recent information about POIs, in addition to still valid historical information.

5   Conclusions

This paper presents a mathematical and an ontological formalisation for describing geographically bounded social awareness streams through the concept of "localised forums". It also presents an approach for deriving transient categorisation of Points of Interest through the use of concepts derived from the social, semantic knowledge source DBpedia. To illustrate the use of localised forums for location profiling we presented the *Topica* Travel Mashup. *Topica* uses social stream data as an information source, to provide knowledge-based, business logic that meets the travel information needs of different end users. It associates the retrieval of POIs with topical information, based on the end user's expressed and implicit interests, that characterise POIs in the user's physical location.

Future work includes the modelling of relevance decay functions for the latent features of POIs. Future perspectives of the dynamic representation of a location include the real-time integration of composite services able to adapt and react on the fly to these changes. We envisage that research in the field will continue to result in improvements to existing services, as well as the development of new Semantic Web-enabled services to meet the expectations of the modern, ubiquitous technology user.

Acknowledgements. A.E. Cano is funded by CONACyT, grant 175203. A.-S. Dadzie is currently funded by the UK MRC project Time to Change (129941). G. Burel is currently funded by the EC-FP7 project Robust (257859).

References

1. Abel, F., Gao, Q., Houben, G.-J., Tao, K.: Semantic enrichment of twitter posts for user profile construction on the social web. In: Antoniou, G., Grobelnik, M., Simperl, E., Parsia, B., Plexousakis, D., De Leenheer, P., Pan, J. (eds.) ESWC 2011, Part II. LNCS, vol. 6644, pp. 375–389. Springer, Heidelberg (2011)
2. Auer, S., Lehmann, J., Hellmann, S.: LinkedGeoData: Adding a spatial dimension to the Web of Data. In: Bernstein, A., Karger, D.R., Heath, T., Feigenbaum, L., Maynard, D., Motta, E., Thirunarayan, K. (eds.) ISWC 2009. LNCS, vol. 5823, pp. 731–746. Springer, Heidelberg (2009)
3. Barkhuus, L., Brown, B., Bell, M., Sherwood, S., Hall, M., Chalmers, M.: From awareness to repartee: sharing location within social groups. In: CHI 2008: Proc., 26th Annual SIGCHI Conference on Human Factors in Computing Systems, pp. 497–506 (2008)
4. Becker, H., Iter, D., Naaman, M., Gravano, L.: Identifying content for planned events across social media sites. In: WSDM 2012: Proc., 5th ACM International Conference on Web Search and Data Mining, pp. 533–542 (2012)

5. Braun, M., Schmeiß, D., Scherp, A., Staab, S.: Stevie–collaborative creation and exchange of events and POIs on a mobile phone. In: EiMM 2010: Proc., 2nd ACM International Workshop on Events in Multimedia, pp. 35–40 (2010)
6. Brown, B., Chalmers, M.: Tourism and mobile technology. In: ECSCW 2003: Proc., 8th Conference on European Conference on Computer Supported Cooperative Work, pp. 335–354 (2003)
7. Cano, A.E., Tucker, S., Ciravegna, F.: Follow me: Capturing entity-based semantics emerging from personal awareness streams. In: Proc., 1st Workshop on Making Sense of Microposts, #MSM 2011, pp. 33–44 (2011)
8. Cheng, Z., Caverlee, J., Lee, K.: You are where you tweet: a content-based approach to geo-locating twitter users. In: Proc., 19th ACM International Conference on Information and Knowledge Management, CIKM 2010, pp. 759–768 (2010)
9. Cheng, Z., Caverlee, J., Lee, K., Sui, D.Z.: Exploring millions of footprints in location sharing services. In: Proc., 5th International Conference on Weblogs and Social Media, ICWSM (2011)
10. Freyne, J., Brennan, A., Smyth, B., Byrne, D., Smeaton, A., Jones, G.: Automated murmurs: The social mobile tourist application. In: CSE 2009: Proc., International Conference on 2009 Computational Science and Engineering, pp. 1021–1026 (2009)
11. Hao, Q., Cai, R., Wang, C., Xiao, R., Yang, J.-M., Pang, Y., Zhang, L.: Equip tourists with knowledge mined from travelogues. In: WWW 2010: Proc., 19th International Conference on World Wide Web, pp. 401–410 (2010)
12. Hightower, J.: From position to place. In: Proc., Workshop on Location-Aware Computing, pp. 10–12 (2003)
13. Hornecker, E., Swindells, S., Dunlop, M.: A mobile guide for serendipitous exploration of cities. In: MobileHCI 2011: Proc., 13th International Conference on Human-Computer Interaction with Mobile Devices and Services, pp. 557–562 (2011)
14. Laurier, E.: Why people say where they are during mobile phone calls. Environment and Planning D: Society and Space 19(4), 485–504 (2000)
15. Schegloff, E.: Notes on a conversational practice: formulating place. In: Sudnow, D. (ed.) Studies in Social Interaction, pp. 75–119. Free Press (1972)
16. Sheth, A.: Citizen sensing, social signals, and enriching human experience. IEEE Internet Computing 13(4), 87–92 (2009)
17. Sheth, A., Thomas, C., Mehra, P.: Continuous semantics to analyze real-time data. IEEE Internet Computing 14, 84–89 (2010)
18. Stadler, C., Lehmann, J., Höffner, K., Auer, S.: Linkedgeodata: A core for a web of spatial open data. Semantic Web 3(4), 333–354 (2011)
19. Tintarev, N., Flores, A., Amatriain, X.: Off the beaten track: a mobile field study exploring the long tail of tourist recommendations. In: MobileHCI 2010: Proc., 12th International Conference on Human-Computer Interaction with Mobile Devices and Services, pp. 209–218 (2010)
20. Wagner, C., Strohmaier, M.: The wisdom in tweetonomies: Acquiring latent conceptual structures from social awareness streams. In: Proc., Semantic Search 2010 Workshop, SemSearch 2010 (2010)
21. Weilenmann, A.: "i can't talk now, i'm in a fitting room": formulating availability and location in mobile-phone conversations. Environment and Planning A 35(9), 1589–1605 (2003)
22. Womer, M.: Points of Interest core – W3C working draft (May 2011), http://www.w3.org/TR/poi-core

A Community-Driven Approach to Development of an Ontology-Based Application Management Framework

Marut Buranarach, Ye Myat Thein, and Thepchai Supnithi

Language and Semantic Technology Laboratory
National Electronics and Computer Technology Center (NECTEC), Thailand
{marut.bur,thepchai.sup}@nectec.or.th, yemyatthein@gmail.com

Abstract. Although the semantic web standards are established, applications and uses of the data are relatively limited. This is partly due to high learning curve and efforts demanded in building semantic web and ontology-based applications. In this paper, we describe an ontology application management framework that aims to simplify creation and adoption of a semantic web application. The framework supports application development in ontology- database mapping, recommendation rule management and application templates focusing on semantic search and recommender system applications. We present some case studies that adopted our application framework in their projects. Evolution of the software tool significantly profited from the semantic web research community in Thailand who has contributed both in terms of the tool development and adoption support.

Keywords: semantic web application framework, ontology application development tool, software tool development model.

1 Introduction

With the Semantic Web data standards being established, some organizations and initiatives started creating and sharing their data in the RDF format, aka. Linked data initiative [1]. Further, domain knowledge in different areas has been increasingly captured in ontology form that can be shared as OWL data that can be linked with the RDF data. Although creation of the Semantic Web data rapidly grows, e.g. the Linked data cloud [1], applications and uses of the data are relatively limited. This is partly due to high learning curve and efforts demanded in building semantic web and ontology-based applications. To facilitate development of such applications, development tools should allow application developers to focus more on domain problems and knowledge rather than implementation details. Put another way, application development tools should not only be designed for technologists but also researchers or domain experts who are non-technology experts.

In this paper, we introduce the Ontology-based Application Management (OAM) framework, a development platform for simplifying creation and adoption of a seman-

H. Takeda et al. (Eds.): JIST 2012, LNCS 7774, pp. 306–312, 2013.

tic web application. The framework is primarily built on top of some existing tools and frameworks, including the Jena framework [2] and D2RQ [3]. OAM allows the user to interactively define mapping between an existing database schema with an OWL ontology to produce the RDF data based on the mapping. It also provides some application templates that support RDF data processing focusing on semantic search and recommender system applications. Our framework is different from existing semantic web application frameworks in that it does not require user's programming skill in building a semantic web application prototype. Thus, researchers spend less time and effort in adopting the semantic web technologies and can focus more on higher-level application logics. The framework also provides application-level APIs and Web service interface to support a more advanced application development.

A community-driven approach was applied to the development of our software tool. In this approach, research community can contribute to evolution of a software tool. We have organized several activities for a semantic web research community in Thailand, which consists of interested users, developers, students, teachers, domain experts, and researchers, to support deployment of our tools. The activities include both adoption and development supporting activities. The adoption supporting activities include trainings and workshops where the research community can learn about ontology development and application development using the tool. The development supporting activities include student projects and theses and coding marathon event. The community-driven development approach has contributed to the tool evolution by means of user feedbacks, requirement gathering, and collaborative design, coding and testing.

The paper is organized as follows. Section 2 introduces a community-driven software tool development model which has been used in the development of our software tool. Section 3 focuses on some unique approaches and functions of our application framework. Section 4 present two case studies that adopted our application framework as a research tool in their projects. Section 5 discusses current status of the tool and some future development directions.

2   A Community-Driven Software Tool Development Model

A community-driven approach was applied to the development of our software tool. The approach can be illustrated as a software development cycle as shown in Fig. 1. In this model, contributions from research community are significant to the

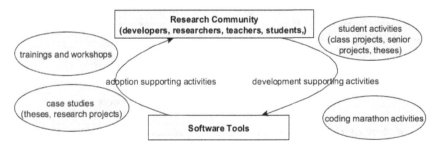

Fig. 1. A community-driven software tool development cycle

development and adoption of a software tool. In supporting tool adoption, training and workshop activities can be organized to introduce the tool to the research community. In addition, some case studies are needed to demonstrate application potentials of the tool. Related student theses or research projects should be set up and adopt the tool as a research tool. The community users who adopted the tool can help to provide feedbacks, testing and evaluation results, which can contribute to gathering additional requirements for improving the tool. In supporting tool development, university teachers can integrate some parts of the tool development as student assignments for class projects, senior projects or theses. In addition, coding marathon activity, which is popularly adopted in opensource software development, can be organized to promote collaborative efforts between researchers and developers in improving design and implementation of the tool.

Since the initial release of our software tool in 2010, we have conducted user training sessions to introduce the tool to the research community. Each session comprises of two hand-on trainings: ontology development[1] and ontology application development. Feedbacks from the participated users, who were developers, researchers, university teachers and students, were gathered as user requirements to guide the tool improvement. In addition, some university teachers have taken these requirements and assigned them as class and senior projects for their students. In addition, we organized a two-nights coding marathon activity which involved nearly 60 students, professional developers and researchers. These supporting activities have collectively contributed to improving the tool both in terms of functionalities and user interface designs. Further, several student projects and theses were set up and adopted the tool, which helped to demonstrate applications of the tool.

3   Ontology-Based Application Management Framework

The Ontology-based Application Management (OAM) framework[2] is an application development platform aims to simplify creation and adoption of a semantic web application. Our application framework differs from the existing tools in two main aspects. First, the framework provides common application templates that can process

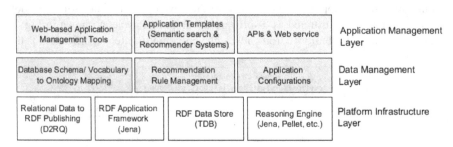

Fig. 2. A layered architecture of the OAM framework

---

[1] Hozo ontology editor (http://www.hozo.jp/) was used in the ontology development trainings.

[2] OAM Homepage – http://text.hlt.nectec.or.th/ontology/

the user's published RDF data. Thus, it does not require user's programming skill in building an application. Second, it is an integrated platform that supports both RDF data publishing from databases and processing of the published data in ontology-based applications, i.e. semantic search, recommender system applications.

Fig. 2 shows a layered architecture of the OAM framework. The framework was implemented on top of existing semantic web data and application platforms, i.e., Jena, D2RQ, RDF data storage and a reasoning engine. OAM added some data and application management functions, including database schema and vocabulary to ontology mapping, recommendation rule and application configuration management. The user can create and manage ontology-based applications using web-based application templates and management tools. OAM also provides APIs and Web service interfaces to support a more advanced application development.

3.1   Database Schema and Vocabulary to Ontology Mapping Management

There are typically two methods in creating instances for ontology classes. The first method is to manually construct an instance and define its attribute values based on a class. This is typically done using instance editor provided in ontology development tool. The second method is to create instances from some existing information sources, such as database records. This normally requires mapping process between the existing database schema and ontology structure [4]. After the mapping process, a database record can be transformed into a class instance. This method is most suitable when an organization already stored the data in some databases. Our framework focuses on the latter approach in creating instance data from existing databases.

OAM's data mapping management tool supports both schema mapping and vocabulary mapping between OWL ontology and a relational database source. In schema mapping, the user can define mapping between ontology classes and database tables. For each class, the user can define mapping between each property, i.e. either datatype property or object property, with a database column of any table with an optional join condition. The property mapping supports one-to-one, one-to-many and many-to-many relationship types in databases. Based on the mapping configuration, the tool can generate instances of the ontology classes as an RDF/XML file. In vocabulary mapping, the user can choose a table column and assign each attribute value as label of an ontology class. This will allow synonymous terms to be mapped with a class in ontology that would allow semantic-based processing in applications.

3.2   Recommendation Rule Management

OAM also focuses on simplifying creation and management of recommendation rules. It provides a recommendation rule management tool that supports two processes: create recommendation and link recommendation. Creating recommendation will create a recommendation class instance where the user can define conditions of class instances to be associated with the recommendation. For example, user can create a car recommendation instance and associate it with car model instances that match a condition of "a Japanese brand and priced under $20,000". Linking recommendation allows the user to define conditions of class instances to which the recommendation is assigned, e.g., customer instances that match a condition of "young adults with Asian nationality". Fig. 3 shows an example of the resulted recommendation rules created in the Jena rule syntax. The tool facilitates the user to create such business logics using a form-based user interface and hides complexity of the rule syntax to be processed by reasoning engine.

```
[Create_car_rec_1: (?x rdf:type ns:CarModel) (?x ns:has_brand ?y) (?y rdf:type ns:JapaneseBrand)
                   (?x ns:has_price ?z) lessThan(?z, 20000) -> (comp1:car_rec_1 rdf:type ns:CarRecommend)
                   (comp1:car_rec_1 ns:has_rec_id '1') (comp1:car_rec_1 ns:has_car_model ?x)]

[Link_car_rec_1: (?x rdf:type ns:Customer) (?x ns:has_nation ?y) (?y rdf:type ns:Asian)
                 (?x ns:has_age_group ?z) (?z rdf:type ns:YoungAdult) (?a rdf:type ns:CarRecommend)
                 (?a ns:has_rec_id '1') -> (?x ns:has_car_model_recommendation ?a)]
```

Fig. 3. An example of the resulted recommendation rules in the Jena rule syntax

3.3   Application Templates and Management

OAM allows the user to create a semantic web application using a provided application template. Using application template, the user only needs to define application configuration and does not need programming skill in building an application. This is suitable for researchers who want to experiment on research ideas that can be realized by means of the semantic web technology. Application template is typically ideal for rapid prototyping and hypotheses testing.

Currently, the framework provides two application templates: semantic search and recommender system applications. The semantic search application template provides a faceted search [5] interface. Using the provided form-based interface, the end-user can select a class in the ontology to search for its instance data and define some search property conditions. A search allows value comparison using both string and number comparators and semantic-based comparators, i.e. IS-A. It also allows the user to customize properties in displayed search results. The user's faceted search condition is automatically transformed to a SPARQL query for retrieving the instance data from an RDF database. The recommender system application template extends the semantic search template to support viewing of recommendation rules and results.

4   Case Studies

This section discusses some projects that adopted our framework in their research. One project was development of a food recommendation system [6] based on user's health status and nutrition goals. In another project, OAM was used to support human activity recognition task in smart home domain [7]. The ontology-database mapping tool helped to simplify mapping process between the domain ontology and the database storing the sensor-based data and to produce the RDF data. The recommendation rule management component helped to facilitate defining rules for the human activity recognition task. Finally, the semantic search application template allows for rapid prototyping and evaluating the recommendation results. Fig. 4 shows an example use of the application framework for activity recognition task in smart home domain.

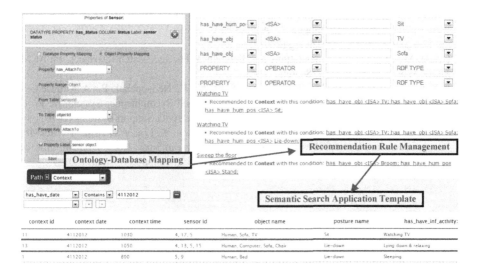

Fig. 4. Example use of the application framework for activity recognition in smart home domain

5   Discussion

In this paper, we describe OAM, an application framework aimed to simplify development of ontology-based applications that use the semantic web technology. Our tool development approach relied significantly on contributions from the research community. Adoption supporting activities, such as trainings and workshops, have contributed to promoting deployment of the software tool among the community. In addition, development supporting activities, i.e. student projects and coding marathon activity, have allowed the research community to contribute to improving design and implementation of the tools.

Based on download statistics in 2012, OAM has approximately 45 downloads monthly. Approximately 200 users have completed the user trainings. Some future development includes adding support for more application templates such as decision support system and NLP applications. We also aim to improve compatibility with various ontology tools and to add support for Linked data interoperability.

References

1. Bizer, C., Heath, T., Berners-Lee, T.: Linked Data - The Story So Far. International Journal on Semantic Web and Information Systems 5, 1–22 (2009)
2. Carroll, J.J., Reynolds, D., Dickinson, I., Seaborne, A., Dollin, C., Wilkinson, K.: Jena: Implementing the Semantic Web Recommendations. In: Proc. of the 13th International World Wide Web Conference on Alternate Track Papers & Posters, pp. 74–83 (2004)
3. Bizer, C.: D2R MAP - A Database to RDF Mapping Language. In: Proc. of the 12th International World Wide Web Conference, WWW 2003 (2003)

4. Kozaki, K., Hayashi, Y., Sasajima, M., Tarumi, S., Mizoguchi, R.: Understanding Semantic Web Applications. In: Domingue, J., Anutariya, C. (eds.) ASWC 2008. LNCS, vol. 5367, pp. 524–539. Springer, Heidelberg (2008)
5. Hearst, M.: Design Recommendations for Hierarchical Faceted Search Interfaces. In: ACM SIGIR Workshop on Faceted Search (2006)
6. Suksom, N., Buranarach, M., Thein, Y.M., Supnithi, T., Netisopakul, P.: A Knowledge-based Framework for Development of Personalized Food Recommender System. In: Proc. of the 5th Int. Conf. on Knowledge, Information and Creativity Support Systems (2010)
7. Wongpatikaseree, K., Ikeda, M., Buranarach, M., Supnithi, T., Lim, A.O., Tan, Y.: Activity Recognition using Context-Aware Infrastructure Ontology in Smart Home Domain. Proc. of the 7th Int. Conf. on Knowledge, Information and Creativity Support Systems (2012)

When Mommy Blogs Are Semantically Tagged

Jinan El-Hachem and Volker Haarslev

Engineering and Computer Science Department
Concordia University, Montreal, Quebec, Canada
ji_elhac@encs.concordia.ca

Abstract. OWL 2-supported Semantic Tagging is a non compulsory yet decisive and highly influential component of a multidisciplinary knowledge architecture framework which synergetically combines the Semantic and the Social Webs. The facility consists of a semantic tagging layer based on OWL 2 axioms and expressions enticing social network users, typically mommy bloggers, to annotate their chaos of textual data with natural language verbalized versions of ontological elements. This paper provides a comprehensive short summary of the overall framework along with its backbone metamodel and its parenting analysis and surveillance ontology ParOnt, laying a particular emphasis on its semantic expression-based tagging feature, and accordingly highlighting the attained gains and improvements in terms of effective results, services and recommendations, all falling in the scope of public parenting orientation and awareness.

Keywords: Semantic Web, Social Web, Social Network Sites (SNS), Mommy Blog, Natural Language Processing (NLP),Natural Language Generation (NLG), Web Ontology Language (OWL), Description Logics (DL), Recommender Systems, Parenting, Public Awareness.

1 Introduction

The emergence of tagging with Web 2.0 sites, and hence the increasing willingness of Social Network users to provide collaborative tagging make the availability of metadata with significant potentials an undisputable reality. Furthermore, these realized facts lead to the assessment that it would be simply unjustifiable if the opportunity of properly taking advantage of the potentials in question is missed through not fostering appropriate approaches to create and manage tags. Among the variety of accessible Web 2.0 sites, "Mommy blogs" are those in which Social Network users provide tons of information related to children, their problems and behaviors, to parenting in general; and "Mommy bloggers", usually parents, are perceived as extremely active and cooperative users who constantly access or manage this particular type of blogs.

In this work, traditional semantic tagging notions are further developed to include identified rules and well defined OWL 2 ontology axioms. The paper thus briefly introduces a social semantic platform already proposed in the related previous works [5] and [6], and extends its semantic and rules tagging layer

H. Takeda et al. (Eds.): JIST 2012, LNCS 7774, pp. 313–318, 2013.

using ontological formalisms and Natural Language Processing and Generation techniques to bestow effective and efficient ontology population, reasoning and querying facilities. The framework thus relies on highly expressive domain expert ontologies, namely ParOnt, a conceived ontology for parenting cross-sectional analysis. Its backbone metamodel fosters the different methods for semantic web language components' distinctive characteristics support, for instance OWL 2 profiles and sublanguages, providing projection mechanisms for efficient and substantiable performance support.

At the end, it demonstrates that if mommy bloggers are further involved in an efficient process of knowledge tagging, massive and efficient ontology population is easily achieved. Thus, supported by advanced reasoning and querying techniques, different services are offered, including community profiling and segmentation, generic and customized parenting recommendations, as well as alerts on encountered fallacies and misleading notions or beliefs, all in the scope of public parenting orientation and awareness.

The remaining part of this paper proceeds by presenting an overview of the knowledge framework and platform with the details of the incorporated semantic tagging feature. Following that, demonstrating scenarios and experimental examples that endow with recommender systems based on the Parenting Ontology ParOnt are exposed right before the concluding section that comprises a closing discussion along with highlights on some relevant future work.

2   Architectural Framework Overview: Emphasis on Semantic Tagging

Figure 1 below tries to overcome the page number limitations by presenting a simplistic high-level description of the knowledge architectural framework's main components. In this illustration, social mommy blog data are subject to an ontology aware analysis process, with the call of NLP and data mining techniques. Following that, a process of semantic annotations generation takes place, within a multidimensional approach that relies on prerequisite data reduction methods and ontology restrictions based on axioms and expressions' complexity levels. As a result, post-tagging ontology population is achieved. Tag suggestions facilities occur based on ontology experts concepts' annotations and axioms' natural language generation, leading to possible user certification and validation.

The populated ontology is subsequently accessible for the different reasoning engine techniques, including classification and subsumption, satisfiability and instance checking, inference discovery and query answering, rule validation and processing. These techniques are henceforth the means by which decision support systems capabilities are attainable.

The different parts of the process rely on the backbone repository. This repository encloses the domain ontologies as well as the metamodel encompassing "meta-semantics" structures to allow the recognition of sublanguages along with suitable reasoning services. Meta-semantics also play a role in the verbalization process of the different constructs prior to the user tagging phase, which additional details are explored next.

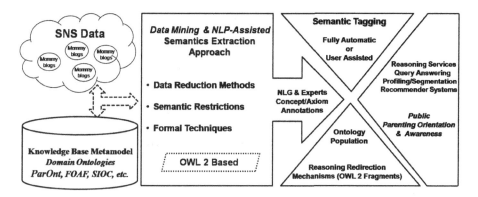

Fig. 1. Knowledge-based Architecture and Modeling Platform General Overview

The primary formal definition of a tag was first provided in [7]. The definition was based on a tripartite model relying on an actor (a user), a concept (a tag or keyword), and an instance (annotated resource):

$$T \subseteq A \times C \times I \tag{1}$$

where A is for Actor, C for Concept and I for Instance.

Later on in [1], the tripartite model definition was extended to a quadripartite one, after adding a local semantic meaning to each tag, obtained by a URI:

$$Tagging \subseteq User \times Resource \times Tag \times Meaning \tag{2}$$

In this work, we further extend the above definition, by assigning a more granular element to the definition, using Description Logics (DL) to denote constructs, axioms and expressions.

Our definition can thus be denoted by the following:

$$Tagging \subseteq User \times Resource \times Tag \times DL_{SEMANTICS} \tag{3}$$

where $DL_{SEMANTICS}$ are OWL 2 Constructs, Axioms and Expressions, in other words DL Building blocks forming OWL 2 fragments and languages, such as ALC and $SHOIN(D)$ [4,3].

The framework thus promotes a syntactic and more formalized approach, benefiting from its metamodel repository's already briefly introduced meta-semantics structures that allow the distinguishing of constructs such as existential restrictions, class conjunctions, disjunctions and negations, cardinality restrictions, ranges and datatypes, nominals, role properties (inverse, transitive, hierarchical, and so on).

The motivation behind attaining this level of granularity is to overcome data mining and NLP limitations by reusing ontology definitions and rules tagging, enforced by a possible user cooperation, thus cutting down complex algorithms and compensating for them through reasoning.

Being aware of the fact that non-ontology experts will surely face difficulties trying to read and understand ontology elements, natural language definitions of classes and axioms are made available based on ontology and metamodel prepared annotations on one hand, and on NLG-based techniques (such as OntoVerbal[1]) on the other. The SN user is thus faced with verbalised naturalistic versions of the formal semantics.

3   Proof of Concept, Experimental Application Scenarios

The first step leading to the platform's realization is the design of its metamodel, with the meta-semantics structures, along with all inferred patterns and configurations. The main ontology conceived for the purpose of parents' orientation and awareness is OWL 2's ParOnt (Parenting Ontology), an upgraded integrated version of COPE(Childhood Obesity Prevention [Knowledge] Enterprise) [2,6], with source information derived from trustful data sources such as RAMQ[2], Canadian Community Health Survey (CCHS)[3] (population health database), CARTaGENE[4].

ParOnt's major subdomains include baby development, behavior, nutrition, health, safety, activities, etc.

Useful Social Network Sites (SNS) data sources typically beneficial for our parenting domain are "Mommy Bloggers", such as Babycenter[5] (which alone counts more than 20 million users), Canada Moms Blog[6], Raising Children Network[7], Asian Mommy[8], among others. The study reported in this paper is based on 2000 blogs and replies collected from these SNS, falling within the same period interval (between 2011 and 2012). A group of almost 60 taggers was in charge of regularly annotating textual blogs with DL semantics, submitted to them in the form of short true false questionnaires.

While the full strategy to process blog data (based on semantics for domain and rule dedicated search on one hand, and filtering criteria for data reduction on the other [5]) are not the focus of this paper, a flat straightforward workflow was adopted to provide a minimum level of support to show the preliminary advantage of a semi-automatic ontology population process.

Figure 2 provides sample ontology axioms along with their verbalized definitions, as well as a comparative graphical illustration denoting, based on OWL DL ontology subsets and sample filtered blog extracts, the advantage of a semi-automatic user assisted population process. Users are offered a facility through which the enrichment of the ontology with instances goes beyond automatic

---

[1] http://swatproject.org/demos.asp
[2] Régie de l'assurance maladie du Québec: www.ramq.gouv.qc.ca/index_en.shtml
[3] www.statcan.gc.ca/concepts/health-sante/index-eng.htm
[4] www.cartagene.qc.ca/index.php?lang=english
[5] www.babycenter.com
[6] www.canadamomsblog.com
[7] www.raisingchildren.net.au
[8] www.asianmommy.com

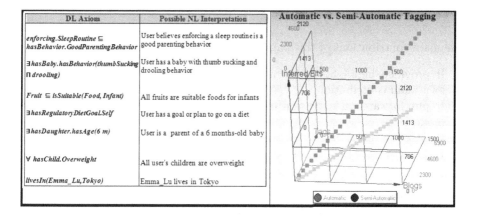

DL Axiom	Possible NL Interpretation	Automatic vs. Semi-Automatic Tagging
enforcing.SleepRoutine ⊑ hasBehavior.GoodParentingBehavior	User believes enforcing a sleep routine is a good parenting behavior	
∃hasBaby.hasBehavior(thumbSucking ⊓ drooling)	User has a baby with thumb sucking and drooling behavior	
Fruit ⊑ IsSuitable(Food, Infant)	All fruits are suitable foods for infants	
∃hasRegulatoryDietGoal.Self	User has a goal or plan to go on a diet	
∃hasDaughter.hasAge(6 m)	User is a parent of a 6 months-old baby	
∀ hasChild.Overweight	All user's children are overweight	
livesIn(Emma_Lu,Tokyo)	Emma_Lu lives in Tokyo	

Fig. 2. To the left, examples of DL constructs, along with their possible Natural Language interpretation; to the right, a comparative graph highlighting the difference between automatic and semi-automatic approaches for tagging/ ontology population. Note that the more we add complexity, the more NLP and mining techniques become limited, and the more the user tagging role is highlighted and rendered crucial.

NLP and data mining capabilities. This naturally increases the number of inferred elements, which represent query answering, inference discovery, instance checking and association to additional ontology classes (taking into consideration the multidimensional nature of an ontology). It is worth noting that the estimations provided (for inferred elements) are not static or fixed numbers. The idea is to prove that for a given number of analyzed blogs, the number of tags easily doubles between an automatic and a semi-automatic approach, proportionally with the number of inferred new knowledge, of services, etc. Furthermore, the more the level of expressivity and complexity increases, the steeper the slope will be, as extra automatic tagging restrictions are reinforced: with more complex semantics, accuracy levels are aggravated; as a consequence, the positive input brought in by the user's tagging cooperation is exponentially valued.

Once this infrastructure is achieved, reasoning procedures can be applied in order to attain the required services for our parenting awareness and orientation systems. Redirection mechanisms, based on the projected languages and fragments, target advanced and powerful reasoners, query and rule engines.

Useful retrieved information, mostly based on conjunctive queries, bring in recommender services, profiling, segmentation, opinion mining capabilities; more concrete examples include: the retrieval of the list of frequent undesired behaviors, of health problems for children per age in certain locations, of sets of best practice recommendations for parenting in general or in specific situations, of detected bad parents' behaviors, misconceptions, and so on.

4   Conclusions

The main contributions of this paper consist in the following:

- It provided highlights on an already proposed framework surrounding its own main contribution
- It put forward an extension of the tagging semantic model quadripartite model
- It introduced ParOnt, a newly conceived Parenting Ontology
- It demonstrated the overall advantages and services of getting SN users more involved in a formalized tagging process, all falling under the scope of public parenting orientation and awareness

In terms of future work, we plan to further develop implementation and verification tools, looking for the incorporation of maximized sets of rules and Description Logics-based languages.

References

1. Passant, A., Laublet, P.: Meaning of a Tag: A collaborative approach to bridge the gap between tagging and Linked Data. In: Proceedings of the WWW 2008 Workshop Linked Data on the Web, LDOW 2008 (April 2008)
2. Shaban-Nejad, A., Buckeridge, D.L., Dubé, L.: COPE: Childhood Obesity Prevention [Knowledge] Enterprise. In: Peleg, M., Lavrač, N., Combi, C. (eds.) AIME 2011. LNCS, vol. 6747, pp. 225–229. Springer, Heidelberg (2011)
3. Cuenca Grau, B., Horrocks, I., Motik, B., Parsia, B., Patel-Schneider, P., Sattler, U.: OWL 2: The next step for OWL. Journal of Web Semantics: Science, Services and Agents on the World Wide Web 6(4), 309–322 (2008)
4. Baader, F., Calvanese, D., McGuinness, D., Nardi, D., Patel-Schneider, P. (eds.): The Description Logic Handbook: Theory, Implementations and Applications. Cambridge University Press (2007)
5. El-Hachem, J., Haarslev, V.: A User and NLP-Assisted Strategic Workflow for a Social Semantic OWL 2-Based Knowledge Platform. In: Semantic Analysis in Social Media, at EACL 2012 (April 2012)
6. El-Hachem, J., Shaban-Nejad, A., Haarslev, V., Dube, L., Buckeridge, D.: An OWL 2-Based Knowledge Platform Combining the Social and Semantic Webs for an Ambient Childhood Obesity Prevention System. Procedia Computer Science 10, 110–119 (2012)
7. Mika, P.: Ontologies are us: A unified model of social networks and semantics. Journal of Web Semantics 5(1), 5–15 (2007)

Applying Semantic Technologies to Public Sector: A Case Study in Fraud Detection

Bo Hu, Nuno Carvalho, Loredana Laera, Vivian Lee,
Takahide Matsutsuka, Roger Menday, and Aisha Naseer

Fujitsu Laboratories Europe, Ltd.
bo.hu@uk.fujitsu.com

Abstract. Fraudulent claims cost both the public and private sectors an enormous amount of money each year. The existence of data silos is considered one of the main barriers to cross-region, cross-department, and cross-domain data analysis that can detect abnormalities not easily seen when focusing on single data sources. An evident advantage of leveraging Linked Data and semantic technologies is the smooth integration of distributed data sets. This paper reports a proof-of-concept study in the benefit fraud detection area. We believe that the design considerations, study outcomes, and learnt lessons can help making decisions of how one should adopt semantic technologies in similar contexts.

1 Introduction

Fraudulent claims account for a significant portion of claims received by private finance providers as well as public welfare authorities. It is estimated that fraudulent benefit and tax credit claims cost the UK government at least £1.6 Billion in 2011 and is expected to increase significantly in 2012 [3]. Challenges of similar magnitude are observed in many other industrialised countries [7,8]. Thus far, benefit fraud detection is largely performed within individual councils and is labour intensive. Though some ICT support is in place to relieve civil servants of manual data comparison and anomaly detection, given the scale and complexity of the problem, such support is very limited. Our studies highlighted the following barriers and inefficiencies in the current best practice:

1. Semantic discrepancy: even within one local council, semantic mismatches are evident across different governmental data repositories which are normally created and curated by different departments and/or outsourced to different ICT vendors each only has a fragmented view of the domain of discourse. Syntactical heterogeneity also widely presents across independently modelled data schemata.
2. Silo'ed information sources: regardless the efforts made in e-governance, progress in integrating public sector data is rather discouraging. Data warehousing is highly appraised but not widely practiced due to security and privacy as well as ownership concerns.

H. Takeda et al. (Eds.): JIST 2012, LNCS 7774, pp. 319–325, 2013.

3. Model rigidity: with the rapid pace in the change of fraudulent minds, extensibility is always a challenge. This calls for flexibility in data schemata, anomaly patterns, and data sources. Newly confirmed fraudulent cases should be analysed, whose patterns should be easily incorporated.
4. Immaturity of well defined business intelligence model: thus far, most of the tools are not designed and developed for social benefit fraud detection domain. They are generic OLAP (On Line Analytical Processing) tools [9] and business intelligence tools lacking intuitive features for the specific domain.

The relational database has recognised limitations as a solution basis for scenarios where data is highly distributed, sizable and where model structures are evolving and de-centralised. New paradigms in data management, collected under the label "Big Data", offer alternative solutions able to process increasing amounts of available data. For fraud detection the challenge is efficiently pinpointing small anomalies in Big Data. This is often based on patterns of relationships between data. Thus, here we report on a case study in Public Sector—applying connected Big data and semantic web technologies to make it easier to detect fraud. The main motivation of our work is not to develop top-notch fraud detection algorithms or tools, but to demonstrate how semantic technologies and linked data can solve a typical analytical task in real-life settings.

2   The Fraud Detection Case

In the past a few years, fraudulent benefit claims have attracted more attention due to both political and technical reasons. On the one hand, the austerity measures exercised in many developed countries emphasise more on cutting unnecessary public spending, leading to rigid benefit claim evaluation and review. On the other hand, the US and UK government's initiative to publish government data online [5] has fuelled better utilisation and transparency of data. Theoretically, cross-sector and cross-council references are made possible. The positive aspect of this movement is that by integrating multiple data sources, new fraud patterns can be discovered. The down side is equally evident: it obsoletes current ICT supports and calls for "smart" solutions.

Essentially, benefit fraud detection is a semantic alignment and pattern matching problem. Our studies have revealed several common heuristics that civil servants practice to identify suspicious claims in identify theft, housing benefit fraud, and council tax exemption fraud. Some exemplar clues are "one mobile number registered for housing benefit claim in multiple councils" and "the distance between home address and school address of the claimant's children is more than a predefined threshold". Normally, there is also a blacklist of known perpetrators of known fraudulent claims. Upon receiving a new claim, the typical workflow is: matching the claimant against those in the blacklist, pooling data from multiple repositories (e.g. DVLA for vehicle registration, EduBase for school addresses and name disambiguation, and Land Registry for property addresses/coordinates), aligning the claimant with all those from approved claims, identifying and highlighting potential fraud claims, and investigating candidate claims with high confidence. It is evident that semantic technology can play an important role in such a process. For instance, comparing attributes of new claimants with those from the blacklist can gain better performance if enhanced with semantics. Reconciling new and existing claims is tantamount to semantic alignment and integration across multiple data sources. SPARQL queries and rules help to increase the coverage through semantic query rewriting and transformation. Detected anomalies can be explicated with semantic equivalency relationships (e.g. ⟨owl : sameAs⟩ or ⟨skos : relatedMatch⟩). In a nutshell, by leveraging semantic technology, organisations are able to dynamically describe new fraud cases and facilitate the integration, analysis, and visualisation of disparate and heterogeneous data from multiple sources. Also by using the semantic technology and hence generating *semantic* fraud detection rules, we offer to convert labour intensive tasks into (semi-)automated processes.

3 Semantic Technology "Inside"

The real challenge of this pilot study is to find the balance between new and existing technology to achieve the best cost-effective outcomes, avoiding putting more pressure on the over-stretched public budget. For the social benefit fraud detection use case, we have extracted the following design requirements and steered the system design accordingly:

1. Graph is the fundamental data structure. It allows utilising graph analysis algorithms to discover new fraud patterns and rules. This is not readily available in conventional tabular data structures.
2. Extensibility manifests itself at different dimensions. Data repositories from different councils might be warehoused when deemed useful; different data schemata might be aligned; a particular column (of current tabulate data structure) might be selected or deselected; detection rules might be added or updated. It is highly likely that all these changes are made on the fly when users are experimenting with different scenarios.
3. Data models from different councils only present simple semantic and syntactical discrepancies. Lightweight and supervised reconciliation methods based on string comparison algorithms are enough for model alignment. Structure-based methods are excluded due to its complexity, which might lead to significant delay when large amount of claims arrive simultaneously.
4. There are two types of inputs: the incoming flow of new claims and existing claims. The underlying storage needs to accommodate both efficiently.
5. Instead of incorporating many (potentially useful) data sources, we exercise an on-demand approach to maintain the balance between overhead and benefit: only use the most necessary pubic data repositories to minimise the cost of data cleansing and conversion. Due to this consideration, many "fancier" data sets (e.g. London Gazette, http://www.london-gazette.co.uk/ and news stream) are opted out.

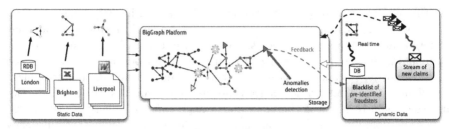

Fig. 1. System architecture

The architecture of the pilot system is illustrated in Figure 1. Our proposal combines common practice in the community with unique domain specific characteristics. We duplicate all the data in a graph data storage. This is to observe the integrity and ownership of the original data sources. Following the example of other systems targeting similar use cases, we offer an interface to triplify data from existing relational databases. This is particularly desirable and useful to query data from third-party RDBs and file-based systems. Data is triplified against a lightweight ontology manually crafted based on existing models from different councils. Many open source tools are available for RDF to RDB format conversion (e.g. Google Refine with RDF extension). New claims continuously arrive at the system. We exercise an eventually consistent approach when updating the original data sources: new claims are processed by the fraud detector first; they are then periodically delivered to the external storage as batch jobs.

The triplified data for benefit fraud detection can be potentially very large. They are stored in a distributed graph storage. We pipelined Jena Graph with a distributed Key-value Store (KVS) to enjoy the benefits from both sides: scalability and transaction-oriented data operation offered by KVS and graph operations and RDF-specific reasoning through Jena. Jena exposes the claim data as semantic graphs, allowing us to easily navigate through the data space by traversing along property and instantiation references.

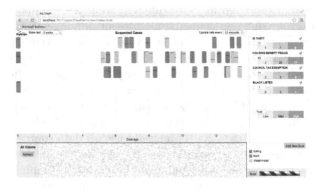

Fig. 2. Fraud Detection System

Fraud detection is facilitated with rule-based inferences. Known fraud patterns are solicited from domain experts and coded as SPARQL rules. When matching rule patterns against the semantic data graph, we employ multiple pattern relaxation methods to maximise the recall. For instance, semantic alignment is performed to avoid ruling out potential matches; third-party ontologies (e.g. GeoName ontology) are consulted for terminology knowledge. Figure 2 presents the everyday web-based user interface for council workers.

4   Discussions

In this pilot study, we have encountered many issues commonly known to the community (*c.f.* [2] and [6]). There are also very specific issues that have not been raised previously. We believe these issues worth further discussions.

Dashboard data visualisation. Data visualisation reveals the stories hidden behind data sets for sense-making and for easy communication. In the context of LOD, graph is the most intuitive visualisation that highlights and in many cases allows easy navigation through the relationships among data entities [4]. Graphs, however, are less desirable when the data size grows. Even though visual tricks can be employed to maximise the encoded information, there is a limit that our eyes can discern and our brain can understand [10]. On the other hand, for users of the fraud detection system, dashboards and tabular views are the day-to-day norm. They seldom trudge through the data space unless a suspicious case requires further scrutiny. We tried to offer flexibility by combining dashboard panel with navigation restricted to manageable subgraphs. The dashboard presents a concise and accurate summary of the data to show trends/statistics in individual councils, highlights of most recent fraudulent candidates, summary of different types of offends, fraud-specific key risk indicators, etc. High risk claims will trigger visual alert on the GUI. Users are then able to drill into the subgraph containing the suspicious claim together with neighbouring graph resources contributing to identifying the claim. Comparing to the entire graph, this subgraph is much smaller in size and easier to navigate. It is possible to enrich dashboard with other popular LOD visualisation methods, e.g. *map overlay.*

OLTP versus OLAP. In semantic applications the differences between OLTP and OLAP are not well presented and are not as well defined as in the database community. This (possibly) intentional blurring makes it difficult to effectively communicate with people not well exposed to semantic technologies. In fraud detection, OLTP is seen when traversing through the data sets (e.g. inserting and updating a new claim). The OLTP characteristics become more evident if we extend the use case to general fraud detection where credit card transactions need to be processed in a few milliseconds. OLAP is demonstrated through periodical discovery and review of fraud patterns. We try to accommodate the requirements of OLTP by combining NoSQL KVS and RDF semantic models. OLAP is supported through range queries over triple prefixes. There is, however, a lack of OLTP/OLAP-specific benchmarking on semantic data stores. It is our contentions that a community-wide effort on this direction will allow us better aligning with industrial practice and put forward a more convincing argument underpinned by numeric figures.

Focus small to aim big. Instead of draining the limited monetary and human resources with large scale projects, we focus on a very specific problem from the public sector domain. We further restrict ourselves to a specific aspect of the problem, i.e. reconciling new claims to emerge potential frauds. Comparing to the so-called "killer" applications, we are not aiming to fundamentally revamp the practice, but improve productivity with minimum disturbance and manageable changes. Thus far, this moderate approach has been effective in gaining the initial trust and gradually enlarge the project scope.

5   Conclusion

Recently, public sectors' interests in LOD and other semantic technologies are fuelled by national and international directives on sharing and reusing governmental data (*c.f.* [1]). Linking multiple data sets has demonstrated significant merits of revealing patterns and correlations that are otherwise locked in data silos. In this paper, we report a case study of applying semantic technologies to social benefit fraud detection. We present the unique requirement from the domain and corresponding design considerations and challenges during the execution of the project.

This paper aim to demonstrate that semantic technology could be the means for public institutions to best manage the vast amounts and variety of data. It provides at least an alternative to harnessing the true value of data in a fraud detection case where speed in analytics and insight is increasingly critical. Semantic technology can help to reduce the level of fraud and to tackle the problem of wasteful expenditure in the public sector.

References

1. The re-use of public sector information regulations (2005), http://www.legislation.gov.uk/uksi/2005/1515/pdfs/uksi_20051515_en.pdf
2. Alani, H., Dupplaw, D., Sheridan, J., O'Hara, K., Darlington, J., Shadbolt, N., Tullo, C.: Unlocking the potential of public sector information with semantic web technology. In: Aberer, K., Choi, K.-S., Noy, N., Allemang, D., Lee, K.-I., Nixon, L.J.B., Golbeck, J., Mika, P., Maynard, D., Mizoguchi, R., Schreiber, G., Cudré-Mauroux, P. (eds.) ISWC/ASWC 2007. LNCS, vol. 4825, pp. 708–721. Springer, Heidelberg (2007)
3. National Fraud Authority: Annual fraud indicator (March 2012)
4. Dadzie, A.S., Rowe, M.: Approaches to visualising linked data: A survey. Semantic Web 2(2), 89–124 (2011)
5. DataGov: The principles of open public data (June 2010), http://data.gov.uk/blog/new-public-sector-transparency-board-and-public-datatransparency-principles
6. Hu, B., Svensson, G.: A case study of linked enterprise data. In: Patel-Schneider, P.F., Pan, Y., Hitzler, P., Mika, P., Zhang, L., Pan, J.Z., Horrocks, I., Glimm, B. (eds.) ISWC 2010, Part II. LNCS, vol. 6497, pp. 129–144. Springer, Heidelberg (2010)

7. Mie, A.: Government to crack down on welfare fraud as payouts balloon (June 2012), http://www.japantimes.co.jp/text/nn20120606a6.html
8. Prenzler, T.: Welfare fraud in Australia: Dimensions and issues. In: Trends & Issues in Crime and Criminal Justice. Australian Institute of Criminology
9. Thomsen, E.: OLAP Solutions: Building Multidimensional Information Systems. John Wiley & Sons (1997)
10. Ware, C.: Information Visualization: Perception for Design (Interactive Technologies), 1st edn. Morgan Kaufmann (February 2000)

Location-Based Concept in Activity Log Ontology for Activity Recognition in Smart Home Domain

Konlakorn Wongpatikaseree[1], Mitsuru Ikeda[2], Marut Buranarach[3],
Thepchai Supnithi[3], Azman Osman Lim[1], and Yasuo Tan[1]

[1] School of Information Science, JAIST, Ishikawa, Japan
[2] School of Knowledge Science, JAIST, Ishikawa, Japan
{w.konlakorn,ikeda,aolim,ytan}@jaist.ac.jp
[3] Language and Semantic Technology Laboratory, NECTEC, Pathumthani, Thailand
{marut.buranarach,thepchai.supnithi}@nectec.or.th

Abstract. Activity recognition plays an important role in several researches. Nevertheless, the existing researches suffer various kinds of problems when human has a different lifestyle. To address these shortcomings, this paper proposes the activity log in the context-aware infrastructure ontology in order to interlink the history user's context and current user's context. In this approach, the location-based concept is built into the activity log for producing the description logic (DL) rules. The relationship between activities in the same location is investigated for making the result of activity recognition more accurately. We also conduct the semantic ontology search (SOS) system for evaluating the effectiveness of our proposed ideas. The semantic data can be retrieved through SOS system, including, human activity and activity of daily living (ADL). The results from SOS system showed the advantage overcome the existing system when uses the location-based concept in activity log ontology.

Keywords: Activity recognition, Activity log, Location-based concept, Description logic rules, The semantic ontology search, Activity of daily living.

1 Introduction

Nowadays, various kinds of healthcare systems have been proposed because people taking an interest in health is more increase continually. Numerous techniques are developed under Smart Home (SH) concept [1] to support the healthcare system more precisely. Activity recognition is a one of sub-system in the healthcare system. The results from the activity recognition are extremely useful because they can be used in many directions. For example, activity of daily living (ADL) is necessary information to know what human does in each day. Physician can use the ADL information to diagnose patient correctly.

Recently, the ontology concept has been developed in several activity recognition researches. The knowledge engineering is used to describe a semantic of

H. Takeda et al. (Eds.): JIST 2012, LNCS 7774, pp. 326–331, 2013.

context-aware infrastructure in the SH domain [2, 3]. Even though, the existing systems can recognize the human activity, they still need to improve recognition's ability in some ambiguous cases [4]. The main problem of the activity ambiguous problem is an input data. Mostly, snapshot data from the sensors is used as the input data for the activity recognition system. An interval time for receiving data from sensors is set depending on experimental environment. The system can identify the activity only when the interval time comes. However, the concept of snapshot information may not suitable for the activity recognition system. Since, It can be a lack of information for recognition in some cases.

According to the former problem, primary goal of this research is to classify the human activity more accurately. 13 activities of daily living are emphasized such as "Sleeping", "Cooking", "Watching TV", or "Eating or Drinking". There are three goals as following. First, we propose the activity log in the context-aware infrastructure ontology in the SH domain. The semantic of user's context can be obtained through this ontology. The activity log can also improve the ability of activity recognition. The reusability of knowledge is used to improve the generation for the new knowledge. Therefore, the relationship between activities is considered and leads to the hypothesis of this research, which is "activities in a specific location relate each other". Second, the location-based concept is developed in the activity log and DL rules. This concept makes the system knows what the activities were done by the user at the current user's location. It makes different result with the existing research [5]. For instance, the user does not need to do the "Drinking" activity after "Preparing drinking" activity immediately. It is possible that the user may do another activity before. We cannot use the recent activity for recognition in some cases. Finally, we demonstrate the advantage of proposed ideas through the semantic ontology search (SOS) system. We also present the ADL information, which gathers the activities in one day.

2   Proposed Ideas

2.1   Context-Aware Infrastructure Ontology with Activity Log

Ontology model uses the knowledge engineering to define a semantic of context information, to explicit and formalize specification of a shared conceptualization. At present time, the ontology is adopted to various areas of researches, including the activity recognition. The context-aware information in SH domain has been introduced in several ontology models. [6] proposed a new user's context, which added the human posture in the context-aware information to reduce the ambiguous case problem in the activity recognition system. Although the benefit of user's context is exhibited, it still needs to improve the relationship between semantic data. In this paper, we introduce the activity log to analyze the relationship between the current activity and last activity. The connection between current activated object and last activated object is also described for solving the "activity pattern ordering" problem.

Hozo application is used to build the ontology in this research. Fig. 1 illustrates the context-aware infrastructure ontology with the activity log. Based on

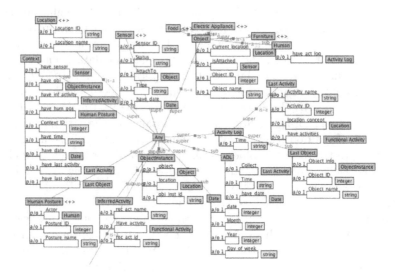

Fig. 1. Context-aware infrastructure ontology with activity log

our ontology modeling is centered on *Context* class, which relates to location information and surrounding entities such as sensor, object, and human posture. The data from *Context* class is operated as an input data because it can link to all of the semantic data. *Activity log* class, moreover, is proposed to aggregate a sequence of history of activities and object activations. It is necessary for collecting the log data because it is difficult to guarantee the accuracy when we used only data at a specific time point. To reach the benefit of the activity log, the location-based concept is presented in the activity log to make the results of activity recognition system more accurately. *ADL* class is also presented to express what activities human does in each day.

2.2   Location-based Concept in Activity Log and DL Rules

The location information in the *Activity log* class can be referred through two properties: "location_concept" and "Object_info". In the *last activity* class, the "location_concept" property refers to the activities, which user performed in the current user's location. For example, if the user is doing some activities in the kitchen, the activities that user performed in the kitchen will be considered in the activity log for classifying the current activity. The relationship between activities in the same place is investigated. For example, "Wash dishes" activity will happen if and only if the "Cooking" and "Eating and Drinking" activities occurred before. Hence, we can ignore the activities, which are performed out of the current user's location. As described from the ontology in Fig 1, we can construct various kinds of semantic data from *last object* class, including the location information. It is reasonable to use the "Object_info" property to monitor the object, which has dynamic location. For example, user can use the "Broom" object to sweep the floor in any rooms. In addition, the "last object" can figure out the lack of data problem because the snapshot data at the specific time point maybe not enough for recognition. For example, the system perceives that the "Cupboard" object is being used. It cannot refer to any activities. However, if we know the object used by user before such as refrigerator, the system may recognize the current activity as "Making a cold drink".

The idea to create the DL rules is available at this stage. The existing researches have been proposed the basic DL rules for recognition the current human activity, described in the introduction section. Normally, the ontology concept does not support the temporal reasoning, so it is not an easy task to analyze the relationship between the recent activity and the current activity. In this paper, the external program is implemented to capture the temporal reasoning. The DL rules are established based on the location-based concept and used the external program to link with the temporal reasoning. The example below indicates the DL rule for "Wash dishes" activity. The example rule used the relationship between activities in the same location through "LastActivity.Kitchen Activity" property.

Wash dishes $\sqsubseteq$ *Functional Activity*
$\sqcap$ *use(Object.Furniture(Sink))*
$\sqcap$ *Object.Human.Current_location(kitchen)*
$\sqcap$ *HumanPosture(Stand)*
$\sqcap$ *LastActivity.Kitchen Activity(Eating or drinking)*

3   Experiments and Results

In order to validate our approach, we have implemented the SOS system, with NECTECs ontology search framework [7], to recognize the human activity in the SH domain. For setting up the input data, the context-aware information in one day is fed into the system. We have also developed the external Java application to capture the temporal reasoning.

Figure 2 shows example results of the SOS system on 18th Aug 2012. We obtain good results when we use the reusability knowledge to promote the new knowledge. Context_id 29 is an outstanding example to explain the strong point of our proposed ideas. In this section, we will describe the usefulness of proposed ideas in two directions.

First, we compare between the systems that use activity log and does not use activity log. In general, "Sink" object in the kitchen can be activated in several purposes such as "Wash hands" activity, or "Wash dishes" activity. Thus, the

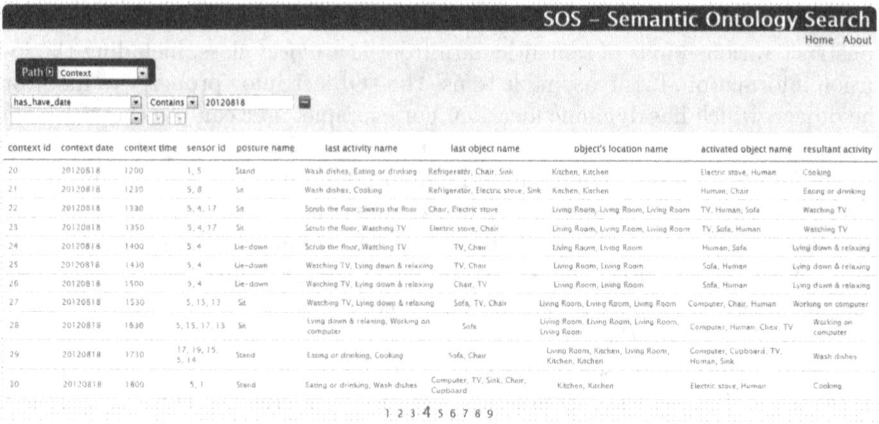

SOS – Semantic Ontology Search

Home About

Fig. 2. Result of SOS system at 18th August 2012

system, which does not have the activity log, cannot get any supported ideas to ensure the resultant activity. On the other hand, the system with activity log considers the relationship between activities to achieve the correct resultant activity. For example, the "Eating or drinking" activity follows the "Cooking" activity. "Wash dishes" activity comes after "Eating or drinking" or "Cooking" activity. Furthermore, the system, which uses the activity log, also solves the "activity pattern ordering" problem. Since, each human has own lifestyle to complete the task, even the same task. The aggregated data from the history of activated objects can make the reasonable information, although the user has different ordering. Nonetheless, the sequence of activities may not be always arranged in the same order. It is depended on the human behavior.

Second, we compare the performance of two methods. One of them uses only the activity log to classify the activity, whereas another one method applies both the location-based concept and the activity log. The challenging point of the activity log is that it is difficult to keep track of activities, which one should consider with current activity, in some cases. The experimental results in Fig. 2 indicate the advantageous of the location-based concept in the activity log. If we consider in a column "last activity name" in context_id 28 carefully, the last activities are "Working on computer" and "Lying down & relaxing" activities. It is difficult to connect the relationship between these two activities and the user's context in context_id 29. Thus, the system uses the location-based concept and retrieves the last activities that user performed in the current location (context_id 20 = "Cooking" and context_id 21 = "Eating or drinking") to classify the resultant activity of this context. Truly, the resultant activity in context_id 29 can be other activities, but the "Wash dishes" is a high probability that the user may be performing because we have the supported conditions as "Cooking" and "Eating or drinking" to verify this result.

4   Conclusion and Future Works

We have proposed the technique to improve the ability of activity recognition in the SH domain. In this paper, we presented the activity log in context-aware infrastructure ontology to interconnect between history user's context and current user's context. The result shows the advantage overcome the existing systems when we add the activity log into the ontology. Furthermore, we also enhanced the activity log by including the location-based concept. The activities, which performed in the same location, are considered to decide the current activity. The experimental results also indicate how the importance of location-based concept is. With location information, system will scan the recent activities only in the interesting area. It makes the system classify the human activity more reasonable. In this paper, the SOS system has also introduced to evaluate the proposed ideas. In this system, various kinds of semantic data can be searched through the proposed ontology. The ADL information is recorded and formed in term of last activity – separated each ADL by date as shown in Fig. 1. Hence, we can retrieve the ADL information from SOS system for further processing.

The future directions of this research can be developed in many ways. For example, it can be established more specific rules to cover more activities in the SH domain or analyzed the human activity in each day, ADL, in order to perceive the human behavior. Then, analyze the human behavior to provide various kinds of services such as healthcare service, household appliances service, or entertainment service.

References

1. Chan, M., Estve, D., Escriba, C., Campo, E.: A review of smart homes-Present state and future challenges. Journal of Computer Methods and Programs in Biomedicine 91, 55–81 (2008)
2. Chen, L., Okeyo, G., Wang, H., Sterritt, R., Nugent, C.: A Systematic Approach to Adaptive Activity Modeling and Discovery in Smart Homes. In: 4th International Conference on Biomedical Engineering and Informatics, Shanghai, pp. 2192–2196 (2011)
3. Riboni, D., Bettini, C.: Context-aware activity recognition through a combination of ontological and statistical reasoning. In: Zhang, D., Portmann, M., Tan, A.-H., Indulska, J. (eds.) UIC 2009. LNCS, vol. 5585, pp. 39–53. Springer, Heidelberg (2009)
4. Henricksen, K., Indulska, J.: Modelling and using imperfect context information. In: 2nd IEEE Annual Conference on Pervasive Computing and Communications Workshops, Washington, DC, pp. 33–37 (2004)
5. Riboni, D., Pareschi, L., Radaelli, L., Bettini, C.: Is Ontology-based Activity Recognition Really Effective? In: 8th IEEE Workshop on Context Modeling and Reasoning, Seattle, WA, pp. 427–431 (2011)
6. Wongpatikaseree, K., Ikeda, M., Buranarach, M., Supnithi, T., Lim, A.O., Tan, Y.: Activity Recognition using Context-Aware Infrastructure Ontology in Smart Home Domain. In: International Conference on Knowledge, Information and Creativity Support Systems, Melbourne, pp. 50–57 (2012)
7. National Electronics and Computer Technology Center, Language and Semantic Technology Laboratory, http://text.hlt.nectec.or.th/ontology/

Improving the Performance of the DL-Learner SPARQL Component for Semantic Web Applications

Didier Cherix, Sebastian Hellmann, and Jens Lehmann

Universität Leipzig, IFI/BIS/AKSW, D-04109 Leipzig, Germany
{cherix,hellmann,lehmann}@informatik.uni-leipzig.de
http://aksw.org

Abstract. The vision of the Semantic Web is to make use of semantic representations on the largest possible scale - the Web. Large knowledge bases such as DBpedia, OpenCyc, GovTrack are emerging and freely available as Linked Data and SPARQL endpoints. Exploring and analysing such knowledge bases is a significant hurdle for Semantic Web research and practice. As one possible direction for tackling this problem, we present an approach for obtaining complex class expressions from objects in knowledge bases by using Machine Learning techniques. We describe in detail how they leverage existing techniques to achieve scalability on large knowledge bases available as SPARQL endpoints or Linked Data. The algorithms are made available in the open source DL-Learner project and we present several real-life scenarios in which they can be used by Semantic Web applications. Because of the wide usage of the method in several well-known tools, we optimized and benchmarked the existing algorithms and show that we achieve an approximately 3-fold increase in speed, in addition to a more robust implementation.

1 Introduction

The vision of the Semantic Web aims to make use of semantic representations on the largest possible scale - the Web. Large knowledge bases such as DBpedia, OpenCyc, GovTrack are emerging and freely available as Linked Data and SPARQL endpoints. Due to their sheer size, however, users of large Semantic Web knowledge bases are often facing the problem, that they can hardly know which identifiers are used and are available for the construction of queries. Furthermore, domain experts might not be able to express their queries in a structured form at all, but they often have a very precise imagination what kind of results they would like to retrieve. A historian, for example, searching in DBpedia for ancient Greek law philosophers influenced by Plato can easily name some examples and if presented a selection of prospective results he will be able to quickly identify false results.

However, he might not be able to efficiently construct a formal query adhering to the large DBpedia knowledge base a priori. The construction of queries asking for objects of a certain kind contained in an ontology, such as in the previous

H. Takeda et al. (Eds.): JIST 2012, LNCS 7774, pp. 332–337, 2013.
© Springer-Verlag Berlin Heidelberg 2013

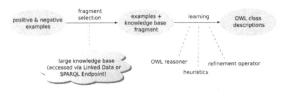

Fig. 1. Process illustration: In a first step, a fragment is selected based on instances from a knowledge source and in a second step the learning process is started on this fragment and the given examples. [1]

example, can be understood as a class construction problem: We are searching for a class expression which subsumes exactly those objects adhering to our informal query (e.g. ancient Greek law philosophers influenced by Plato). Recently, several methods have been proposed for constructing ontology classes by means of Machine Learning techniques from positive and negative examples [1]. These techniques are tailored for small and medium size knowledge bases, while they cannot be directly applied to large knowledge bases (such as the initially mentioned ones) due to their dependency on reasoning methods. The scalability of the algorithms is ensured by reasoning only over "interesting parts" of a knowledge base for a given task. As a result users of large knowledge bases are empowered to construct queries by iteratively providing positive and negative examples to be contained in the prospective result set.

In this paper, we will motivate the usefulness of the method first presented in [1] by describing the algorithm and how it is employed in several well-known applications(HANNE, ORE, DL-Learner, Tiger Corpus Navigator, etc.). Then we will introduce the changes we have made to improve performance of the data acquisition method. Although the basic problem was introduced several years ago[2], we will discuss lessons learned and actual problems in the last section.

2   Summary of the Existing SPARQL Component

In this paper, we focus on the knowledge source component, which is responsible for downloading a "relevant" fragment of the available knowledge base via SPARQL. At the beginning, the component receives a list of positive and negative examples needed for supervised machine learning. As the separation into positive and negative examples is irrelevant for data acquisition, a list of all (seed) examples is created by merging. We also assume that the given examples must be instances of an OWL class. Based on a given recursion depth parameter, the old component retrieved all information of the seed examples naively, by traversing the graph with SPARQL as shown in Figure 2 and without making any distinction between A- and T-Box (i.e. treating OWL axioms as triples) . Reasoning on the resulting RDF fragment is sound, because of the monotonicity of DL, but naturally incomplete, since we restricted the available knowledge. Nevertheless, Hellmann et. al. [1] have shown that results are comparable to learning with the complete data. Actually, better results can be achieved in the same time and

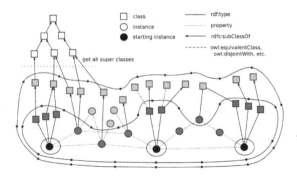

Fig. 2. Extraction with three starting example instances. The circles represent different recursion depths. The circles around the starting instances signify recursion depth 0. The larger inner circle represents the fragment with recursion depth 1 and the largest outer circle with recursion depth 2. [1]

with less memory, as unnecessary data is not loaded into the reasoner. Next, we briefly introduce the applications that use a SPARQL component.

HANNE (**H**olistic **A**pplication for **N**avigational K**N**owledge **E**ngineering) [3] enables users and domain experts to navigate through knowledge bases by selecting examples. From these examples, formal OWL class expressions are learned on the fly by the approach presented in this article. When saved by users, these class expressions form an expressive OWL ontology, which can be exploited in numerous ways: as navigation suggestions for users, as a hierarchy for browsing, and as input for a team of ontology editors.

The Tiger Corpus Navigator was the predecessor of HANNE and can profit from the improving [4].

Another interesting use case for the proposed fragment selection approach is the debugging and maintenance of large scale ontologies. For this reason, the fragmentation approach has been integrated in the **ORE (ontology repair and enrichment) tool**. ORE [5] uses the approach to scale to larger knowledge bases such as DBpedia and OpenCyc. ORE can detect inconsistencies and unsatisfiable classes within large knowledge bases by continuously loading fragments of increasing size. Furthermore, it can also learn new definitions of classes as described previously.

AutoSPARQL [6] is a question answering support system, which uses fragment extraction for an active learning algorithm. The algorithm allows to refine results obtained from a question answering system.

DL-Learner is a framework to learn concepts in description logics [2,7], the source code for the component is available through this project.

3   The New Version of the SPARQL Component

The three tasks of the SPARQL component that are illustrated in Figure 1 are now realized in four separate steps. 1.) The T-Box of the ontology is loaded and indexed (T-Box index). 2.) All outgoing properties of seed examples and related objects and literals are retrieved. 3.) All asserted classes are retrieved via SPARQL. 4.) The T-Box index is used to infer the class hierarchy.

Step 1: Indexing the T-Box. The previous component traversed the T-Box syntactically based on the retrieved triples during fragment extraction. Our improved method first retrieves all T-Box axioms from the SPARQL endpoint and can, optionally, also directly load an ontology file if available, which we did in out experiments. The ontology is loaded into a reasoner and queried once to materialize the transitive closure of the subclass hierarchy. This subclass hierarchy is then stored in an index, which allows retrieval of superclasses in constant time $O(1)$.

Step 2: A-Box queries. Given a certain recursion depth, the algorithm now traverses the A-Box part of the RDF graph based on the SPARQL template given in Listing 1.1. EX1,..,EXn are the seed examples. After each recursion step, EX1,...,EXn are replaced by the new objects (?o) which have not yet been queried. The SPARQL 1.1 "IN" feature is used.[1] As a last optional step, manual filtering is used to remove data irrelevant for a learning processs.

Listing 1.1. SPARQL query for the A-Box

```
1 CONSTRUCT {?s ?p ?o} { ?s ?p ?o }
2 #list of all seed examples
3 FILTER (?s IN (<EX1>, <EX2>,...,<EXn>))
4 FILTER (!(?p=rdf:type))
5 #other filters, here excluding one specific property and all properties
        of the foaf namespace
6 FILTER ( ?p!=<http://dbpedia.org/property/wikiPageUsesTemplate> $$ !
        regex(str(?p), '^http://xmlns.com/foaf/0.1/') && ...)
```

Step 3: Typing retrieved instances. Based on the A-Box data retrieved in Step 2, we are able to query all types of the found instances (again using the SPARQL "IN" construct) and include them in the fragment as well. To be able to correctly type all resources, we implemented the rules introduced by Bechhofer and Volz [8]. The SPARQL query is given here:

Listing 1.2. SPARQL query for the T-Box

```
1 CONSTRUCT { ?ex a ?class . } {?ex a ?class . }
2 #list of all objects
3 FILTER (?s IN (<EX1>, <EX2>,...,<EXn>))
4 #other filters, here excluding all classes of the yago namespace
5 FILTER ( !regex(str(?class), '^http://dbpedia.org/class/yago/')) . }
```

Step 4: T-Box Index. To complete the fragment, we iterate over all retrieved and included classes and query the T-Box index for all subclassof axioms and include them in the fragment.

4   Perfomance Evaluation

The new component is compared to the old one regarding two aspects: execution time and correctness. For this, we designed an experiment using DB-

---

[1] http://www.w3.org/TR/sparql11-query/#func-in

Table 1. Performance evaluation showing runtimes in ms.(rounded)

Iteration	new component			old component		
	Min	Total	Avg	Min	Total	Avg
1	205	1,642,678	6545	3359	5,999,248	23901
2	136	744,953	4756	2295	3,707,947	19337
3	132	984,861	3924	1703	2,956,095	16817
	158	1,124,165	5,075	2452	4,221,097	20019

pedia and DL-Learner. Specifically, we used DL-Learner to generate defini-
tions for the DBpedia classes. The experiment is a learning problem with
positive and negative examples. Figure 1 explains how DL-Learner is working. We created one learning task for each class in the DBpedia ontology[2].
Starting with the DBpedia T-Box description, we extracted all its classes.
For each of class, we retrieved all instances until a limit of 10000 is reached.
Then, we randomly choose 30 instances and use those as positive examples.
For the negatives examples we use one sister class. A sister class is a class
that has the same superclass as the one we are processing, but naturally not
the class itself. For each sister class we randomly picked 30 examples from its
instances. In the case where the class to learn or its sister have less than 30
instances, all instances are used as positive and negative examples, respectively.
As endpoint we use `http://live.dbpedia.org/sparql`. The recursion depth
for each learning problem (a DBpedia class is one learning problem) is set to 1.
We excluded the following properties from the fragment: `dbo:wikiPageUsesTemplate`,
`dbo:wikiPageExternalLink`, `dbo:wordnet_type`, `<http://www.w3.org/2002/07/owl#sameAs>`
and classes having those prefixes:
`http://dbpedia.org/class/yago/` or `http://dbpedia.org/resource/Category:`.

The experiments are repeated three times to reduce the effect of the network
and caching from SPARQL-endpoint. To measure the needed time we use the
JAMon framework[3]. We measure the execution time of the SPARQL-component
for each class to learn. After each experiment (when all classes are learned), we
compute the average time for this process. Results are shown in Table 1.

The experiments ran on a 64 bit 2.26 GHZ dual core processor and 8GB
RAM. For the old component we use the cache database. The cache has be
cleared between each class, because if a class is learned after his superclass to
ensure a fair experiment.

5   Discussion and Conclusion

Table 1 shows that the new implementation improves performance significantly.
The old component has a maximum query time of 171245 ms, the new one of
68800 ms. As explained above, a cache is not implemented (and possible not

---

[2] `http://downloads.dbpedia.org/3.6/dbpedia_3.6.owl`
[3] `http://jamonapi.sourceforge.net/`

necessary) in the new component. Still, the runtime improves by factor 4 on average.

In the distribution of the F-measure, we see that the new component has a little bit better results as the old one. One possible explanation for this is that the queries of the new component can be executed faster and, thus, the endpoint returns more results, which improves the approximations in the machine learning processes. Another explanation and probably the bigger effect is that DL-Learner with the new component returns an answer for much more classes compared to the previous component. With the new, DL-Learner found definitions for 248 classes, whereas it did only for 158 previously.

Reasons that for DL-Learner not returning definitions are Java exceptions, which are usually due to errors in the retrieved data. The new component fired 61 exceptions until the experiment. But only three of those caused an interruption of the algorithm and cause the algorithm to return no answer. All of those are `com.hp.hpl.jena.shared.BadUriExceptions` resulting from malformed output from the SPARQL endpoint. The others exceptions are all parse exceptions due to errors in the XML output of the endpoint. Those exceptions cause a loss of some retrieved triples when they occur, but the algorithm can still recover and terminate. The behavior is more stable than and able to recover from more types of errors than the previous component.

Overall, we presented an improved fragment extraction component for machine learning over large SPARQL endpoints, which is more efficient and more stable than previously published methods.

References

1. Hellmann, S., Lehmann, J., Auer, S.: Learning of OWL class descriptions on very large knowledge bases. IJSWIS 5(2), 25–48 (2009)
2. Lehmann, J.: DL-Learner: learning concepts in description logics. The Journal of Machine Learning Research 10, 2639–2642 (2009)
3. Hellmann, S., Unbehauen, J., Lehmann, J.: Hanne - a holistic application for navigational knowledge engineering. In: Posters and Demos of ISWC (2010)
4. Hellmann, S., Unbehauen, J., Chiarcos, C., Ngonga Ngomo, A.C.: The tiger corpus navigator. In: Proceedings of the Ninth International Workshop on Treebanks and Linguistic Theories, TLT9. NEALT Proceeding Series (2010)
5. Lehmann, J., Bühmann, L.: ORE - a tool for repairing and enriching knowledge bases. In: Patel-Schneider, P.F., Pan, Y., Hitzler, P., Mika, P., Zhang, L., Pan, J.Z., Horrocks, I., Glimm, B. (eds.) ISWC 2010, Part II. LNCS, vol. 6497, pp. 177–193. Springer, Heidelberg (2010)
6. Lehmann, J., Bühmann, L.: AutoSPARQL: Let users query your knowledge base. In: Antoniou, G., Grobelnik, M., Simperl, E., Parsia, B., Plexousakis, D., De Leenheer, P., Pan, J. (eds.) ESWC 2011, Part I. LNCS, vol. 6643, pp. 63–79. Springer, Heidelberg (2011)
7. Lehmann, J., Hitzler, P.: Concept learning in description logics using refinement operators. Machine Learning Journal 78(1-2), 203–250 (2010)
8. Bechhofer, S., Volz, R.: Patching syntax in OWL ontologies. In: McIlraith, S.A., Plexousakis, D., van Harmelen, F. (eds.) ISWC 2004. LNCS, vol. 3298, pp. 668–682. Springer, Heidelberg (2004)

DashSearch LD: Exploratory Search for Linked Data

Takayuki Goto[1], Hideaki Takeda[1,2], and Masahiro Hamasaki[3]

[1] National Institute of Informatics
tygoto@nii.ac.jp, takeda@nii.ac.jp
[2] The Graduate University for Advanced Studies
[3] National Institute of Advanced Industrial Science and Technology (AIST)
masahiro.hamasaki@aist.go.jp

Abstract. Although a large number of datasets gathered as Linked Open Data (LOD) is better for data sharing and re-using, the datasets themselves become more difficult to understand. Since each dataset has its own data structure, we need to understand datasets individually. In addition, since the entities in datasets are interconnected, we need to understand the interconnections between datasets. In other words, understanding the data is crucial for exploiting LOD. In this paper, we show a novel system called DashSearch LD to understand and use LOD with an exploratory search approach. The user interactively explores datasets by viewing and selecting entities in the datasets. Specifically, the user manipulates widgets on the screen by moving and overlapping them with a mouse to check entities, draw detail data on them, and obtain other entities linked by the widgets.

Keywords: Linked Open Data, SPARQL, Exploratory Search, Search Interface, Facet Search, Human-Computer Interaction.

1 Introduction

Linked Open Data (LOD) is now rapidly growing in size and variety. LOD has great potential to raise the value of data by sharing and re-using data that has been hidden behind websites.

Although the increasing number of datasets gathered as LOD is better for data sharing and re-using, the datasets themselves become more difficult to understand. Since each dataset has its own data structure, we need to understand datasets individually. In addition, since the entities in datasets are interconnected, we need to understand the interconnections between datasets. In other words, LOD as a whole consists of a huge and complicated database, and understanding the data is crucial for exploiting LOD.

Understanding data is twofold; one aspect is understanding the schema and their relationships. For example, users need to know what kinds of classes exist, how classes are represented, and how they are interconnected. The other aspect is understanding the data distribution. Among the defined classes in a dataset,

H. Takeda et al. (Eds.): JIST 2012, LNCS 7774, pp. 338–343, 2013.

some are heavily used, so many entities exist. Conversely, classes not heavily used have few entities. The same is true for properties: some properties always have values but others often lack values. Such skewed distributions are observed in real datasets.

In this paper, we show a novel system called DashSearch LD to understand and use LOD with an exploratory search approach, which can help users to understand data with respect to their structure and distribution. DashSearch LD is an LOD version of a widget-based desktop search system [2]. The user interactively explores datasets by viewing and selecting entities by manipulating widgets, e.g., moving and overlapping them, on the screen with a mouse, to check entities, draw detail data on them, and obtain other entities linked by the widgets. Through browsing and exploring data, the user can understand data *as is*, i.e., the distribution as well as the structure of the data.

2 Related Work

In LOD, although understanding both the distribution and the structure is important, existing studies mainly focus on one or the other. iSPARQL [6] and NITELIGHT [8] help users to compose SPARQL queries by providing a graph interface. A graph interface is suitable because it conforms to the nature of the Resource Description Framework (RDF) and is intuitive for users. These interfaces rely on pre-registered vocabularies, but many vocabularies can occur when traversing LOD graphs. So, it is difficult to apply graph interfaces to LOD . In addition, understanding the schema is required. DERI Pipes [7], which is a semantic version of Yahoo! Pipes[1], enables rapid development of mashup applications by providing a graphical user interface for flow diagrams. Instead of RSS, DERI Pipes can manipulate RDF triples and SPARQL queries. DERI Pipes is user friendly and convenient for composing applications, but understanding the schema is again required. MashQL [4], although similar to DERI Pipes, is a more sophisticated approach to construct SPARQL queries. In MashQL, the schema is detected by the system and specified graphically by the user. It is an advantage to detect the schema from the data, but its interpretation is still left to the users.

Understanding data is another issue. Deligiannidis et al. proposed a system called Paged Graph Visualization [1], which visualizes RDF triples so that the user can explore them by traversing graphs. This system helps the user to understand both the data and the data structure. However, showing all individual relations is often troublesome and prohibits the user from comprehensive understanding. gFacet [3] is the most relevant work of applying graphical user interface techniques to browse databases. Whereas gFacet enables tree-like exploration of a dataset, DashSearch LD emphasizes the flexibility of the exploration of datasets. DashSearch LD offers more flexible data operations, such as combining the search results from different datasets, clipping search results, as well as providing more extensive interfaces using widget metaphors.

---

[1] `http://pipes.yahoo.com/pipes/`

3   DashSearch Interface

DashSearch LD is a Flash web application (Fig. 1). DashSearch LD provides a
workspace and visual objects called widgets for operations of Linked Data. Widgets include an "endpoint widget" with the function of the SPARQL endpoint
and a "metadata widget" that displays properties and their values.

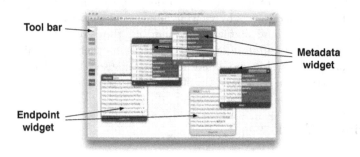

Fig. 1. DashSearch LD interface

Users can retrieve resources by typing strings on the endpoint widget. The
result is a list of resources. Next, users can see all the properties by overlapping
a metadata widget, as shown in Fig. 2. If they select a property in the list, all
the values of the property starting from the retrieved resources are shown with
the frequency of the values. Users can understand the distribution of data by
browsing properties with the property values. Selecting a value of a property
narrows the search results.

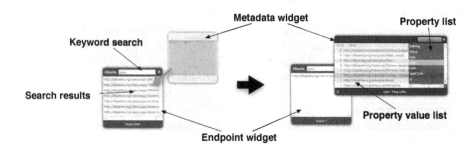

Fig. 2. Property browsing

Users can compose a more complex search condition by combining metadata
widgets. They can compose AND queries easily by attaching multiple metadata
widgets (Fig. 3 Left). The search condition can be removed from the query and
re-used by separating metadata widgets (Fig. 3 Right). The detached metadata

widget still contains the search condition (either a property or a value for a metadata widget) and can be used again. Thus, the addition, deletion, and repetition of search conditions can be facilitated by rearranging widgets by controlling the mouse, and a complex search can be formalized and performed quickly.

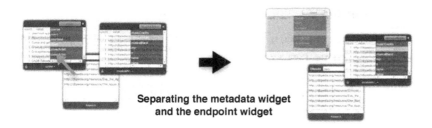

Separating the metadata widget and the endpoint widget

Fig. 3. AND operation by combining widgets

DashSearch LD enables a faceted search by overlaying multiple metadata widgets. First, users stack another metadata widget on one metadata widget. Next, they select the value of the property of the metadata widget, as shown in Fig. 4. Then, the metadata widget's properties and their values on the stack are narrowed with resource narrowing. By employing the facet search, users can understand the values of the properties related to the value of the selected property.

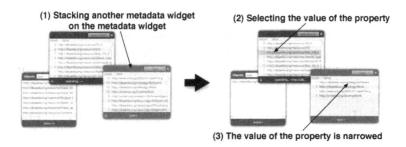

(1) Stacking another metadata widget on the metadata widget

(2) Selecting the value of the property

(3) The value of the property is narrowed

Fig. 4. Faceted search

With DashSearch LD, users can mix multiple datasets by using multiple endpoint widgets. The AND search with multiple endpoint widgets can yield results shared by the datasets. As mentioned, users can easily re-use a search condition of a metadata widget. Re-use of a search condition is also applicable to other endpoint widgets, and so users can apply the same widgets used for browsing other datasets. In addition, users can search different datasets by the same metadata used with another endpoint widget (Fig. 5).

**Search by the value of the property obtained
from the search results**

**Metadata widget is also available
for a different endpoint**

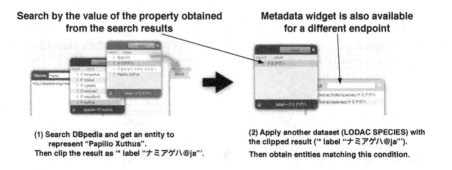

(1) Search DBpedia and get an entity to
represent "Papilio Xuthus".
Then clip the result as "* label "ナミアゲハ@ja"".

(2) Apply another dataset (LODAC SPECIES) with
the clipped result ("* label "ナミアゲハ@ja"").
Then obtain entities matching this condition.

Fig. 5. Search by the same metadata widget

Thus, by using DashSearch LD, users can understand the data structures and data distribution by browsing the metadata.

4   Implementation

The architecture of the system is shown in Fig. 6. The basic flow of the system is 1) acquiring data by SPARQL via SPARQL endpoints, 2) caching the acquired data, 3) filtering the data, and 4) displaying the data. An endpoint widget obtains triples, the subjects of which have a label containing the search query. Querying by SPARQL is done iteratively with consideration of the server's load, which can yield up to 10,000 triples as the result. Then, metadata widgets can filter the set of these triples. A metadata widget obtains RDF data from an endpoint widget with respect to a selected property. If a metadata widget is put on another metadata widget, it obtains RDF data from the metadata widget located below it. Thus, the RDF data is filtered sequentially. The search results of the endpoint widget are narrowed by the AND query that consists of the value of the property specified by each metadata widget. Thus, DashSearch LD enables users to browse data by filtering from various viewpoints. The system is fully implemented and openly accessible[2].

Currently, DBpedia[3], DBpedia Japanese[4], LODAC Museum [5] and LODAC Species are listed as endpoints[5]. A simple query example with these datasets is searching some entities (such as universities) that have the same property and value (such as the homepage URL) in DBpedia and DBpedia Japanese. A more complicated example is exploring species information, such as searching species by name from DBpedia, obtaining all species in the same genus as the search species, selecting and clipping one of the species, then searching LODAC Species with the clipped search condition. All operations are done without typing except for the initial input for the text search (Fig. 5).

---

[2] http://www.ahirulab.com/dashsearch/lod/
[3] http://dbpedia.org/sparql
[4] http://ja.dbpedia.org/sparql
[5] http://lod.ac

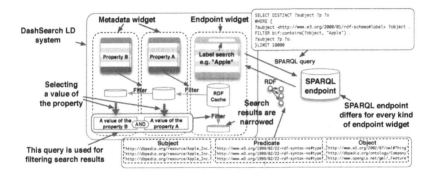

Fig. 6. DashSearch LD system

5 Conclusion

We described a retrieval interface called DashSearch LD that uses widgets to understand the LOD. Our main contribution is to provide a novel exploratory search environment for LOD that enables users to understand LOD through an intuitive interface. Users can interactively explore one or multiple datasets by viewing, selecting, and interacting with the entities in these datasets. In addition, users can just manipulate widgets on the screen to check entities, draw detail data on them, and obtain other entities linked by the widgets. Thus, through exploratory searching, DashSearch LD can help users to understand LOD *as is*, including not only the structure and the distribution of data but also the implicit association between datasets.

References

1. Deligiannidis, L., Kochut, K.J., Sheth, A.P.: RDF data exploration and visualization. In: Proc. of CIMS 2007, pp. 39–46 (2007)
2. Goto, T., Takeda, H., Yasumura, M.: Dash Search: Desktop Widget based Desktop Search for Metadata Exploitation. In: Adjunct Proc. of UIST 2008, pp. 37–38 (2008)
3. Heim, P., Ertl, T., Ziegler, J.: Facet Graphs: Complex Semantic Querying Made Easy. In: Aroyo, L., Antoniou, G., Hyvönen, E., ten Teije, A., Stuckenschmidt, H., Cabral, L., Tudorache, T. (eds.) ESWC 2010, Part I. LNCS, vol. 6088, pp. 288–302. Springer, Heidelberg (2010)
4. Jarrar, M., Dikaiakos, M.D.: A Query Formulation Language for the Data Web. IEEE Transactions on Knowledge and Data Engineering (2011)
5. Kamura, T., Takeda, H., Ohmukai, I., Kato, F., Takahashi, T., Ueda, H.: Study Support and Integration of Cultural Information Resources with Linked Data. In: Proc. of Culture and Computing 2011, pp. 177–178 (2011)
6. Kiefer, C., Bernstein, A., Stocker, M.: The Fundamentals of iSPARQL: A Virtual Triple Approach for Similarity-Based Semantic Web Tasks. In: Aberer, K., Choi, K.-S., Noy, N., Allemang, D., Lee, K.-I., Nixon, L.J.B., Golbeck, J., Mika, P., Maynard, D., Mizoguchi, R., Schreiber, G., Cudré-Mauroux, P. (eds.) ISWC/ASWC 2007. LNCS, vol. 4825, pp. 295–309. Springer, Heidelberg (2007)
7. Phuoc, D.L., Polleres, A., Morbidoni, C., Hauswirth, M., Tummarello, G.: Rapid semantic web mashup development through semantic web pipes. In: WWW 2009 (2009)
8. Russell, A., Smart, P.R., Braines, D., Shadbolt, N.: NITELIGHT: A Graphical Tool for Semantic Query Construction. In: SWUI 2008 (2008)

Entity-Based Semantic Search on Conversational Transcripts Semantic

Search on Hansard

Obinna Onyimadu[1,2], Keiichi Nakata[1], Ying Wang[1,2],
Tony Wilson[2], and Kecheng Liu[1]

[1] Informatics Research Centre, Henley Business School, University of Reading, UK
[2] System Associates Ltd, Maidenhead, UK
obinnao@systemassociates.co.uk, k.nakata@henley.reading.ac.uk

Abstract. This paper describes the implementation of a semantic web search engine on conversation styled transcripts. Our choice of data is Hansard, a publicly available conversation style transcript of parliamentary debates. The current search engine implementation on Hansard is limited to running search queries based on keywords or phrases hence lacks the ability to make semantic inferences from user queries. By making use of knowledge such as the relationship between members of parliament, constituencies, terms of office, as well as topics of debates the search results can be improved in terms of both relevance and coverage. Our contribution is not algorithmic instead we describe how we exploit a collection of external data sources, ontologies, semantic web vocabularies and named entity extraction in the analysis of underlying semantics of user queries as well as the semantic enrichment of the search index thereby improving the quality of results.

Keywords: Hansard, Named Entities, Semantic Search, RDF.

1 Introduction

Information retrieval in a number of domain centric search applications is still vastly limited to keyword based extraction techniques and the current search infrastructure for Hansard is no exception. As a consequence, the semantics and word sense behind a user's information need are not adequately captured in the search result. The importance of Hansard to the parliamentary community as a means for not just recording parliamentary debates but as an avenue for discovering trends and sourcing opinions from commentary made by members of parliament (MPs) makes its current search capability a prime candidate for semantic enhancement. Central to achieving this goal is the enrichment of the search index with contextual and semantic metadata as well as an extensive domain ontology for disambiguating topics, persons, places and events. Our approach entails extracting and indexing named entities using traditional information extraction methods [2,3], semantic metadata from documents which point to concepts in a domain ontology to support named entity (NE) extraction or conceptual disambiguation as well as the ability to make inferences about entities and

H. Takeda et al. (Eds.): JIST 2012, LNCS 7774, pp. 344–349, 2013.

instances in the domain. These features enriches and enhances the usefulness of the index and paves the way for semantically classifying commentaries and documents and finally augments the task of cross referencing content. Also, because we capture temporal and geographical information about MPs such as time in office, changes in position, party or constituency in our ontology means that geographical and temporal queries can be answered and ranked while incorporating information from external sources like the linked data space through a common URI. In the rest of this paper, we describe the characteristics of Hansard (Sec. 2), system architecture and design (Sec. 3), open source technologies used in implementing the design (Sec. 4), while sections 5 and 6 reports on the results, observations and conclusions respectively.

2   Hansard

The document search space of this semantic search is Hansard, the official proceedings of parliamentary debates, which is publicly available via web[1]. The electronic data on Hansard is semi structured and its underlying structure features the identity of a speaker followed by the commentary, within a theme or subject of debate. By transforming each Hansard document into speaker and commentary pairs, we can connect entities to external related content on the web through their URIs and analyze each speaker-commentary pair independently and in the context of sequential speaker-commentary pair embedded within a debate.

The domain specificity of Hansard to parliament means that various knowledge about aspects of the parliament and MPs are assumed, such as MPs and their constituencies, terms of office, and their changes over the years. These are often known to the users but not captured by standard keyword-based searches, which results in mismatches between user expectations and search results. The semantic search engine developed aims to improve both the relevance and coverage of search.

3   Design

3.1   System Architecture

A component based software development approach was employed for the system architecture. This offers rapid and easy development of system resources as well as its wide acceptance in the development community [1]. It consists of three tiers, data tier, analysis tier, and client tier, as illustrated in Fig. 1.

The **Data Tier** consists of two primary components. The *document harvest* component consists of a crawler that acquires all the Hansard web pages. Each document is fed to a document transformer tasked with preprocessing the documents. This involves cleaning unwanted html tags, extracting the required segments of the document and transforming it to a normalized format to enable us deal with the disparity and variation in the style of each document. The resulting transformed documents are

---

[1] http://www.parliament.uk/business/publications/hansard/: Last accessed 070/9/2012.

stored in database from which they are accessed, semantically analyzed in the analysis tier and subsequently indexed. The *document knowledge* component consists of a web data extractor that queries external data sources for parliamentary domain information ranging from personal details to time in office or positions held is organized, converted to triples and through a template in the ontology builder, marked up in RDF. We implement a parliamentary ontology and populate it with the triples after which they are stored in the Ontology Repository. The ontology is accessed through a REST API.

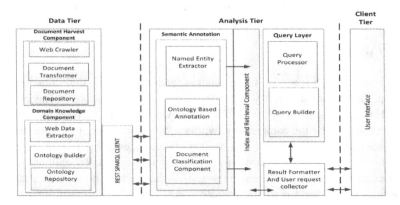

Fig. 1. System Architecture

The **Analysis Tier** encompasses the Semantic Annotation Layer (SAL), the indexing mechanism, a search query layer and request and result collector which basically serves as a bridge between the rest of the analysis tier and the client by receiving client request and returning formatted search result response to the client interface. The SAL consists of a named entity (NE) extractor that includes tokenizer, stemmer, English language POS tagger, a dictionary of parliamentary terms and a syntactic analyzer for extracting entities such as events, names, locations, organizations, dates, financial and numerical figures, dates and addresses. This rich metadata set when combined together provides the foundation for semantic and contextual relevance in the result set. We enhance the metadata derived from the NE extractor by employing the ontology based annotation component in extracting relevant annotations which point to concepts in the domain ontology. The hierarchical structure of the ontology means that we can disambiguate extracted NEs and make inferences. The document classification model is based on the assumption that a set of words occurring together will determine its sense [4, 5] so through the existing classification of concepts and entities in domain ontology we have a credible base from which document classification can be performed by analyzing the document similarity based on extracted NEs.

All extracted metadata and categorization information is indexed and queried via the query layer. Query layer comprises of a query builder and processor. The processor also harnesses NE tools in extracting the key components of the user requests. For instance a query like "How much was spent by the Labour Government on Health in

1998?" The syntactic analyzer would classify the query type as a question. Relations among the entities are derived from the query, i.e., a relationship "on" is formed between "Labour Government" (Type: Organization) and "Health" (Type: Topic). The temporal assertion in the request with the year "1998" identified by the NE extractor provides a temporal constraint. In addition, through the domain ontology, the entities "Labour Government" and "Health" are associated with related entities. These parameters forms a query structure based on which relevant documents are searched.

3.2   Process Flow

There are two primary processes involved in the system (Fig. 2). **Data collection and semantic indexing** (dotted arrows in) is a non-real time scheduled process primarily to support updates to ontology and Hansard documents. **User queries** (solid arrows) trigger a search process. The data collection and semantic indexing process commences with the crawling, transformation and population of transformed documents into the database repository. Domain information is sourced from the web, marked up in RDF and used to populate the ontology repository. The completion of these processes triggers the task of analyzing and extracting named entities from the documents which generates indexes based on semantic metadata. User queries are first processed by the query processor to derive a set of query terms which include query entities, query context, relations in the query and disambiguation of entities. The resulting query parameters are converted to a structured query by the query builder which then queries the index, which returns the results to a result formatter which forwards it to the client interface.

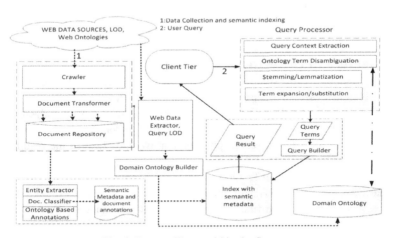

Fig. 2. Process Flows within the System

4   Implementation

In the current implementation, the system crawled Hansard documents over a five year period. Our objective was to obtain an initial sizable corpus that captures periods covering notable transitions in parliament through elections, hand-overs, resignations etc. Preprocessing these documents involved stripping them of unwanted html tags and extracting speaker-commentary pairs and valuable metadata like the original document's URL, date and topic if available. Through this speaker-commentary pairs and other relevant document metadata are obtained. Mongo[2] DB was used in housing the transformed document and its metadata. The choice of Mongo is due to its document oriented structure, scalability, JSON like data structure and popularity. The domain ontology was implemented using Bigdata, an open source semantic web repository which supports SPARQL queries and RDFS inferences. Domain ontology data is sourced primarily from "theyworkforyou.co.uk" web site which maintains a REST API for querying MP and parliamentary information. The resulting parliamentary ontology makes use of popular semantic web vocabularies like FOAF, SKOS and Dublin Core. The data from "theyworkforyou.co.uk" is converted to triples and stored in the ontology. The resulting ontology contains about 2 million triples. NE extraction is carried out using GATE[3], an open source text engineering tool by modifying its Gazetteer and creating new grammar rules to identify and extract additional context like questions, responses to questions, statements. For our topic categorization, we make use of Open Calais[4]'s topic categorization and scoring mechanism through its public API. Commentaries are therefore classified according to a genre such as politics, social issues etc. The resultant entities are indexed in Solr[5], an open source search platform based on Lucene. At query time, terms in the user query are expanded through the domain ontology and disambiguated in the way that we can indicate that a search for the term "Maidenhead" refers to the constituency of Maidenhead, which has as its MP Theresa May. We can also make temporal disambiguation such that Prime Minister between 2010 and 2012 is David Cameron while Prime Minister in 2009 was Gordon Brown, the MP for Kirkcaldy & Cowdenbeath. This sort of disambiguation extends to names, titles and positions.

5   Results and Observations

Using independent assessors, early stage evaluation of our search engine in comparison with the current one focused on relevance of the results. Testers judged and recorded degree of relevance for the first 15 results as "Relevant", "Quite relevant" and "Not relevant" for 6 different search phrases. On average, 32% of the results in our search were classified as very relevant as compared to 11.6% in the current system. Our result also exhibits salient features of semantic expansion. For example, a search on "What David Cameron has said about Health Care between 2001 and 2011" returns two sets of results. The first set features statements made by David Cameron, the MP for Witney prior to his appointment as Prime Minister in 2010, while the second set includes statements made by the Prime Minister David Cameron. Fig. 3 is a screenshot of the outputs from the demonstrator.

---

[2] http://www.mongodb.org/: Last accessed 07/09/2012.

[3] http://Gate.ac.uk: Last accessed 07/09/2012.

[4] http://Opencalais.com: Last accessed 07/09/2012.

[5] http://lucene.apache.org/solr/: Last accessed 07/09/2012.

The combination of domain ontology, the indexing of named entities and contextual information means that we can compose more expansible yet focused queries which fit the user's information need. Our component based approach also means that components can be replaced or substituted. For instance, we can enhance the strength of our domain ontology and disambiguation process by directing queries to Dbpedia to obtain more extensive information on entities. The NE extraction process is not always accurate and there are situations where we are unable to identify some named entities or even misidentify certain named entities.

Fig. 3. A screenshot of search results

6   Conclusion and Future Work

In this paper we have shown that a hybrid approach combining NE extraction resources and domain ontology can serve to implement semantic search for Hansard. We intend to further enrich the domain ontology by accessing additional data sources. Future directions include extending the semantic extract sentiments and opinion in the comments made by MPs so that search results can be enriched by offering both single and collective sentiments on debated subject matters.

References

1. Emmerich, W.W.: Distributed Component Technologies and their Software Engineering Implications. In: Proceedings of the 24th International Conference on Software Engineering, Orlando, Florida, pp. 537–546 (2002)
2. Junhui, Y., Chan, H.: Keywords Weights Improvement and Application of Information Extraction. In: Gaol, F.L., Nguyen, Q.V. (eds.) Proc. of the 2011 2nd International Congress on CACS. AISC, vol. 144, pp. 95–100. Springer, Heidelberg (2012)
3. Lam, M.I., Gong, Z., Muyeba, M.K.: A Method for Web Information Extraction. In: Zhang, Y., Yu, G., Bertino, E., Xu, G. (eds.) APWeb 2008. LNCS, vol. 4976, pp. 383–394. Springer, Heidelberg (2008)
4. Liu, Y., Scheuermann, P., Li, X., Zhu, X.: Using WordNet to Disambiguate Word Senses for Text Classification. In: Shi, Y., van Albada, G.D., Dongarra, J., Sloot, P.M.A. (eds.) ICCS 2007, Part III. LNCS, vol. 4489, pp. 781–789. Springer, Heidelberg (2007)
5. Navigli, R.: Word Sense Disambiguation: A Survey. ACM Computing Surveys 41(2), Article No. 10 (February 2009)

Development of Linked Open Data for Bioresources

Hiroshi Masuya[1,*], Terue Takatsuki[1], Yuko Makita[2], Yuko Yoshida[2],
Yoshiki Mochizuki[2], Norio Kobayashi[2], Atsushi Yoshiki[1], Yukio Nakamura[1],
Tetsuro Toyoda[2], and Yuichi Obata[1]

[1] RIKEN BioResource Center, Tsukuba, Japan
(hmasuya,takatter,yoshiki,yukionak,yobata)@brc.riken.jp
[2] RIKEN BASE, Yokohama, Japan
(makita,yoshida,ym,toyoda)@base.riken.jp

Abstract. The broad dissemination of information is a key issue in improving access to existing bioresources. We attempted to develop Linked Open Data (LOD) for bioresources available at the RIKEN BioResource Center. The LOD consists of standardized, structured data available openly on the World Wide Web, including published bioresource information for 5,000 mouse strains and 3,600 cell lines. The LOD includes links to publically available information, such as genes, alleles, and ontologies, providing phenotypic information through the BioLOD website. As a result, information on mouse strains and cell lines have been connected to various data items in public databases and other project-oriented databases. Thus, through the use of LOD, dispersed efforts to produce different databases can be easily combined. Through these efforts, we expect to contribute to the global improvement of access to bioresources.

Keywords: bioresource, mouse strain, cell line, linked open data.

1   Introduction

Experimental materials, often referred to as "bioresources" or "biological resources", are the most fundamental elements of research in life sciences. The scientific need for the common use of bioresources has been increasing, moving toward the establishment of global interoperability among biological data and knowledge. The improvement of access to existing and newly developed bioresources is a major challenge in the realization of a sustainable infrastructure for biomedical studies.

Information technology is essential to improving access and promoting the common use of bioresources. In particular, Linked Open Data (LOD) is expected to be a next-generation methodology for exposing, sharing, and connecting pieces of data, information, and knowledge on the World Wide Web (WWW) using Uniform Resource Identifier (URI) and Resource Description Framework (RDF) coding. Based on the LOD strategy, distributed (non-centralized) efforts are expected to contribute to the integration of data throughout the entire Internet.

---

* Corresponding author.

H. Takeda et al. (Eds.): JIST 2012, LNCS 7774, pp. 350–355, 2013.

With the goal of contributing to the global improvement of access to bioresources, we have attempted to develop LOD for the bioresources available at the RIKEN BioResource Center, a not-for-profit public organization involved in the deposition, preservation, and distribution of bioresources including living strains of mice and experimental plants, cell lines of human and animal origin, experimental plants, genetic materials (DNA) from humans and animals, and microorganisms [1-3]. Recently, we published bioresource information for several mouse strains and cell lines available at the center with connections to publically available information, such as genomic databases and ontologies, at BioLOD website. In addition, we have been cooperating with international efforts to formalize an inter-relationship schema among different types of data items through the International Mouse Phenotyping Consortium (IMPC) [4] and the Cell Line Ontology (CLO) [5] consortium. In this paper, we describe the development of LOD for bioresources using the infrastructure system provided at the RIKEN BioResource Center and discuss the benefits and future needs of this endeavor.

2   Construction of LOD for Bioresource Data

Re-organization of Bioresource Data Available in the SciNetS System

As the first step in constructing the LOD, we re-organized the data items in the Scientists' Networking System (SciNetS) [6]. The bioresource data for mouse strains [1] and cell lines [2] available at the RIKEN BioResource Center were captured from the BioResource Web Catalog databases of mouse strains [7] and cell lines [8]. In the SciNetS system, the data items were managed as roughly equivalent forms of the RDF and the Web Ontology Language (OWL). When importing the bioresource data to SciNetS, the individual data base records of 5,000 mouse strains and 2,200 cell lines were converted as instances of classes named "RIKEN BRC mouse resource" and "cell line resource", respectively. The data columns and their values in the database tables were then converted to properties (Table 1) equivalent to the RDF:Property and triples. We established a monthly workflow to update the latest data.

Table 1. A fraction of properties defined in the BRC mouse and cell line resources classes

Domain for property	Name of property	Type of property	Explanation of property	What is described by the property?
BRC mouse strain	Taxon	object	A biological species of the strain.	Nature of mouse strain.
	BRC strain type	object	Practical type of strain defined in BRC database.	Nature of mouse strain.
	has allele	object	Link to allele instance, which the strain carries.	Nature of mouse strain.
	Strain origin	object	Original strain(s) from which the mouse strain was derived.	A nature of the mouse strain.
	depositor	object	Person who deposited the mouse strain	Aspect as a common resource.
	Institute depositor	object	Institute that deposited the mouse strain.	Aspect as a common resource.
	BRC mouse strain status	object	Current preservation status of the mouse strain.	Aspect as a common resource.
	BRC mouse strain availability	object	Availability of the strain.	Aspect as a common resource.
BRC cell resource	Taxon	object	A biological species of the cell line.	Nature of cell line.
	BRC cell category	object	Practical type of cell line defined in BRC database.	Nature of cell line.
	Taxon	object	Biological species of cell line.	Nature of cell line.
	derived from	object	Body part from which the cell line was derived.	Nature of cell line.
	donor	object	Individual (person or animal) from which the cell line derived.	Nature of cell line.
	depositor	object	Person who deposited the cell line.	Aspect as a common resource.
	depositor institute	object	Institute that deposited the cell line.	Aspect as a common resource.

In the SciNetS system, data from multiple public databases were imported and implemented to form integrated databases for mammals [9], plants, and omics studies. The implementation of triples to refer to existing data items enabled bioresource data to be utilized in broad networks containing mutual relationships among biological data items (Fig. 1).

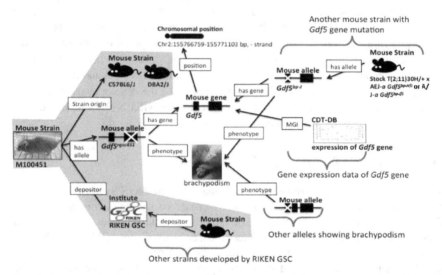

Fig. 1. An example of the data connection within a record for a mouse strain (M100451) in SciNetS. The shaded area represents data items prepared in this study.

Publication of LOD from BioLOD Website

We have developed a fully automated workflow to generate, publish, and provide an SPARQL-based query interface for LOD from databases stored in SciNetS (Fig. 2). In the workflow, imports from newly developed databases in SciNetS are converted into multiple data formats (e.g., RDF/OWL, RDFa, OBO, GFF and plaintext) and are published in an open archive termed "BioLOD" [10]. A SPARQL-based query interface, BioSPARQL is provided for the RDF/OWL formatted data in BioLOD [11]. Using this workflow, we published all the bioresource data from BioLOD [12,13].

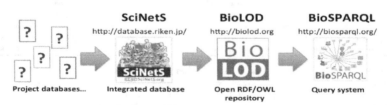

Fig. 2. Integrated workflow for generating, publishing, and providing a query interface for data provided by SciNetS, BioLOD and BioSPARQL

3 Benefits of Developing LOD for Bioresources

To promote the common use of bioresources, webpages containing not only information on the whereabouts and biological characteristics of the bioresources plays the quite important role. However, scientific community often requires provision of "latest" result, which is obtained from bioresources. Therefore, the establishment of cost-effective way or technology to maintain related-information with bioresources is desired. Such a technology will be an efficient means of providing positive feedback and research results obtained using the bioresources. The LOD strategy seems to be one of the best candidates for integrating biomedical information with multiple mutual relationships.

In this study, we attempted to develop LOD for mouse strains and cell lines. Using the LOD strategy, we performed the distributed (non-centralized) development of organized biological data (or an integrated database) together with other developers working on non-bioresource databases, such as for genome biology, at RIKEN. As a result, we have developed rich and well-organized integrated database in which information on mouse strains and cell lines were connected to various data items from public databases and other project-oriented databases (Fig. 1). User interface to browse bioresource-related LOD is provided by SciNetS in which various types of data are organized along common biological knowledge of gene-organism relationship with previously proposed methodology [9]. Thus, within the scope of LOD, dispersed efforts to produce different databases were easily organized and mainained. This methodology may be one of the most cost-effective ways of developing sustainable integrated databases.

The international coordination of knowledge representation using RDF/OWL technologies may be another important component in realizing useful LOD. We have participated in several activities concerning the international coordination of bioresource data. Within the IMPC consortium [4], multiple major mouse research centers are working on a global systematic genome-wide phenotyping project for all (20,000) genes in the mouse. Within this consortium, experimental data on mouse mutant strains are represented in the form of biological ontologies to increase the interoperability with other biomedical databases. The development of an RDF-based LOD may be an extension of the IMPC concept to maximize the reuse of open biological experiment data.

Within the CLO consortium [5], a global methodology involving OWL representation for cell line resources across international resource centers is being discussed. This activity is based on the strategy proposed by the Open Biomedical Ontology (OBO) Consortium and the OBO Foundry initiatives [14], open discussions of the development of ontology, the reuse of existing bio-ontologies for the description of composite knowledge, and the design of logically and well-formed representations of biological entities. We are now in the process of aligning our properties in the BRC Cell Resource project with those of the CLO. When combined with standardized knowledge representation technology, the LOD strategy may be one of the most powerful means of promoting the common use of bioresources.

In addition, the development of an advanced methodology for describing applications and uses of biological materials is also an open issue necessary for the improvement of access to bioresources. To investigate how "bioresources" can be integrated into an upper-ontology, we verified the meaning (what the property describes) of each property defined in the BRC Mouse Resource and BRC Cell Resources classes in the SciNetS. Interestingly, we found that two kinds of properties were present (Table 1). Some properties (e.g., "has allele", "background strain", and "donor") describe the natures of a mouse or a cell. On the other hand, other properties (e.g., "depositor" and "research applications") describe aspects of "common resources" in biomedical studies. Both the biological nature of a living organism and its use as a common resource in the context of scientific studies are important aspects of bioresources. Furthermore, descriptions of the applications or roles of bioresources, which can be diverse depending on the field of study, seem to be extremely important. The methodology for knowledge description of roles and their contexts has been well established using the ontological model of role concepts [15-17]. Further investigation of the roles of bioresources and RDF/OWL-based descriptions should break new ground for the efficient use of LOD for bioresources.

Acknowledgement. This study was partially supported by the Life Science Database Integration Project of Japan Science and Technology Agency (JST).

References

1. Yoshiki, A., Ike, F., Mekada, K., Kitaura, Y., Nakata, H., Hiraiwa, N., Mochida, K., Ijuin, M., Kadota, M., Murakami, A., Ogura, A., Abe, K., Moriwaki, K., Obata, Y.: The mouse resources at the RIKEN BioResource center. Exp. Anim. 58, 85–96 (2009)
2. Nakamura, Y.: Bio-resource of human and animal-derived cell materials. Exp. Anim. 59, 1–7 (2010)
3. Yokoyama, K.K., Murata, T., Pan, J., Nakade, K., Kishikawa, S., Ugai, H., Kimura, M., Kujime, Y., Hirose, M., Masuzaki, S., Yamasaki, T., Kurihara, C., Okubo, M., Nakano, Y., Kusa, Y., Yoshikawa, A., Inabe, K., Ueno, K., Obata, Y.: Genetic materials at the gene engineering division, RIKEN BioResource Center. Exp. Anim. 59, 115–124 (2010)
4. The International Mouse Phenotyping Consortium, http://www.mousephenotype.org
5. Cell Line Ontology, http://bioportal.bioontology.org/ontologies/1314
6. RIKEN adopts Semantic Web as the data-release standard of its database-construction infrastructure. RIKEN Press Release, http://www.riken.jp/engn/r-world/info/release/press/2009/090331_2/
7. RIKEN BRC Mouse Web Catalog, http://www2.brc.riken.jp/lab/animal/search.php
8. RKIEN BRC Cell Line Web Catalog, http://www2.brc.riken.jp/lab/cell/search.php
9. Masuya, H., Makita, Y., Kobayashi, N., Nishikata, K., Yoshida, Y., Mochizuki, Y., Doi, K., Takatsuki, T., Waki, K., Tanaka, N., Ishii, M., Matsushima, A., Takahashi, S., Hijikata, A., Kozaki, K., Furuichi, T., Kawaji, H., Wakana, S., Nakamura, Y., Yoshiki, A., Murata, T., Fukami-Kobayashi, K., Mohan, S., Ohara, O., Hayashizaki, Y., Mizoguchi, R., Obata, Y., Toyoda, T.: The RIKEN integrated database of mammals. Nucleic Acids Res. 39(Database issue), D861–D870 (2011)

10. BioLOD, http://biolod.org
11. BioSPARQL
12. BioLOD for BRC mouse strain, http://biolod.org/class/cria315s1i/BRC_Mouse_Strain
13. BioLOD for BRC cell line, http://biolod.org/class/cria322s1i/BRC_Cell_Line
14. Smith, B., Ashburner, M., Rosse, C., Bard, J., Bug, W., Ceusters, W., Goldberg, L.J., Eilbeck, K., Ireland, A., Mungall, C.J., OBI Consortium, Leontis, N., Rocca-Serra, P., Ruttenberg, A., Sansone, S.A., Scheuermann, R.H., Shah, N., Whetzel, P.L., Lewis, S.: The OBO Foundry: coordinated evolution of ontologies to support biomedical data integration. Nat. Biotechnol. 25, 1251–1255 (2007)
15. Mizoguchi, R.: Tutorial on ontological engineering - Part 2: Ontology development, tools and languages. New Generation Computing 22, 61–96 (2004)
16. Mizoguchi, R.: Tutorial on ontological engineering - Part 3: Advanced course of ontological engineering. New Generation Computing 22, 198–220 (2004)
17. Kozaki, K., Sunagawa, E., Kitamura, Y., Mizoguchi, R.: Rôle Representation Model Using OWL and SWRL. In: Proc. of 2nd Workshop on Roles and Relationships in Object Oriented Programming, Multiagent Systems, and Ontologies, pp. 39–46 (2007)

Towards a Data Hub for Biodiversity with LOD

Yoshitaka Minami[1,*], Hideaki Takeda[1,2], Fumihiro Kato[1], Ikki Ohmukai[1,2],
Noriko Arai[1], Utsugi Jinbo[3], Motomi Ito[4], Satoshi Kobayashi[5], and Shoko Kawamoto[6]

[1] National Institute of Informatics
{minami1105,takeda,fumi,i2k,arai}@nii.ac.jp
[2] Graduate University for Advanced Studies
[3] National Museum of Nature and Science, Tokyo
ujinbo@kahaku.go.jp
[4] Department of General Systems Studies, Graduate School of Arts and Sciences,
The University of Tokyo
cmito@mail.ecc.u-tokyo.ac.jp
[5] Transdisciplinary Research Integration Center, National Institute of Polar Research
kobayashi.satoshi@nipr.ac.jp
[6] Database Center for Life Science, Research Organization of Information and Systems
shoko@dbcls.rois.ac.jp

Abstract. Because of a huge variety of biological studies focused on different
targets, i.e., from molecules to ecosystem, data produced and used in each field
is also managed independently so that it is difficult to know the relationship
among them. We aim to build a data hub with LOD to connect data in different biological fields to enhance search and use of data across the fields. We
build a prototype data hub on taxonomic information on species, which is a key
to retrieve data and link to databases in different fields. We also demonstrate
how the data hub can be used with an application to assist search on other database.

Keywords: Linked Open Data (LOD), biodiversity, taxonomy, data integration.

1 Introduction

Life is still a big target for our science, even physics can explicate elementary particles. In
particular biodiversity [1] becomes a big social problem. Biology consists of many research fields focused on various targets, from molecules to ecosystem. Thus, there are
many biological disciplines, from molecular and cell biology, to ecology, evolution and
taxonomy. Data collected from biological studies are highly diverse in contents and
formats. Each research field can yield and use data for own field, but such data often lacks
information on relationships to one another. As collaborative research projects across
different fields are developing, demands for a data coordination system is increasing.

We aim to build a data hub to connect data collected in various biological fields to
enhance researchers to search and use data across fields.

---

[*] Corresponding author.

H. Takeda et al. (Eds.): JIST 2012, LNCS 7774, pp. 356–361, 2013.
© Springer-Verlag Berlin Heidelberg 2013

When focusing on data, diversity can be categorized in the following three ways.

1. Diversity in subjects of biological researches.   There are different fields depending on hierarchical level of focus ranging from molecular biology to ecology, and analysis using multiscale data is often required.   There are large databases for molecular data (e.g. DDBJ, NCBI) and for specimen and observation data (GBIF), but their relationships are rather weak.   On the other hand, some specialists of specific groups of organisms build their own specific databases.   Such databases contain valuable data but are often independent from each other.
2.   Representation in local languages is needed for wide range of people, for example, governmental people working on biological resource management or biodiversity conservation in individual countries.
3. Diversity by people.   In addition to researchers and governmental people, general people are also looking for biological data for their activities. Recently, Citizen Science programs, namely, researches and studies in collaboration with general people, are emerging and biology is in its forefront.

Therefore, the information infrastructure for biodiversity should be required to treat heterogeneous, multiscale and multilingual data in scattered databases and to provide them for various people. We aim to build a data hub to absorb the above diversity. We focus on taxonomic information on species since they are common and mandatory fields for most biological information. It provides very basic information on each species including classification, and scientific and general (English and Japanese) names. It also provides links to entries on other database such as NCBI (National Center for Biotechnology Information), EOL (Encyclopedia of Life), and DBpedia based on scientific name. We also build an application to expand queries for database search to demonstrate how data can be used.

2   Related Work

In biodiversity informatics [2], ensuring interoperability of the various databases specialized in individual purpose, is one of the most important issues [3].   Several researchers and groups have started to research about Linked Data in biodiversity information but they have not reached standard or consensus fully [4]. Peterson et al. [5] emphasized that the integration of scientific names using linked data approach has a big potential and enables to create rich services that biologist can benefit.   Darwin Core [6] is a well-known standard for metadata for biodiversity information, but Linked Data for Darwin Core is under discussion. TaxonConcept[1] provides ontology and data for species but data is limited to the specific geographical area.

3   Selection of the Datasets

There exist various datasets even just concerning species information. Among them we selected mainly three datasets to examine feasibility of integration as a necessary

---

[1] http://www.taxonconcept.org

condition to publish and to contain species data. One of them covers a general dataset which covers a wide range of organisms, and other two datasets are domain-specific ones.

1. A core dataset for taxa: BDLS

We selected the dataset called BDLS (Building Dictionary for Life Science) for a core dataset for taxa. BDLS[2] is an integrated dictionary for biology which is built from nearly 100 sources which include various illustrated books and specimen dataset in museums, the latter of that are provided by the Science Museum Net[3] in Japan.

2. A domain-specific dataset for taxon names: the Current Checklist of Japanese Butterflies

We selected the Current Checklist of Japanese Butterflies [7] as a domain-specific dataset. It is a checklist (a list of species names) created and authorized by the butterfly taxonomists. It covers all butterfly species which number is 327 found in Japan, and describes each species by scientific and Japanese general names and higher taxa.

3. A domain-specific dataset for specimens: Bryophytes Specimen Collection

We selected the Bryophytes Specimen Collection that National Institute of Polar Research developed and maintained as another domain-specific dataset. This data has 56590 specimen data.

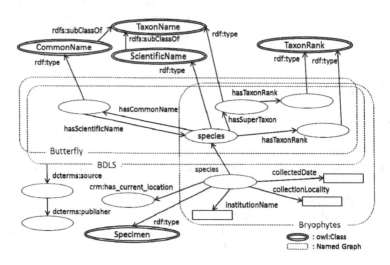

Fig. 1. Data model

---

[2] It is developed by Database Center for Life Science (DBCLS), Research Organization of Information and Systems (ROIS), Japan and available from http://lifesciencedb.jp/bdls/

[3] The network is maintained by the National Museum of Nature and Science, Tokyo. http://science-net.kahaku.go.jp/

4   The Data Model for Species Information

These dataset that we have selected have common information about scientific name and taxon. Then, we made a data model shown in Fig.1 in order to link to entries among the databases through the common information and to be used from outside conveniently. This data model was expressed in Named Graph for the data sources, i.e., the sub datasets in BDLS, Butterflies and Bryophytes.

One of the big issues in species information is treatment on various names. Each species has its scientific name but can be represented differently. One case is caused by different citation forms (e.g., use of abbreviation for genus, omission of authors). The other case is derived from multiple names for one species. The valid species name and the combination of genus and species might be changed as taxonomic studies proceed. Furthermore a species may have multiple general names in local languages. It is a delicate problem in taxonomy to choose a unique valid name for each species and it is beyond our scope[4]. Rather we represent each name as a node and associate nodes by relationship such as synonyms.

A basic model is as follows. We provide a class for taxon name and classes for scientific name and common name as its subclass. All nodes on names are instances of these classes. A node can have a "hasTaxonRank" property of which value is an instance of Class "TaxonRank", i.e., either kingdom, phylum, class, order, family, subfamily, tribe, subtribe, genus, subgenus or species. Another type is a node for specimen which is an instance of Class "Specimen" providing specific properties on specimen. All triples on instances are associated to data source URIs in Named Graph.

The benefit of name-based approach is rapid integration of data from different data sources. Its drawback is complexness of representation since a single specimen is represented as a network but it can be compensated by inference in RDF.

Representation of species name common to three datasets is defined as follows. First, nodes of the ScientificName type and CommonName type are generated for species name and for common name respectively. Next, the hasCommonName property links a node of the ScientificName to a node of the CommonName, and the hasScientificName property vice versa. A node representing taxon is linked to a node of the TaxonRank type by the hasTaxonRank property. And, other items are literal.

Nodes representing specimen in Bryophytes dataset describe ID, collected date, collector, latitude, longitude, floral region, floral subsection, locality, sporophyte, altitude, determiner, and herbarium housed[5]. Nodes representing specimen are defined as follows. First, the node is generated assigning a URI to each specimen because there can multiple specimens for one species. Next, a specimen node is linked to the node of ScientificName by the species property. A specimen node has a link by the crm:has_current_location property to a node representing a facility to represent herbarium housed relationship. And, other Items are literal.

---

[4] The exception in our dataset is Butterflies dataset where a list of valid scientific and common names is authorized by taxonomic experts.

[5] The list of herbarium index is available from Index Herbariorum:
   (http://sciweb.nybg.org/science2/IndexHerbariorum.asp)

Source information in BDLS dataset is represented by name in Named Graph. RDF triples representing data in a data source has a name which has a link by the dcterms:source property to the node representing the data source. Then it has properties such as rdfs:label and dcterms:publisher.

5   The Results

In accordance with the data model, we generated LOD from the selected data. As a result, the number of taxon name is 443,248, scientific name of species is 226,141, common name of species is 219,865, hasScientificName property node is 87160 and hasCommonName property node is 84610.

Our approach is successful in integration. But it causes some problems when the dataset is used. For example, we implemented a simple taxon search interface by name. We can show the results by matching names but we can just a set of taxon names but not show representative names. One of the potential problems of our approach is homonymy, i.e., two taxa may share a name. Though the naming rule of scientific name is not essentially permitted a name sharing two or more taxa, there are some exceptional cases e.g., an animal and a plant can share one genus name. We checked how it is in the real dataset by using the NCBI taxonomy. We found 1797 homonymical names. By using this data, we can distinguish taxon names properly.

6   An Application: Query Expansion for Paper Search

In addition to a data browsing view for the dataset, we also developed an application to demonstrate how the integrated dataset could be used. The application *CiNii-x-Species* is a tiny application which assists search for academic articles with species names. CiNii[6] is a database service which provides academic information of articles, books and journals published in Japan.   *CiNii-x-Species* assists people to search CiNii by extending search keywords variously based on the prototype dataset. If a

Fig. 2. Search example of *CiNii-x-Species*

---

[6] http://ci.nii.ac.jp/

Japanese common name is given, it searches CiNii with the corresponding scientific name (the left column in Fig. 2). It also searches CiNii with names of species belonging to the same genus with the query species, and with its other names (the center and right column Fig. 2). *CiNii-x-Species* uses our SPARQL Endpoint to find corresponding names to an input name in order to expand a query to CiNii.

7   Conclusion

The prototype shows the potential of LOD in biodiversity.   Firstly, biodiversity information can be really linkable among various datasets. We adopt the simple way for integration, i.e., name-based approach, but it can work thanks to relatively stable name convention in biology. It means that various datasets could be interlinked easily. Secondly the integrated dataset can show a new value, in particular associating information in English and local languages are beneficial for ordinary people. Thirdly, as demonstrated in the application, the link structure enables to exploit the dataset in new ways which are difficult in conventional databases.

We have shown the potential value of applying LOD on the biodiversity field. Since the biodiversity field is really varied and now emerging, data is created and published day by day. The evolutionaly nature of LOD is suitable in such an emerging field. Any dataset can be added to the existing LOD without modifying the latter, and any data model can be applied on the LOD.

In the future, we aim to extend data, to make mechanism adding data easily and to develop applications with a user-friendly interface complying with purpose. We expect that this work contributes to build the information infrastructure to solve the social problem about biodiversity conservation.

References

1. Bisby, F.A.: The quiet revolution: biodiversity informatics and the Internet. Science 289(5488), 2309–2312 (2000)
2. Edwards, J.L., Meredith, A.L., Nielsen, E.S.: Interoperability of Biodiversity Databases: Biodiversity Information on Every Desktop. Science 289(5488), 2312–2314 (2000)
3. Jenkins, M.: Prospects in biodiversity. Science 302, 1175–1177 (2003)
4. Global Biodiversity Information Facility: Recommendations for the Use of Knowledge Organisation Systems by GBIF (2011), accessible online at http://links.gbif.org/gbif_kos_whitepaper_v1.pdf
5. Patterson, D.J., Cooper, J., Kirk, P.M., Pyle, R.L., Remsen, D.P.: Names are key to the big new biology. Trends in Ecology and Evolution 25, 686–691 (2010), doi:10.1016/j.tree.2010.09.004
6. Wieczorek, J., Bloom, D., Guralnick, R., Blum, S., Döring, M., et al.: Darwin Core: An Evolving Community-Developed Biodiversity Data Standard. PLoS ONE 7(1), e29715 (2012), doi:10.1371/journal.pone.0029715
7. Inomata, T., Uémura, Y., Yago, M., Jinbo, U., Ueda, K.: The Current Checklist of Japanese Butterflies (2010), http://binran.lepimages.jp/

Linking Open Data Resources for Semantic Enhancement of User–Generated Content*

Dong-Po Deng[1,3], Guan-Shuo Mai[2], Cheng-Hsin Hsu[2],
Chin-Lung Chang[1,4], Tyng-Ruey Chuang[1], and Kwang-Tsao Shao[2]

[1] Institute of Information Science
[2] Biodiversity Research Center
Academia Sinica, Taipei, Taiwan
[3] Faculty of Geo-Information Science and Earth Observation (ITC)
Twente University, Enscahede, The Netherlands
[4] Department of Computer Science and Information Engineering
National Taiwan University of Science and Technology, Taipei, Taiwan

Abstract. This paper describes our experiences in developing a Linking Open Data (LOD) resource for Taiwanese Geographic Names (LOD TGN), extracting Taiwanese place names found in Facebook posts, and linking such place names to the entries in LOD TGN. The aim of this study is to enhance the semantics of User-Generated Content (UGC) through the use of LOD resources, so that for example the content of Facebook posts can be more reusable and discoverable. This study actually is a development of a geospatial semantic annotation method for Facebook posts through the use of LOD resources.

Keywords: Linking Open Data, User-Generated Content, Facebook, Gazetteers, Place Names.

1 Introduction

Social media is software and services for human communication. A key characteristic of social media is to enable people to create and exchange information on the Web. Such information is often called User-Generated Content (UGC). A large number of users voluntarily contribute in situ information on the Web. These users are often considered as human sensors; they actively report what happen in their surroundings [3]. The emergence of social media offers an opportunity for recruiting more volunteers to contribute data for scientific purposes. Engaging with a large number of users can be a way to improve data collection in scientific projects over a large geographic region and a long time span. For example, people often post to online forums what birds they observe and where. However, UGC is often unstructured data, such as in the form of text and photo. The semantics of unstructured UGC is often ambiguous or poor. Although UGC

* This research is supported in part by National Science Council of Taiwan under grant NSC 101-2119-M-001-004 ("A Semantic Web of Place Names: Gazetteers, Ontologies, and Web Services").

H. Takeda et al. (Eds.): JIST 2012, LNCS 7774, pp. 362–367, 2013.

is considered as a potential resource for scientific projects, it is a challenge to process unstructured data for scientific purposes. To efficiently and systematically collect data from UGC sources, there is a need for tools which can clarify the semantics in UGC by connecting UGC to LOD, so that input from users can be more easily handled for scientific works.

Linked data is a new research area which studies how to make data available on the Web, and to interconnect data with the aim of increasing its value for users [1]. In the context of the linked data, there have been about 300 datasets consisting of over 31 billion Resource Description Framework (RDF) triples within LOD projects. Each entry representing a fact in LOD datasets has a Unique Resource Identifier (URI) which is referenceable and linkable on the Web. The high interconnectivity between entries potentially increases discoverability, reusability, and the utility of information [5]. Therefore, if named entities of UGC can be identified and connected to entries of LOD, the semantics of named entities would be disambiguated, so that the UGC could be easier to process.

Take place names as examples. Although the use of name *Amsterdam* mostly means the city of Amsterdam in the Netherlands, it is also the name of some cities in the United States. If each Amsterdam has its own URI on the Web, the use of the name can be distinguished. Besides the name Amsterdam, *A'dam* and *Ams* are also intended to represent the city of the Netherlands. Imagine that three posts from social media respectively contain the three names. The three posts can be linked to the same place Amsterdam. That is, the URI of a place name plays a role as geospatial information hub. As a consequence, if posts from different social media contain place names which all refer to the same place identified by an entry in a LOD resource, the posts can be aggregated via linking to the same entry.

In this paper, we describe our experiences in developing a LOD resource for Taiwanese Geographic Names (LOD TGN), extracting Taiwanese place names found in Facebook posts, and linking such place names to the entries in LOD TGN. This study in addition includes the development of a geospatial semantic annotation method for UGC such as Facebook posts through the use of LOD resources.

2   Extracting Ecological Observations from Facebook Posts

Geometry coordinates do not mean that much to most people. Naming is one of the ways through which space can be given meaning and become place [2]. That is, a place name is associated to knowledge about the place; it is used to refer to a location in space. A place name actually plays a very important role in human communication. Even though nowadays GPS devices are essential components in smart phones and digital cameras, place names are still often used in social media to refer to locations. Figure 1 is a Facebook post reporting an observation

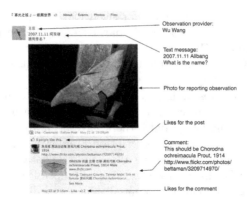

Fig. 1. An ecological observation was posted on the Facebook interest group *Enjoy Moth*. It can be found at http://www.facebook.com/440860179259546/.

of a moth *Chorodna Ochreimacula Prout*. The observation provider indicated the location of the observation through a place name *Alibang* (阿里磅). To transfer the Facebook thread as an ecological observation, there is a need for a tool to extract not only species names but also place names. By using Taiwan species checklists and gazetteers, we develop a Natural Language Processing (NLP) tool for extracting ecological observation information from Facebook posts. The extracted ecological observations are published via LOD principles. To make such data linkable on the Web, we firstly designed an ontology for semantically organizing the extracted data, as shown on Figure 2. Then, we used a D2R server to publish such information on the Web. Figure 3 illustrates an entry of the extracted ecological observation which comes originally from the post shown in Figure 1.

3   LOD Ecology and LOD Taiwan Geographic Names

Linked Open Data of Ecology (LODE) is a validated dataset from a LOD project. LODE integrated 4 previously distributed databases: fire ecology from Taiwan Forestry Research Institute (TFRI) (Firedb), herbarium of TFRI (Taif), insect collection of TFRI (Flyhorse), and Catalogue of Life in Taiwan (TaiBNET, http://taibnet.sinica.edu.tw/). In addition, LODE also uses the metadata document from the TFRI Research Data Catalogue which itself uses Ecological Metadata Language (EML) as the standard to describe Forest Dynamics Plot census of Nanshi Survey (FDP-NS) by Providence University [4]. Figure 4 illustrates that the extracted species name are linked to LODE using *owl:sameAs*.

Geonames.org is an important LOD resource. Although it provides multilingual place names, it is not rich enough to support the use of place names in social media. *Taiwan Gazetteer* is another source of place names in Taiwan. To properly link UGC resources to Taiwan geographic names, it is needed to transform Taiwan Gazetteer to become a LOD resource. Using LOD principles, we transferred Taiwan Gazetteer to a resource called LOD Taiwan Geographic Names

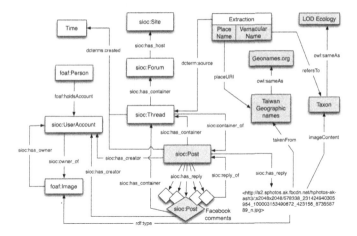

Fig. 2. An ontology is designed for formalizing extracted information of a Facebook thread

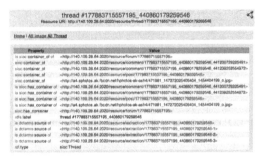

(a) A LOD entry transferred from a Facebook thread.

(b) The extracted species name from the Facebook thread is linked to LOD resources.

(c) The extracted place name from the Facebook thread is linked to LOD resources.

Fig. 3. The Facebook thread listed in Figure 1 is published via a D2R server

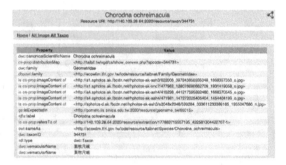

Fig. 4. A taxon of *Theretra Nessus* is the extracted species name in Figure 3. This entry is connected to LODE via *owl:sameAs*.

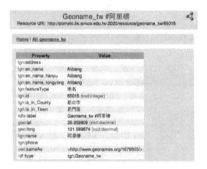

Fig. 5. The entry of LOD TGN transferred from Taiwan Gazetteer. It is linked to geoname.org via *owl:sameAs*.

(LOD TGN). Also, we link the entries in LOD TGN to those in geonames.org via *owl:sameAs*, as shown on Figure 5.

4   Semantic Annotations Using LOD Resources

To improve user input in Facebook groups, we developed a browser extension using JavaScript. The extension actually consists of an NLP engine which can recognize LOD place and species name entities in user input text. If a name is found to match some entity names in LOD resources, the extension would pop up a box to ask the user to identify the exact place or species name entity the user intend to input. As shown in Figure 6, word segmentation and identification of name entities is activated after a user starting to enter text. Since a geographic name *Wu Shan* (烏山) is identified, all geographic entities named *Wu Shan* are listed in the pop-up box. After the user selects the intended geographic name entity, the ID of the geographic name entity is inserted into the text message. This tool thus assists UGC creators to enter structured data.

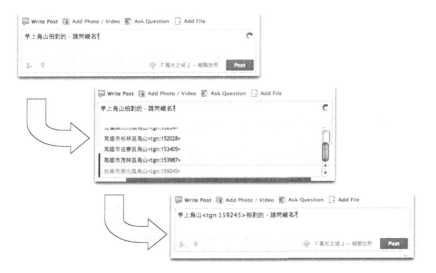

Fig. 6. A semantic annotation plug-in for name disambiguation

5   Conclusion and Feature Works

This study reports our experiences in transferring Facebook threads to ecological observations through information extraction methods and LOD principles. To properly interlink the ecological observations, we developed not only LODE but also LOD TGN. With these information extraction tools and LOD resources, we developed a tool for semantic enhancement of user input. The LOD TGN is an on-going project. In the future, we will consolidate the feature types of the geographic names, and we plan to make the LOD TGN a geospatial semantics reference resource.

References

1. Bizer, C., Heath, T., Berners-Lee, T.: Linked Data-the story so far. Int. J. Semant. Web Inf. 5(3), 1–22 (2009)
2. Cresswell, T.: Place: A Short Introduction. Blackwell Publishing Ltd., Oxford (2004)
3. Goodchild, M.F.: Citizens as sensors: the world of volunteered geography. GeoJournal 69(4), 211–221 (2007)
4. Mai, G.-S., Wang, Y.-H., Hsia, Y.-J., Lu, S.-S., Lin, C.-C.: Linked Open Data of Ecology (LODE): A new approach for Ecological Data Sharing. Taiwan J. of Forest Sci. 26(4), 417–424 (2011)
5. Mendes, P.N., Jakob, M., García-Silva, A., Bizer, C.: DBpedia Spotlight: Shedding Light on the Web of Documents. In: I-Semantics (2011)

Korean Linked Data on the Web: Text to RDF

Martín Rezk[1], Jungyeul Park[2], Yoon Yongun[1], Kyungtae Lim[1], John Larsen[1],
YoungGyun Hahm[1], and Key-Sun Choi[1]

[1] Semantic Web Research Centre, KAIST, Daejeon, South Korea
{mrezk,yoon,kyungtaelim,jlarsen,hahmyg,kschoi}@kaist.ac.kr
[2] Les Editions an Amzer Vak, Lannion, France
park@amzer-vak.fr

Abstract. Interlinking data coming from different sources has been a
long standing goal [4] aiming to increase reusability, discoverability, and
as a result the usefulness of information. Nowadays, Linked Open Data
(LOD) tackles this issue in the context of semantic web. However, currently most of the web data is stored in relational databases and published as unstructured text. This triggers the need of *(i)* combining the
current semantic technologies with relational databases; *(ii)* processing
text integrating several NLP tools, and being able to query the outcome using the standard semantic web query language: SPARQL; and
(iii) linking the outcome with the LOD cloud. The work presented here
shows a solution for the needs listed above in the context of Korean
language, but our approach can be adapted to other languages as well.

Keywords: NLP2RDF, Linked Open Data, Korean, RDF, Reasoning.

1 Introduction

The Web of Linked Open Data (LOD) is developing rapidly, and with it the
number of resources described, which are represented by RDF statements. However, still most of the web data is stored in relational databases and published
as unstructured text. Moreover, the number of links between the resources that
are already published is low compared with the amount of data published in the
LOD cloud—less than 2% at the moment. A way to overcome this gap between
the traditional *Web of Documents* and the LOD is to extract facts and links
from unstructured text using and chaining the different available NLP tools. In
order to allow interoperability between different NLP tools it is desirable to have
ontologies that define and establish the vocabulary to be used, and the relation
among the different terms in it. However, when this approach is applied to non-
Latin languages, such as Korean, there are further issues that need to be solved
regarding internationalization (*i18n*) of NLP tools, ontologies, and standards—
such as URI vs IRI. Furthermore, it is necessary to minimize the performance
gap between relational and RDF data management. This can be done using existing technologies to access and reason with ontological data that is stored in
relational databases.

H. Takeda et al. (Eds.): JIST 2012, LNCS 7774, pp. 368–374, 2013.
© Springer-Verlag Berlin Heidelberg 2013

The first task towards linking Korean data with the LOD cloud is to identify which resources and which properties we want to describe. The resources we are interested in this paper are *morphemes, words (eojeols)* and *sentences* in Korean. We are also interested in modelling linguistic properties such as part-of-speech (POS) information and grammatical roles among others. Our final goal is to chain several Korean NLP tools and be able to efficiently publish the outcome on the web linking Korean text with Korean DBpedia [3]. Summarizing, the issues we need to solve are:

1. **Linguistic Modelling:** We need to model the outcome of the different Korean NLP tools—such as POS–with reference language-independent concepts that allow: *(i)* interoperability with other NLP tools, and *(ii)* conceptual interoperability with other linguistic annotations. To this end, we need ontologies.
2. **Producing RDF Triples and Accessing the Data:** We need to align the outcome with the different ontologies, produce the RDF triples, and be able to query and *reason* with this data enhanced by the ontology.
3. **Linking the Resources with the LOD Cloud:** The final critical step is to link these resources with existing elements in Korean DBpedia.

Our main focus in this paper is to show how we have solved Items 1 and 2 and our first approach towards solving Item 3. In the current work we are linking Korean entities with Wikipedia, and in a follow-up paper we will extend this work to link these entities with the most specific resource in the LOD cloud exploiting the linguistic information provided by the NLP tools. We show preliminary results evaluating this first attempt tackling Item 3.

The contribution of this work is many-fold: *(i)* Ontologies for Korean linguistic annotations; *(ii)* i18n of the NLP Interchange Format; *(iii)* An application (available online) that allows processing Korean text, producing RDF triples, and efficiently querying and *reasoning* with the outcome; *(iv)* Links connecting Korean entities with Korean Wikipedia and preliminary results evaluating this approach. Observe that the method applied here to connect entities with Wikipedia can be applied also to connect them to DBpedia. However, currently Korean DBpedia is under migration tasks.

2   Formats and Ontologies for Korean Annotations

To allow interoperability between different NLP tools, the outcome of these tools must be modelled with formal conceptual descriptions and linked with language-independent reference concepts. To this end, we need ontologies defining these concepts, and specifying the relation among them. Since we focus on describing the linguistic properties of eojeols and Korean sentences; the first step was to identify and model these Strings and then the Sejong tagset—containing POS tags for Korean—and grammatical roles into an OWL ontology (Sejong Ontology) for linguistic annotations.

To describe the resources, that is, eojeols and sentences, we relay on the NLP Interchange Format (NIF) [2]. NIF is an RDF/OWL-based format that can represent Strings as RDF resources. NIF relies on a Linked Data enabled URI scheme and defines two ontologies (String and SSO ontologies) that do not need further modifications to be applied to Korean text. The NIF format is used to *(i)* standardize the input/output of the different tools to ease to connection among them, and to *(ii)* uniquely identify (parts of) text, entities and relationships. Further details can be found in [2]. To identify the Strings in a text, NIF provides two URI schemes: The *offset* and *context-hash* based schemes. We opt in our application for the latter one since it has several advantages regarding stability of the URI. The Hash-based URIs have five components: *(i)* The word "hash"; *(ii)* the *context* length—that is, a predefined number of characters surrounding the String to left and right; *(iii)* the overall length of the String; *(iv)* a 32 character md5 hash created of the String and the context; and *(v)* a human readable part consisting of the first 20 characters of the referenced string.

Apart for Item *(i)*, this specification cannot be applied straightforwardly to Korean. An eojeol is composed of several Hangul syllables. One syllable is composed of two to four Hangul alphabet symbols. , for instance "금" is one syllable composed of the symbols: ㄱ, ㅡ and ㅁ. Since not every combination of Hangul alphabet symbols form a syllable, it is desirable to keep the syllables atomic and make one Hangul syllable correspond to one character in Items *(iii)-(v)*. In addition, URIs specification do not support Hangul. Korean DBpedia solve this problem by using the percent-encoding of the Korean Strings. However, such encoding is not readable by humans. Thus, we propose to extend the NIF standard to support Hangul alphabet symbols (i18n) and use syllables instead of characters to define the context in the case of Korean Language. In our prototype we use and support both: percent encoding and Hangul alphabet symbols. All the issues mentioned above also increase the difficulty of linking Korean entities with the LOD cloud. This will be explained in Section 4.

To model the linguistic properties of eojeols and sentences, we categorized the Sejong tagset into twenty one tag classes for linguistic annotation—such as ProperNoun, CaseMarker, Determiner—together with their respective hierarchy. In particular, we carefully defined case markers and verbal endings—present in Korean and other non-Latin languages but not in English—in the class "Particle" where significant information concerning syntactic structure is expressed. Furthermore, we added classes such as "LikelyNoun" for particular tags (c.f. Figure 1) which, to the best of our knowledge, do not exist in English tagsets, such as the Penn tagset. Once the Sejong ontology was well-defined, we used the *Ontologies of Linguistic Annotation* (OLiA) [1] to link the ontological concepts from the Sejong ontology with language-independent reference concepts. The OLiA consist of three different ontologies:

1. The OLiA reference model: specifies the common terminology that different annotation schemes can refer to.

2. The OLiA annotation model[1]:formalizes the annotation scheme and the tagsets. In our case, this is the Sejong ontology.
3. The OLiA linking model: defines the inclusion relationships between concepts and properties in the Annotation Model and the Reference Model.

In Figure 1 we show the correspondence between Sejong tags and concepts and the concepts in the OLiA reference model.

Tag			Sejong	OLiA
superclass			LinguisticAnnotation/Tag/	LinguisticConcept/MorphosyntacticCategory/
MA	MA		Adverb	Adverb
	MAJ		Adverb/ConjunctiveAdverb	Adverb and Conjunction/CoordinatingConjunction
	MAG		Adverb/GeneralAdverb	Adverb
SN, XN			CardinalNumber	Quantifier/Numeral
MM			Determiner	PronounOrDeterminer/Determiner
SH, SL			ForeignWord	Residual/Foreign
IC			Interjection	Interjection
XR			Noun/BaseMorpheme	Noun/CommonNoun
NN	NN		Noun	Noun
	NNB, NNG		Noun/CommonNoun	Noun/CommonNoun
	NNP		Noun/ProperNoun	Noun/ProperNoun
	NA, NF		LikelyNoun	Noun
NP			Pronoun	PronounOrDeterminer/Pronoun
SE, SF, SO, SP, SS			Symbol	Punctuation
V	NV, V		Verb	Verb
	VA		Verb/Adjective	Verb and Adjective/PredicativeAdjective
	VX		Verb/AuxiliaryPredicate	Verb/AuxiliaryVerb
	VC, VCN, VCP		Verb/Copula	Verb/FiniteVerb
	VV		Verb/VerbalPredicate	Verb
E, JK, XP, XS	E, JK, XP, XS		Particle	MorphologicalCategory/morpheme/
	JC, JX		Particle/AuxiliaryPostposition	MorphologicalCategory/morpheme/Morphological Particle
	JKB, JKC, JKG, JKO, JKQ, JKS, JKV		Particle/CaseMarker	MorphologicalCategory/morpheme/Morphological Particle
	XPN		Particle/Prefix	MorphologicalCategory/morpheme/prefix
	XSA, XSN, XSV		Particle/Suffix	MorphologicalCategory/morpheme/suffix
	EC, EF, EP, ETM, ETN		Particle/VerbalEnding	MorphologicalCategory/morpheme/suffix

Fig. 1. Correspondence between Sejong tags and the concepts in OLiA

3   Implementing NLP2RDF

Our prototype[2] (Figure 2(a)) takes as input a Korean sentence, runs a number of Korean NLP tools, and displays the result. The outcome of these NLP tools is stored in a relational database (DB). The DB might not be needed now since we only parse one sentence at the time, but in the future we will need it for parsing large amounts of data simultaneously. To make this data available for other NLP tools—that takes an NIF input—or by any Semantic Web application; a user must follow an *ETL*-like process, that is: (E)xtract the data from the sources, (T)ransform it into RDF triples, and (L)oad it into a query answering system [5]. However, as it is, this process has several drawbacks such as generating duplicated data—the parsed sentences and words are in the RDF triple store and in the relational database—decreasing the performance of the system and introducing the problem of synchronizing the DB and the triple store in each update. To avoid these problems our system relies on the Ontology Based Data Access (OBDA) model from the OnTop framework [5]. Our OBDA model

---

[1] The annotation model might consists of several ontologies.
[2] http://semanticweb.kaist.ac.kr/nlp2rdf

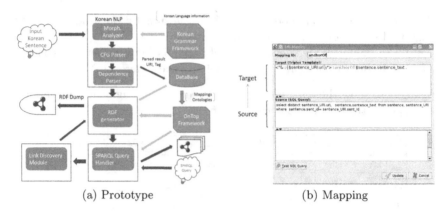

(a) Prototype   (b) Mapping

Fig. 2.

is composed by *(i)* the relational DB definition (database, usr, pswd); and *(ii)* a set of mappings representing the relationship between our relational database and the NIF and OLiA ontologies. Figure 2(b) illustrates these concepts.

The `Target` states a class or a property in our ontologies to be defined— such as the NIF property *anchorOf*—and `Source` is a SQL query defining the members/domain-range of such class/property. For instance, in the case of the *anchorOf* property, we need to join the tables containing the IRIs (for the domain) of the word/sentences and the ones containing the String itself (for the range). In this way, we do not need to materialize the set of triples to answer SPARQL queries, although we can also do this materialization if it is needed. It is worth noticing that when we materialize the RDF triples describing the Strings, we only give the triples with the most specific superclass, thus our approach is more efficient space-wise. The rest of the triples can be obtained using **reasoning** and SPARQL queries. For instance, if x has been tagged as a Common Noun, and the ontology states that every Common Noun is a Noun, we do not provide two triples stating that z has $rdf : type$ Noun and CommonNoun, but only one, that is $< z\ \text{rdf:type}\ :\text{CommonNoun} >$. However, if we query:

$$\text{Select }?x\text{ where }\{?x\quad\text{rdf:type}\quad\text{Noun}\}$$

onTop will use our mappings to reason, and rewrite the SPARQL query into a SQL query in such a way that z will be in the answer. Since SQL is used, we can profit from all the existing optimizations for these tools, and then answer back RDF triples. This closes the *performance gap* between relational and RDF data management. An important feature of this approach in the context in Linked Data, is that our system is aware of the provenance of the data and it also keeps structural information of the data that is lost if we triplify the DB.

To the best of our knowledge there are no other approaches implementing NLP2RDF for Korean, however, we are aware of similar implementations of NLP2RDF for English, for instance, the StandfordCore NIF wrapper.[3] Although

---

[3] `http://nlp2rdf.org/implementations/stanford-corenlp`

the StandfordCore NIF wrapper and our application are similar in nature, the StandfordCore NIF wrapper cannot answer SPARQL queries nor reason as we do. As a consequence, they always produce all the RDF triples that can be derived from the data and the ontologies. It is worth noticing that most of the RDF stores, do not provide reasoning features for query-answering.

4   Towards Linking Korean LRs with the LOD Cloud

In this section we tackle the problem of creating links between the entities discovered by a given NLP tool (which output is in NIF) and Korean Wikipedia. This approach can be adapted straightforwardly to link resources with DBpedia; however, at the moment Koeran DBpedia is under migration tasks and it is unstable. Our final goal is to link these resources (words *and* sentences) with the *most* specific DBpedia resource.

As many well known approaches for Link Discovery, such as LIMES[4], we rely on string-based metrics to measure the similarity between entities. Our approach first accesses the ontological data using SPARQL and obtains all the nouns. Observe that since the vocabulary is given by the OLiA reference model, it is language and Tagset independent; and moreover, since we allow reasoning we just query the super-class Noun without worrying about the substructure below it. IRIs obtained can represent simple or compound entities. Then using the NIF data property *anchorOf* we get the Strings referenced by those IRIs and check if there is a Korean Wikipedia article which title has Levenshtein distance equal zero (that is, exact matching) with such string. If such article exists, we create the link.

We manually evaluate our approach using sentences picked randomly from news articles. This table shows that our parser tagged more nouns than there actually were originally in each sentence. However with the simple linking method we could link a large portion of the tagged entities with the Korean Wikipedia. Further details are available online[5].

# of Sentences	# of Nouns	# of Tagged Nouns	# of Nouns Linked correctly
16	164	191	119

5   Conclusions and Future Work

In this paper we have presented a solution to publish Korean Strings on the web and a first approach towards linking these resources with the LOD. In order to process Korean text and to produce an RDF output that can be re-used by other NLP tools, we provided ontologies for Korean linguistic annotations, and we suggested an internationalization of the URI scheme of the NLP Interchange Format. We presented a prototype (available online) that allows processing Korean text, producing RDF triples, efficiently querying and *reasoning* with the

---

[4] http://aksw.org/projects/limes
[5] http://semanticweb.kaist.ac.kr/nlp2rdf/link.pdf

outcome, and connecting Korean entities with Korean Wikipedia. In addition, we provide preliminary results evaluating this first approach.

Acknowledgments. We thank the anonymous reviewers for useful feedback. This research was supported by the Industrial Technology International Cooperation Program (FT-1102, Creating Knowledge out of Interlinked Data) of MKE/KIAT.

References

1. Chiarcos, C.: Ontologies of linguistic annotation: Survey and perspectives. In: Proceedings of LREC 2012, Istanbul, Turkey (May 2012)
2. Hellmann, S., Lehmann, J., Auer, S.: Towards an ontology for representing strings. In: EKAW 2012. LNCS (LNAI). Springer (2012)
3. Kim, E.-K., Weidl, M., Choi, K.-S., Auer, S.: Towards a Korean dbpedia and an approach for complementing the Korean wikipedia. In: OKCon 2010 (2010)
4. Loomis, M.E.S.: The 78 codasyl database model: a comparison with preceding specifications. In: Proceedings of SIGMOD 1980, New York, NY, USA (1980)
5. Rodriguez-Muro, M., Calvanese, D.: Quest, an OWL 2 QL reasoner for ontology-based data access. In: Proceedings of OWLED 2012 (2012)

Issues for Linking Geographical Open Data of GeoNames and Wikipedia

Masaharu Yoshioka[1] and Noriko Kando[2]

[1] Graduate School of Information Science and Technology, Hokkaido University N-14 W-9, Kita-ku, Sapporo 060-0814, Japan
yoshioka@ist.hokudai.ac.jp
[2] National Institute of Informatics, 2-1-2 Hitotsubashi, Chiyoda-ku, Tokyo 101-8430, Japan

Abstract. It is now possible to use various geographical open data sources such as GeoNames and Wikipedia to construct geographic information systems. In addition, these open data sources are integrated by the concept of Linked Open Data. There have been several attempts to identify links between existing data, but few studies have focused on the quality of such links. In this paper, we introduce an automatic link discovery method for identifying the correspondences between GeoNames entries and Wikipedia pages, based on Wikipedia category information. This method finds not only appropriate links but also inconsistencies between two databases. Based on this integration results, we discuss the type of inconsistencies for making consistent Linked Open Data.

1 Introduction

Recently, the number of collaboratively edited open data Websites has increased, such as Wikipedia[1] and GeoNames[2]. These databases are connected by the principle of Linked Open Data [1]. GeoNames and Wikipedia are commonly used as linking hubs for geographical open data.

There have been several attempts to utilize these open databases by discovering links among them. For example, YAGO2 (Yet Another Great Ontology) [2], which is one of the largest ontologies for open use, expanded their Wikipedia-based ontology using GeoNames information. GeoNames users can also use similar techniques to find links[3]. Based on these efforts, the number of links between Wikipedia and GeoNames has increased from around 148,983 (April 21, 2011) to 385,754 (May 7, 2012).

In earlier work, we have described a method for finding links between Wikipedia and GeoNames by using Wikipedia's category information to estimate classes and locations [3]. However, this method has problems for finding inappropriate links

---

[1] http://www.wikipedia.org/

[2] http://www.geonames.org/ Most of the data are imported from different geographical database such as CIA - The World Factbook.

[3] Based on personal communication with Dr. Marc Wick at GeoNames.

H. Takeda et al. (Eds.): JIST 2012, LNCS 7774, pp. 375–381, 2013.

that can be easily checked by using distance information used in YAGO2. In this paper, we prpose to use distance information for evaluating the quality of links.

However, since the nature of GeoNames and Wikipedia means that both databases contain differing amounts of errors, information obtained by distance may not be so reliable. In addition, there are several cases that are difficult to judge appropriateness of the links. We also classify the problems on finding links between GeoNames and Wikipedia.

2   Integration of Wikipedia and GeoNames

2.1   Integration Method Based on Wikipedia Category Information

We have already proposed an integration method by using Wikipedia categories to estimate the feature class of a Wikipedia page and its location information (e.g., country and administrative division information) [3]. This method generally works well to find out appropriate links, but there are several inappropriate links that can be easily checked by comapring the coordinate information of GeoNames and Wikipedia.

In addition, the previous method used category hierarchy information of "Geography of <country|aministrative area>" for selecting candidate pages and extracting location information. It is not good enough to use common style of Wikipedia category description.

Wikipedia categories use a common style to represent category information, which makes it easy to extract their semantic and attribute information [4]. In the categories of geographic entities, for example, we can find many categories with the following format; "<semantic class> (of|in) <location information>".

In this paper, we use this style for extracting semantic class and location information. Typical examples are "States of the United States" and "populated places in Henry County, Iowa." We can use these categories to infer that the Wikipedia page "Alabama" with the category "States of the United States" is one of the "States" of the "United States."

Figure 1 shows the basic procedure for finding pairs in Wikipedia and GeoNames. The details of each step will be explained later.

1. Comparison of Wikipedia page names and GeoNames names

Wikipedia uses certain conventions to identify information and ensure disambiguation. For example, disambiguation information is added just after the name of the geographic entry to discriminate different geographic entries for "Rome" (e.g., "Rome, Italy," "Rome, Iowa"). Disambiguation information is not included in GeoNames data, so we delete the disambiguation information (i.e., we delete all information after the first ",") during comparisons.

For each Wikipedia page, we select the GeoNames entry that shares the Wikipedia name or an alternative name as candidate pairs to make links.

2. Information extraction from categories

We extract the semantic class and location information by comparing the Wikipedia category with a simple template "<semantic class> (of—in) <location

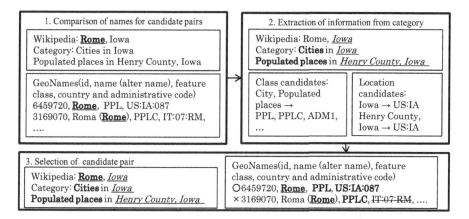

Fig. 1. Procedure for finding candidate pairs in Wikipedia and GeoNames entries

information>." To make a match between the semantic class extracted from a category and feature class in GeoNames, we manually prepare a corresponding list using predefined links between GeoNames and Wikipedia. Several groups of feature classes have similar semantic meanings, which make it difficult to discriminate using category information alone, so we prepare a matching list based on n to m correspondences. Table 1 shows examples of this matching approach. We exclude Wikipedia pages without any corresponding feature classes by using the matching list of candidate pairs.

Table 1. Correspondence between Wikipedia category semantic classes and GeoNames feature classes (example)

General description	Semantic class	Feature class
Administrative division and towns	Populated places, villages, communes, Neighborhoods, ⋯	ADM1,ADM2,ADM3, PPL,PPLA,PPLA2,⋯
Airport	Air force, air base, airfields, air bases, airfield, airports	AIRQ,AIRB,AIRF, AIRP,AIRH
Bay and gulfs	Lochs, gulfs, coves, bays	BAY,BAYS,COVE,GULF

The location information extracted from Wikipedia categories is applied to the selection of candidate pairs, using lists of country and administrative codes as location information. The production of such lists simply requires that we compare the location string with the GeoNames entry with a feature class that described the administrative division and/or seat of such a division. Disambiguation information is also excluded from matching when making this comparison. We use predefined links between GeoNames and Wikipedia to disambiguate GeoNames entries with the same name.

3. Selection of candidate pairs

We select appropriate pairs from the candidates using feature class information estimated from the semantic class string, and the country and administrative codes obtained from location strings and the disambiguation information.

We also use the coordinate information from the Wikipedia page. In addition to the coordinate information extracted by DBPedia [5] [4], we extract coordinate information from GeoHack, which is usually displayed in the right top corner (the top of the infobox).

In addition, there are several groups of pairs with n to m correspondences among the GeoNames entries and Wikipedia pages. The following additional method is used to exclude inappropriate pairs and maintain consistency among candidate pairs.

- Multiple Wikipedia pages for a single GeoNames entry
  We assume that different Wikipedia pages represent different geographical entities, so inappropriate pairs are included in these pairs. We exclude these paris from the candidates for keeping good precision.
- Multiple GeoNames entries for a single Wikipedia page
  If there are multiple GeoNames entries, we calculate the distances between all of the GeoNames entries. If a pair of GeoNames entries are more than 5 km apart (the same as the distance criterion used by YAGO2 [2]), this pair is considered inappropriate according to the 5 km rule of excluding pairs based on the distance from the Wikipedia page. Therefore, we exclude all Wikipedia and GeoNames entry pairs that are more than 5 km apart.

2.2   Integration Results

Based on the procedures explained in the previous section, we conducted experiments to find new links between Wikipedia pages and GeoNames entries.

We used a database containing an English Wikipedia page dump on April 3, 2012 and a GeoNames database download on May 7, 2012. In the GeoNames database, there were 385,754 links between Wikipedia and GeoNames. In this experiment, we aimed to find new links from Wikipedia pages that did not have corresponding GeoNames entries in the database.

Our comparison of Wikipedia names and GeoNames entry names found 1,227,097 Wikipedia pages with at least one GeoNames entry with the same name. After using the Wikipedia category and disambiguation information, however, we only found pairs for 52,016 Wikipedia pages.

We classified these pairs into three groups using geographical coordinate information from Wikipedia, i.e., pairs located within a small distance (less than 5 km: nearby pairs), pairs located within a longer distance (more than 5 km: distant pairs), and pairs with no distance information. We excluded inconsistent pairs from the list that were located a long distance apart and pairs with no distance information (i.e., GeoNames with multiple Wikipedia pages and Wikipedia pages with multiple GeoNames entries located more than 5 km apart).

---

[4] Geographic coordinate of DBPedia 3.7 `http://wiki.dbpedia.org/Downloads37`

As a result, we found the Wikipedia page pairs shown in Table 2. We performed a manual evaluation of the quality of the data extracted using 200 Wikipedia pages, which were selected randomly from each group.

Table 2. Number of Wikipedia pages with corresponding pairs

Types of pairs	Pages	Manual evaluation
Nearby pairs	26,047	200/200
Distant pairs	4,333	180/200
Pairs with no distance information	14,200	190/200

In this evaluation, there were no errors for nearby pairs while we also achieved high precision in other cases. These pairs were not included in the GeoNames link database, so this result is useful for enriching the links between Wikipedia and GeoNames.

Based on a failure analysis, we classified the errors into the following two types.

1. Variations in names
   The names of entities might not be represented in English in GeoNames, which makes it difficult to find their match with a Wikipedia page. We excluded mismatched pairs by applying inconsistency analysis to this type of entry.
2. Failure to estimate the appropriate administrative code
   In several cases, our method failed to select the correct administrative codes for location strings. These errors led to a mismatch between Wikipedia pages and GeoNames entries with different administrative codes. In addition, the location string information was insufficient for selecting the detailed Wikipedia page administrative code for some countries (e.g., few South African categories had location strings with Wikipedia administrative category codes).

After analyzing 200 distant pairs, we found many errors in Wikipedia pages and GeoNames entries, i.e., 12 infobox errors and nine coordinate information errors in Wikipedia pages, five errors in GeoNames, and four inconsistent data pairs (we could not determine which was correct)).

These results suggested that a comparison of different data resources that use Linked Open Data may be helpful for maintaining the quality of databases. We discuss this issue in the next section.

3 Discussion

3.1 Inconsistent Geographical Information

Our proposed method does not rely on geographic coordinate information, so we can use this linking method to evaluate the consistency of geographic coordinate information in different databases.

There were three types of inconsistent pairs according to our method. The number given after the explanation of each type is the number of pairs detected in our experiments.

1. Inconsistent geographic information for appropriate pairs (150/200)
   This type of inconsistency may occur when a geographic entity covers a large area (e.g., mountain ranges, lakes, and countries).
2. Errors in Wikipedia and/or GeoNames (30/200)
   The Wikipedia and GeoNames databases are maintained by humans, so there is a possibility that these databases will contain errors.
3. Errors due to our link detection method (20/200)
   Our method may generate inconsistent pairs.

These results confirmed that our automatic link discovery method may be useful for finding errors in these databases.

3.2 Type of Links between Wikipedia and GeoNames

During the link discovery process, we found that many pairs could not be represented using owl:SameAs [6] links, which are used in DBPedia. We found several cases where a single Wikipedia page corresponded to multiple GeoNames entries.

1. Geographical entities with multiple points
   A typical example of this type of entry is a river. In Wikipedia pages, the infobox of a river contains its source and mouth coordinates. GeoNames also contains different points on a river with the same name.
2. Geographical entities with multiple feature classes
   A single GeoNames entry corresponds to one feature class. As a result, it is necessary to use two GeoNames entries with the same name for different feature classes if the name of an administrative division is also used as the name of a town, e.g., "Milolii, Hawaii" has two corresponding GeoNames entities (5851041: administrative division and 5851402: populated place).
3. Wikipedia pages with multiple geographical entities
   Several Wikipedia pages contained information about multiple geographical entities. For example, some university pages contained different campus information while some mountain range pages contained information about several mountains in the range.

In all three cases, it is acceptable to set links from GeoNames to Wikipedia as "See also," although this may cause problems when constructing SameAs networks [6]. It is necessary to consider this issue when using links to Linked Open Data.

4   Conclusion

In this paper, we proposed a method for integrating the geographical information found in Wikipedia and GeoNames based on Wikipedia category information. This method detected a large number of new links between GeoNames and

Wikipedia. We also demonstrated that this linking method could be used to find inconsistent geographical information in Wikipedia and GeoNames. Based on this result, we discuss the type of inconsistencies for making consistent Linked Open Data.

References

1. Bizer, C., Heath, T., Berners-Lee, T.: Linked data - the story so far. International Journal on Semantic Web and Information Systems 5(3), 1–22 (2009)
2. Hoffart, J., Suchanek, F., Berberich, K., Weikum, G.: YAGO2: A spatially and temporally enhanced knowledge base from wikipedia. Artificial Intelligence (2012) (to appear), http://www.mpi-inf.mpg.de/yago-naga/yago/publications/aij.pdf
3. Yoshioka, M., Liu, Y., Kando, N.: Discovery and maintenance of links between wikipedia and geonames by using wikipedia category. IPSJ Transactions on Databases (TOD) 5(3), 141–148 (2012) (in Japanese)
4. Nastase, V., Strube, M.: Decoding wikipedia categories for knowledge acquisition. In: Proceedings of the 23rd National Conference on Artificial Intelligence, AAAI 2008, vol. 2, pp. 1219–1224. AAAI Press (2008)
5. Bizer, C., Lehmann, J., Kobilarov, G., Auer, S., Becker, C., Cyganiak, R., Hellmann, S.: DBpedia - a crystallization point for the web of data. Web Semantics: Science, Services and Agents on the World Wide Web 7(3), 154–165 (2009)
6. Ding, L., Shinavier, J., Shangguan, Z., McGuinness, D.L.: SameAs networks and beyond: Analyzing deployment status and implications of owl:sameAs in linked data. In: Patel-Schneider, P.F., Pan, Y., Hitzler, P., Mika, P., Zhang, L., Pan, J.Z., Horrocks, I., Glimm, B. (eds.) ISWC 2010, Part I. LNCS, vol. 6496, pp. 145–160. Springer, Heidelberg (2010)

Interlinking Korean Resources on the Web

Soon Gill Hong[1], Saemi Jang[1], Young Ho Chung[1],
Mun Yong Yi[1], and Key-Sun Choi[2]

[1] Department of Knowledge Service Engineering, KAIST, Republic of Korea
{hsoongil,sammyjang,nowespy}@gmail.com,
munyi@kaist.ac.kr
[2] Department of Computer Science, KAIST, Republic of Korea
kschoi@kaist.ac.kr

Abstract. LOD (Linked Open Data) is an international endeavor to interlink structured data on the Web and create the Web of Data on a global level. In this paper, we report about our experience of applying existing LOD frameworks, most of which are designed to run only in European language environments, to Korean resources to build linked data. Through the localization of Silk, we identified localized similarity measures as essential for interlinking Korean resources. Specifically, we built new algorithms to measure distance between Korean strings and to measure distance between transliterated Korean strings. A series of empirical tests have found that the new measures substantially improve the performance of Silk with high precision for matching Korean strings and with high recall for matching transliterated Korean strings. We expect the localization issues described in this paper to be applicable to many non-Western countries.

Keywords: LOD, Silk, Distance measure, Localization, Transliteration.

1   Introduction

One serious drawback of the current Web is its limited support for sharing and interlinking online resources at the data level. Linked Open Data (LOD) is an international endeavor to overcome this limitation of the current Web and create the Web of Data on a global level. Since Web 2.0 emerged, a great amount of data has been released to the LOD cloud in the structured-data format so that computers can understand and interlink them. Many tools and frameworks have been developed to support the transition and successfully deployed in a wide number of areas. However, as the proportion of non-Western data is comparatively small, a couple of projects have been only recently initiated to extend the boundary of LOD over non-Western language resources.

The goal of this research is to report about our experience of applying existing linked open data frameworks, most of which are designed to run only in European language environments, to Korean resources to build linked data. In addition to inherent multi-byte issues related to non-European languages [1] [2], we identified key localization issues that should be taken into account when building links among

H. Takeda et al. (Eds.): JIST 2012, LNCS 7774, pp. 382–387, 2013.

multilingual resources. In particular, we have developed two new Korean similarity metrics and implemented those metrics into Silk, a tool specifically designed for linking LOD resources [3], and compared the performance of the new metrics to Levenshtein Distance. Through a series of empirical tests, we have confirmed that the new metrics offer several advantages over the existing metrics.

2   Related Work

The Korean alphabet system, Hangul, is the native alphabet of the Korean language, consisting of fourteen consonant letters (i.e., 'ㄱ', 'ㄴ', 'ㄷ', 'ㄹ', 'ㅁ', 'ㅂ', 'ㅅ', 'ㅇ', 'ㅈ', 'ㅊ', 'ㅋ', 'ㅌ', 'ㅍ', and 'ㅎ') and ten vowel letters (i.e., 'ㅏ', 'ㅑ', 'ㅓ', 'ㅕ', 'ㅗ', 'ㅛ', 'ㅜ', 'ㅠ', 'ㅡ', and 'ㅣ'). Two consonant letters can be combined to create consonant digraphs (i.e., 'ㄲ' or 'ㄼ'), and two or three vowel letters can be combined to create vowel digraphs or trigraphs (i.e., 'ㅘ' or 'ㅙ'). Syllables are composed by combining one consonant (letter or digraph), one vowel (letter, digraph, or trigraph), and one optional consonant (letter or digraph). The current Unicode system contains 11,172 syllables, which can cover all of the modern Korean words.

Levenshtein Distance, which is also known as 'edit distance', is a popular metric to measure distance between two Latin Alphabet strings and highly recommended to use in Silk. Levenshtein distance is defined as the minimum number of insertion, deletion, or substitution operation of a single character needed to transform one string into the other. Soundex is a widely used phonetic similarity measure for English to score the distance between strings based on not letters but sound.

In the Unicode system, the unit of comparison of English alphabet is one letter because each English letter is assigned to one code point. In Korean, however, a combination of 2 or 3 letters (diagraphs or trigraphs, hereafter letter), representing a syllable in Korean, is assigned to one code point. Thus, even though Levenshtein Distance says that the distance is one, it could mean one, two, or three different letters in Korean. Thus, research has been conducted to develop localized similarity measures for Korean strings. For example, KorED computes distance between two strings by calculating the number of necessary syllable and phoneme operations of insertion, deletion, or substitution to make them identical. GrpSim and OneDSim2 have similar approaches except assigning different weights based on the sound or the location of the phonemes [4].

Transliteration converts letters from one writing system into another and doesn't concern representing original phonemes. For example, one of the Korean popular food "칼국수" (knife-cut Korean noodles in translation) can be transliterated as "Kalguksu" in English. To the best of our knowledge, there are no such metrics to measure similarity between transliterated Korean strings. Instead, there are several studies on transliteration of English to Korean and back-transliteration of Korean to English [5] [6] and there is a study on a similarity measure for Korean transliteration of foreign words using algorithms similar to Soundex [7].

3   Linking Korean Resources with New Phonemic and Phonetic Measures

3.1   New Korean Similarity Measure by Phonemic Distribution

Our approach is based on the idea that the more the phonemes are distributed across the syllables, the less the possibility there is that the strings have the same meaning. Using this new algorithm, we can control the range of the target string more precisely, especially for those string pairs that have only one or two phonemes different from only one syllable. For example, if we specify a threshold of 2 (phonemes), then search for "호랑이" ("tiger" in English and "horangi" in English transliteration) would retrieve its dialect "호래이" ("tiger" in English and "horaei" in English transliteration) as a candidate correctly but not retrieve "오라이" ("OK" in English and "orai" in English transliteration).

This is how our proposed algorithm works; it calculates the number of different syllables. Then it chooses one random syllable with the least number of different phonemes and then regards the number of different phonemes in that syllable as 'α'. The number of different syllables minus one is regarded as 'β'. We can get the final phoneme distance by the formula shown in Eq. (1).

$$\text{Korean Phoneme Distance} = \begin{cases} (sD - 1) * 3 + \min[pD_n] \ if \ sD > 0 \\ 0 \ else \end{cases} \tag{1}$$

In Eq. (1), sD means syllable distance, and pD_n is a list of phoneme distances of syllables. The multiplier 3 is a weighting factor for a syllable because the most common number of phonemes in one Korean syllable is 3.

We tested the proposed algorithm using all of the Korean strings that appear in the English DBpedia. Assuming that the number of relevant records by Levenshtein Distance is the actual number of relevant records, we obtained the performance results summarized in Figure 1. With threshold 1, the recall score of Korean Phoneme Distance (KoPhoDist in the figure, hereafter Phoneme Distance) was almost the same as that of Levenshtein Distance (99.24% vs. 100%), but the precision score was substantially higher than that of Levenshtein Distance (81.23% vs. 21.40%).

	Precision(%)	Recall(%)	F-Score	Retrieved	Relevant	Ret. & Rel.
KoPhoDist[1]	81.23	99.24	0.8934	8,308		6,749
Levenshtein[1]	21.40	100.00	0.3525	31,786	6,801	6,801

[1] threshold=1

Fig. 1. Performance comparison for DBpedia data

From the results, we can find that, with the Phoneme Distance measure the F-score is 0.8934, and with the Levenshtein Distance measure the F-score is 0.3525, showing that the newly developed measure is about two-and-a-half times more effective in finding correct links. This test revealed that Levenshtein measure retrieved a few more correct links at the cost of a formidably large number of incorrect links.

We can further see the benefits of Phoneme Distance by incrementally changing the search scope as shown in Figure 2. Phoneme and Levenshtein Distance produce the same results when the threshold is set to 0. When the threshold is changed to 1, Levenshtein Distance retrieves an extra of 25,100, of which 115 records (0.46%) are relevant and 24,985 records (99.54%) are irrelevant. The figure shows that this formidable amount of the irrelevant records can be effectively controlled by using Phoneme Distance. Phoneme Distance retrieves an extra of 1,662 (63 relevant and 1,559 irrelevant records) with the threshold of 1, an extra of 8,188 (22 relevant and 8,166 irrelevant records) with the threshold of 2, and an extra of 15,290 (30 relevant and 15,260 irrelevant records) with the threshold of 3. It should be noted that the portion of the relevant records is decreasing (3.79% to 0.27% to 0.20%) as the threshold value is incrementally increasing (1 to 2 to 3), clearly indicating that the incentive to include additional range of search is diminished rapidly.

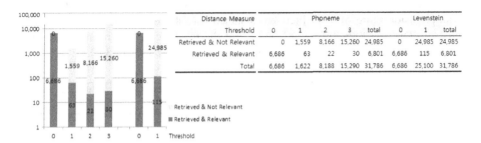

Distance Measure	Phoneme					Levenstein		
Threshold	0	1	2	3	total	0	1	total
Retrieved & Not Relevant	0	1,559	8,166	15,260	24,985	0	24,985	24,985
Retrieved & Relevant	6,686	63	22	30	6,801	6,686	115	6,801
Total	6,686	1,622	8,188	15,290	31,786	6,686	25,100	31,786

Fig. 2. Comparative Search Results - Phoneme Distance vs. Levenshtein Distance

3.2   Korean Transliteration Similarity Measure by Phonetic Features

The best approach to measure distance between transliterated strings would be employing back transliteration of those transliterated strings into the original language and then applying string similarity metrics. This approach, however, is impractical for Korean because a transliterated Korean string could be transliterated back into several possible Korean strings. Due to this difficulty, we decided to take a simpler but more practical approach for measuring similarity between transliterated Korean strings. Our new algorithm replaces 'k' with 'g', 't' with 'd', 'p' with 'b', 'r' with 'l', and then applies Levenshtein Distance on the converted strings. The biggest difference between Soundex and our scheme of Transliterated Korean Distance (hereafter Transliterated Distance) is that we don't eliminate vowels, the other consonants, and any duplicates. And we don't limit the number of letters for comparison.

We tested two cases with each different threshold for Revised Romanization Transliteration (the current standard in Korea). As summarized in Figure 3, with threshold 0 the precision score of Transliterated Distance (KoTrlitDist in the figure) was almost the same as that of Levenshtein Distance (99.86% vs. 99.98%), but the recall score was higher than that of Levenshtein Distance (85.76% vs. 79.88%). With threshold 1 the precision score of Transliterated Distance was a little bit lower than that of Levenshtein Distance (82.20% vs. 83.94%), but the recall score of Transliterated Distance was higher than that of Levenshtein Distance (95.11% vs. 92.37%).

Revised Roman	Precision(%)	Recall(%)	F-Score
KoTrlitDist^0	99.86	85.76	0.9227
Levenshtein^0	99.98	79.88	0.8881
^{0} Threshold = 0			

Revised Roman	Precision(%)	Recall(%)	F-Score
KoTrlitDist^1	82.20	95.11	0.8819
Levenshtein^1	83.94	92.37	0.8795
^{1} Threshold = 1			

Fig. 3. Performance comparison with Revised Romanization Transliteration

With threshold 0, the F-score for Transliterated Distance is 0.9227 and the F-score for Levenshtein Distance 0.8881, and with threshold 1, the F-score for Transliterated Distance is 0.8819 and the F-score for Levenshtein Distance is 0.8795, consistently showing that the newly developed measure is more effective in finding correct links.

We also tested two cases with each different threshold for McCune-Reischauer Transliteration (an old standard). As summarized in Figure 4, with threshold 0 the precision score of Transliterated Distance (KoTrlitDist in the figure) was almost the same as that of Levenshtein Distance (99.91% vs. 99.97%), but the recall score was much higher than that of Levenshtein Distance (66.61% vs. 46.21%). With threshold 1 the precision score of Transliterated Distance was almost the same as that of Levenshtein Distance (85.00% vs. 86.56%), but the recall score of Transliterated Distance was higher than that of Levenshtein Distance (80.90% vs. 71.93%).

McCune-Reichaur	Precision(%)	Recall(%)	F-Score
KoTrlitDist^0	99.91	66.61	0.7993
Levenshtein^0	99.97	46.21	0.6320
^{0} Threshold = 0			

McCune-Reichaur	Precision(%)	Recall(%)	F-Score
KoTrlitDist^1	85.00	80.90	0.8290
Levenshtein^1	85.56	71.93	0.7816
^{1} Threshold = 1			

Fig. 4. Performance Comparison with McCune-Reischauer Transliteration

Taken together, the results consistently show that the proposed scheme of Transliterated Distance outperforms Levenshtein Distance for recall and is similar to Levenshtein Distance for precision. With threshold 0, the F-score for Transliterated Distance is 0.7993 and the F-score for Levenshtein Distance is 0.6320, and with threshold 1, the F-score for Transliterated Distance is 0.8290 and the F-score for Levenshtein Distance is 0.7816, consistently showing that the newly developed measure is more effective in finding correct links.

4   Summary

The Phoneme Distance measure takes the distribution of phonemes in syllables into account to calculate a distance between two Korean strings. Through multiple empirical testing, we have found out that Korean Phoneme Distance is more useful than Levenshtein Distance in interlinking Korean resources because Korean Phoneme Distance reflects the characteristics of encoding scheme of Korean writing system. This measure is especially effective in enhancing precision by reducing the number of irrelevant records. Through multiple empirical testing, we have also confirmed that the proposed Transliterated Distance measure is much more useful than the Levenshtein Distance measure for interlinking transliterated Korean resources because Transliterated Korean Distance reflects the characteristics of phonetics of Korean. This measure is especially effective in enhancing recall keeping precision almost intact, thereby contributing to obtaining a higher number of correct links. The two proposed metrics, Phoneme Distance and Transliterated Distance are original, first defined and explained in this paper. The two measurement approaches can be extended to fuse multi-lingual resources as well.

With the new measures, we could control more precisely the range of the target strings while navigating and generating more quality links with regard to Korean resources. While the localization issues described in this paper needs further empirical validation in different language settings, we expect the ideas implemented through those measures to be applicable to many non-Western countries.

Acknowledgements. This research was conducted by the International Collaborative Research and Development Program (Creating Knowledge out of Interlinked Data) and funded by the Korean Ministry of Knowledge Economy.

References

1. Auer, S., Weidl, M., Lehmann, J., Zaveri, A.J., Choi, K.-S.: I18n of Semantic Web Applications. In: Patel-Schneider, P.F., Pan, Y., Hitzler, P., Mika, P., Zhang, L., Pan, J.Z., Horrocks, I., Glimm, B. (eds.) ISWC 2010, Part II. LNCS, vol. 6497, pp. 1–16. Springer, Heidelberg (2010)
2. Kim, E., Weidl, M., Choi, K.S., Soren, A.: Towards a Korean DBpedia and an Approach for Complementing the Korean Wikipedia based on DBpedia. In: Proceedings of the 5th Open Knowledge Conference 2010, pp. 1–10 (2010)
3. Volz, J., Bizer, C., Gaedke, M.: Silk – A Link Discovery Framework for the Web of Data. In: WWW 2009 Workshop on Linked Data on the Web, LDOW (2009)
4. Roh, K., Park, K., Cho, H.G., Chang, S.: Similarity and Edit Distance Algorithms for the Korean Alphabet using One-Dimensional Array of Phonemes. The Korean Institute of Information Scientists and Engineers 17, 519–526 (2011)
5. Kang, B., Choi, K.: Automatic Transliteration and Back-Transliteration by Decision Tree Learning. In: LREC 2000 Second International Conference on Language Resources and Evaluation Proceedings, Athens, Greece, pp. 1135–1411 (2000)
6. Jeong, K.S., Myaeng, S.H., Lee, J.S., Choi, K.S.: Automatic Identification and Back-Transliteration of Foreign Words for Information Retrieval. Information Processing & Management 35, 523–540 (1999)
7. Kang, B., Lee, J., Choi, K.S.: Phonetic Similarity Measure for Korean Transliterations of Foreign Words. Journal of Korean Information Science Society 26, 1143–1259 (1999)

Author Index